INSIDER TRADING

INSIDER TRADING

HOW MORTUARIES, MEDICINE AND MONEY HAVE BUILT A GLOBAL MARKET IN HUMAN CADAVER PARTS

NAOMI PFEFFER

YALE UNIVERSITY PRESS
NEW HAVEN AND LONDON

Copyright © 2017 Naomi Pfeffer

All rights reserved. This book may not be reproduced in whole or in part, in any form (beyond that copying permitted by Sections 107 and 108 of the U.S. Copyright Law and except by reviewers for the public press) without written permission from the publishers.

For information about this and other Yale University Press publications, please contact:

U.S. Office: sales.press@yale.edu   yalebooks.com

Europe Office: sales@yaleup.co.uk   yalebooks.co.uk

Set in Adobe Caslon Pro by IDSUK (DataConnection) Ltd
Printed in Great Britain by Gomer Press Ltd, Llandysul, Ceredigion, Wales

Library of Congress Cataloging-in-Publication Data

Names: Pfeffer, Naomi, 1946- author.
Title: Insider trading : how mortuaries, medicine and money have built a global market in human cadaver parts / Naomi Pfeffer.
Description: New Haven : Yale University Press, [2017] | Includes bibliographical references and index.
Identifiers: LCCN 2017013302 | ISBN 9780300118551 (cl : alk. paper)
Subjects: | MESH: Tissue and Organ Procurement--legislation & jurisprudence | Tissue and Organ Harvesting—ethics | Mortuary Practice—ethics | Bioprospecting—ethics | Cadaver | Tissue Banks—ethics | United States | United Kingdom
Classification: LCC RD129.5 | NLM WO 690 | DDC 174.2/97954—dc23
LC record available at https://lccn.loc.gov/2017013302

A catalogue record for this book is available from the British Library.

10 9 8 7 6 5 4 3 2 1

CONTENTS

	Acknowledgements	vii
	Introduction	ix
1	Skin Donors-on-the-Hoof	1
2	Pioneers of 'Eye Banking'	8
3	'Doctor, I see you!': Marketing Corpse Philanthropy	16
4	Bioprospecting in Mortuaries	21
5	The Doctrinal Tyranny of Skin	27
6	Growth Hormone Soup	34
7	The American Market for a Growth-promoting Substance	41
8	Civilian Burns: Prevention and Treatment	48
9	Extending Shelf Life After Death	56
10	Cadaver Eyes, Death Denial and the National Health Service	66
11	Whose Corpse Is It?	73
12	Collecting British Cadaver Pituitary Glands	78
13	Lionizing American Eye Banks	86
14	A Gland Lost is a Gland Wasted	92
15	Who's in the Mortuary?	101
16	Representational Dilemmas in Marketing Eye Pledges	107
17	Banking British Cadaver Skin	119

18	The Burn-prone Society	128
19	Harvesting the Dead	139
20	Horse-trading in the Mortuary	146
21	Value for Money in American Mortuaries	154
22	Financing High-volume Eye Banks	163
23	Regulation is Necessary, but How?	171
24	The Blind Eye Act	176
25	Creating American Hybrid Extractors of Cadaver Stuff	183
26	Sharing Pledges and Cadaver Stuff	190
27	Iatrogenesis: Disregarding Risk in Plain Sight	199
28	Ask, or Don't Ask: Inconsistencies in Collecting Sites	207
29	Climbing up the Value Chain	215
30	Contagious Corpses	225
31	British Prions	235
32	Compassion and Commerce	244
33	A Roadmap for the Future	252
34	Repairing the Past	259
35	Globalizing the Gift	266
36	Consolidation Without Cooperation	274
37	From Mortuary to Shopping Cart?	282
	Notes	287
	Bibliography	322
	Figure References	346
	Index	348

ACKNOWLEDGEMENTS

Completing this project, which has taken many years, would have been impossible without the help and forbearance of many people. I am particularly indebted to A. Scott Bruebaker and Ruth M. Warwick for opening doors, arranging introductions, and answering numerous, often naive, questions. Thanks are due to the people who sacrificed time to talk to me about their work; in particular, Patricia Aiken O'Neill, John Armitage, Paul Bowerman, Tom Cochrane, Gerald J. Cole, David Easty, Deidre Fehily, Rollin W. Franks, Nancy Gallo, Frederick N. Griffith, Nancy Stanton Houston, Regina Hunt, Heather Jacobs, Stephen Kaye, John Kearney, Malcolm Keefe, Pamela Keely, Frank Larkin, James H. Leimkuhler, Bob Marchant, Paul Martin, Jean Mowe, Jim Ostrander, David Pegg, Gene Polgar, Michael Preece, James Quirk, Robert Rigney, Michael Roberts, Frank Rollins, Tony Ryan, Andrea Rowe, Khilan Shah, Peter Shakespeare, Heather Shearer and Robert E. Stevenson.

Historians are dependants of people who fetch boxes and files ordered from archives, or allow them to browse through filing cabinets, or direct them to sources, or provide unpublished material. Particular thanks for this kind of help are due to Daniel Barbiero, David Body, James Carson, Robin Chandler, Adele Clarke, Janice Goldblum, Robert Kleinfelder, Deborah O'Malley, Mary Jane O'Neill, Nancy M. Rockafellar and Tom Shakespeare.

I am a beneficiary of the insights of political scientists, anthropologists, sociologists, bioethicists, lawyers, historians and literary scholars who have illuminated the themes explored in this book. My ideas and analysis were developed and tested in the course of conversations, provocations and discussions with many different people – far more than I can adequately acknowledge here – sometimes with and sometimes without their knowledge. I am particularly indebted to Nik Brown, Jane Caplan, Jocelyn Cornwell, Renée Fox, Eddie Higgs, Sarah Hodges, Klaus Høyer, Leonie Kellaher, Julie Kent, Susan Kerrison, Ilana Löwy, Ann Maddox, Luc Noël, Eileen O'Keefe, Allyson Pollock, Belinda Pratten, Marilyn Strathern and Hugh Whittall.

I am grateful to the Wellcome Trust for its generous support, without which this project would have been impossible. Thanks to everyone at Yale University Press London who steered this book's production, especially Robert Baldock for his patience.

Much love and thanks to friends for their encouragement, and providing essential distractions from thinking about death and mortuaries.

The book is dedicated to the countless nameless corpses that haunt its pages.

INTRODUCTION

It is safe to hazard that very few of the many millions of American and British people who have volunteered as posthumous organ donors have envisaged skin being peeled off their corpse, processed, sliced, packed in plastic sleeves like American cheese and retailed as a branded dressing of chronic leg ulcers. Yet this might happen. It is also safe to hazard that few of the millions of people whose head of femur has been replaced are aware that sometimes the prosthesis is secured with putty incorporating ground cadaver bone.

Almost everyone has heard of organ transplantation, where diseased kidneys, hearts, lungs, livers and pancreases are replaced with relatively healthy ones explanted out of fresh corpses. However, cadaver organ recipients are relatively few on the ground, whereas each year many thousands of people undergo a medical procedure incorporating other cadaver stuff, ranging from rescuing people on the brink of death following a drastic burn to penis enlargement. *Insider Trading* is neither the history of people undergoing these procedures nor that of their health-care professionals, although they are frequently mentioned. It is a history of the industries founded in mortuaries, and of the countless fresh corpses that have provisioned their activities.

Language bears some responsibility for why so little is known about these industries. If mentioned at all, these activities are referred to as 'tissue banking', although neither 'tissue' nor 'banking' provides any clue as to what actually happens, how or why.

'Tissue' lacks specificity; it sounds inconsequential when, in practice, following extraction of everything that might be repurposed in medicine, disassembled and substantially incomplete corpses are placed in graves or crematory.

'Tissue' in this context is a medical term, adopted around the turn of the nineteenth century by French physician François-Xavier Bichat to describe twenty-one different types of stuff – mucous, fibrous, cartilaginous, and so on – which he had identified in corpses, each of which, he believed, possesses a unique vital property, such as sensibility and contractility.[1] Bichat's scheme was radical and secular; it reconfigured understandings of the relationship of anatomy and physiology, and at the same time provided medicine with a word that has the capacity to discharge any inconvenient cultural freight carried by various body/corpse fragments: ocular tissue does not sound like eyes, the windows of the soul; liquid tissue does not conjure up blood, the powerful metaphor of life and kinship, although that is what Bichat called it.

So-called tissue banks have no truck with 'tissue'; they deal in a variety of named cadaver stuff: bone, veins, arteries, cartilage, dura mater and so on. *Insider Trading* focuses on eyes/corneas, the cadaver parts medicine most frequently repurposes as replacements; skin, the corpse's least tractable organ, which science has recently disciplined to serve as the raw material of a wide range of branded cadaver products; cadaver pituitary glands, which without the public's knowledge and agreement were extirpated out of the corpses of millions of people (extirpate is the technical term describing complete removal of an organ from body or corpse). These corpse parts are without the capacity to excite, even incite, public attention or social and political action; indeed, thinking and reading about them can be uncomfortable. Nonetheless, their extraction and repurposing have attracted considerable ingenuity and money, both within and outside medicine.

Telling their histories is this book's basic task. It also engages with the wider history of social attitudes towards, and law and policy concerning repurposing of, the material remains of dead people. Perforce, *Insider Trading* confronts and exposes the category error of confusing live and dead persons, bodies and corpses, biopolitics and necropolitics by insisting throughout that while living people can refuse or agree to stuff being extracted from their bodies, irrespective of any direction they might have made during their life, in the mortuary, dead people are without agency; they are capable of neither movement nor speech; they are 'sources' not 'donors', and decisions about whether or not stuff can be extracted are taken by whoever is in legal, physical and/or emotional possession of the corpse, and, lately, is responsible for safeguarding recipients' health: it is their motivations that count, and which are explored.

Endemic death denial, a characteristic of modern death, is responsible for the category error. It has been facilitated by sequestration of corpses into inaccessible and forbidding places: mortuaries situated in hospital bowels, funeral homes and coroners'/medical examiners' premises, places entered through a door typically emblazoned with a notice prohibiting public access.

Everyone dies, but not in the same circumstances. Bichat lived in violent revolutionary times. The guillotine provisioned his investigations, and so many French citizens lost their head that in six months he was able to anatomize 600 corpses.[2] *Insider Trading* engages with modern death, where people typically die following years of poor health and, sometimes, a lengthy period of social isolation, and where the death of young people is exceptional, and is mostly caused by accident, sudden illness or even murder.

No society is indifferent to how its dead are treated. However, what the corpse represents varies in time and place. The modern corpse, which emerged during the nineteenth century, has multiple identities distributed across different sites: in the heads and hearts of kinfolk, the dead person's close circle which may include family members and friends, it is the physical remains of a complex individual person; in certain confessions it is human remains awaiting resurrection; in the funeral parlour it is a business opportunity; in the coroner/medical examiner's office it is a source of forensic evidence; across government departments it is a source of raw data of official statistics; in a hospital mortuary it is a record of an individual's medical and social history; in the anatomy room of medical schools it is teaching material; in public health it is a rapidly putrefying 'nuisance' and potential health hazard.

The identity of the modern corpse as a therapeutic resource is a relative newcomer; it began taking shape during the Second World War through 'cultural work', shorthand for placating opposition, sidelining rival identities, convincing policymakers of the rightness and promoting among the general public what might be called 'corpse philanthropy', a unique expression of posthumous altruism, in which living people volunteer something of their cadaver to medicine.[3]

Cultural work is undertaken mostly in the courts of public opinion and law. It might claim to have some bearing on extractive work, but in practice it follows a separate trajectory. This is because death is deaf to its rhetoric; it sets its own priorities and timetable; its unpredictability subverts the plans of would-be corpse philanthropists and taunts extractors of cadaver stuff.

'Banking' conveys nothing about dealings with death, or of mortuary work, or of how cadaver stuff is processed and distributed. Its first appearance in a medical context was in 1937, when physician Bernard Fantus adopted it to describe his method of securing blood for transfusion, which he had introduced into Cook County Hospital, Chicago.[4] However, the analogy with Main Street financial establishments was misleading, even ironic: Fantus was trying to avoid having to buy blood because neither Cook County, a non-profit public hospital devoted to providing health care to all, nor its patients, could afford to pay professional donors who were the principal source at the time. But the evocation of money is not easily discharged. Indeed, in 1944, New York state regulators refused to incorporate the world's first 'eye bank' on the grounds that the public might mistake it for a financial organization; they acquiesced when its architect R. Townley Paton inserted a hyphen between Eye and Bank, and added 'For Sight Restoration'.[5] Moreover, money did and continues to enter the mortuary and beyond. There are no pockets in shrouds; only living people can engage in financial transactions. The questions explored in *Insider Trading* are: what purpose does money serve, whose is it, and what does its presence reveal.

The comparison of the United States of America and the United Kingdom originated as homage to *The Gift Relationship* (1971), social policy theorist Richard Titmuss's influential analysis of how blood was collected for transfusion medicine on both sides of the Atlantic.[6] Titmuss concluded that the redistributive ethos of the British National Health Service encouraged social solidarity and collective responsibility for the welfare of fellow Britons, which is why its blood transfusion service was keeping pace with growing demand arising from a combination of more people receiving treatment free at the point of delivery, and medical interventions becoming more heroic. In comparison, the record of American blood banks looked dismal, and moreover, Titmuss argued, its blood supply was more likely to be contaminated with the pathogens responsible for yellow jaundice, because refusing blood extracted in exchange for cash by commercial entities operating on skid rows was deemed illegal under American legislation prohibiting anti-competitive business practices.

The Gift Relationship was published during what with hindsight looks like the golden age of welfare states. Politicians have since radically restructured organization and financial arrangements of both American and British health-care systems. Nonetheless, Titmuss's attention to safety has become more relevant: other pathogens have joined forces with those responsible for yellow jaundice and together exert significant influence on how corpse philanthropy is realized and on organizational and business strategies.

Insider Trading follows a loose chronology. Chapters on specific phases in the history of the three cadaver parts in the two sovereign nations are interwoven into a patchwork that confirms how death, corpses and cadaver stuff elude rationalization and generalization. A wide variety of sources is cited: reports of official inquiries, archives, peer-reviewed journal articles, medical textbooks, court hearings, newspapers and magazines, informal interviews, observations during visits to mortuaries and so-called tissue banks, and presentations to professional, trade association and regulatory body meetings. It is also worth noting that much of what happens behind the closed doors of mortuaries emerges only in the context of scandal. The question with which to end is: what kind of entities do British and American mortuaries currently support?

CHAPTER 1

Skin Donors-on-the-Hoof

In 1881, the *New York Times* reported John H. Girdner's (1856–1933) claim to be the first surgeon to have given fresh cadaver skin a new lease of life in a medical procedure.[1] Its source was the corpse of a young man who had cut his throat and had died six hours following admission to Bellevue Hospital. The patient was a ten-year-old boy whose arm and shoulder had been burned when he had been struck by lightning. His condition improved immediately. The lease was short: two weeks later the cadaver skin melted away, but the boy's wounds eventually healed completely.

The *New York Times* article opened with a warning that Girdner's operation 'would strike most persons as revolting in its details, however successful its results'. The editors had good reason for anticipating that readers might experience a visceral reaction. Skin is the human body's largest and most complex organ. Spread out flat, the skin of an average adult measures around 18 square feet (1.67 square metres), every inch of which includes about a yard/metre of blood vessels, four yards (3.66 metres) of nerves, a hundred sweat glands and more than three million cells. Yet skin is so tightly held on the body that it is difficult to envisage it as separate; as English literature scholar Steven Connor observed, 'Where a leg, or a liver, or a heart remain what they are once removed from the body and may be imagined as continuing to function apart from the body which has formed them, the skin itself is no longer a skin once it is detached.'[2]

Another unwelcome thought is repurposing cadaver skin conceptually treats corpses as carcasses, the source of leather, bringing to mind troubling images of lampshades made out of, and books bound in, skin of people murdered in Nazi concentration camps. Indeed, some of the techniques used to process cadaver skin were appropriated from the tanning and leather industries. Perhaps that explains why, for nearly seventy years following Girdner's announcement, there were numerous newspaper articles applauding the bravery of skin donors-on-the-hoof – living people who volunteer body stuff, first used to describe sources of blood for transfusion medicine – but none commending surgeons who had repurposed the skin of cadavers.[3]

Skin can be damaged in many ways. But medicine originally repurposed it to treat wounds caused by a drastic burn or scald. These are complex wounds; their severity is a mixture of extent of damage, injury site, existing medical problems, and cause – flames, chemicals, electricity or cold – and can be exacerbated by inappropriate or inadequate treatment or infection. Victims' plight is dreadful. Indeed, until the mid-twentieth century, treating a drastic burn or scald was regarded as futile, and medical staff would administer a massive dose of morphine and either send victims home or place them in a bed at the end of a hospital ward and hope death would arrive shortly. As a surgeon who admitted to this common practice explained, 'That's not said in a callous way, but there was nothing you could do.'[4] Skin regenerates from the healthy margins of an open wound; the deeper and more extensive the damage the longer it takes to heal, allowing dangerous pathogens ample time in which to feast and multiply. Survivors of the immediate aftermath often succumbed to infection following weeks of pain and misery. Their presence on a hospital ward was frequently announced by the stench of burnt flesh mixed with pus, which is why they were often sequestered in the septic 'dirty' wing alongside patients admitted to have abscesses and carbuncles drained.

The procedure Girdner performed is known as *free* skin grafting.[5] *Free* distinguishes it from the pedicle graft where a thick slice of healthy skin partially severed from one part of the patient's body is attached to the damaged site; the flap is disconnected when it has 'taken', a process that typically lasts several weeks. Whereas the ambition of pioneers of free skin grafting was to replace skin that had been destroyed, the pedicle's purpose is reconstruction. It had been used for centuries in India to replace a nose lost in a fight, cut off as punishment or destroyed by disease – a disfigured or sunken nose is the hallmark of a child born to a syphilitic mother. When it was imported into Europe in the eighteenth century it was sometimes called 'à la Hindu', and it became more widely used following the Napoleonic Wars (1799–1815).[6]

The horticultural connotations of grafting can be misleading. In remaining attached to its source until it 'takes', the pedicle is analogous to propagation by layering where a branch is wounded, the wounded section buried in a medium that encourages root development, and when roots have developed the branch is severed from its source and planted elsewhere. Layering creates new plants identical to their source. Grafting typically is used in order to combine different properties, for instance, a twig cut from a sweet-smelling but slow-growing rose might be grafted on to the rootstock of a vigorous but unattractive variety, which fuse and grow together.

Early skin grafters laboured under the belief that fragments of healthy skin could 'take root' and grow in a fresh or chronic open wound. Different techniques were developed. In 1869, Swiss surgeon Jacques-Louis Reverdin (1842–1929) innovated the 'pinch graft', where a fragment of healthy skin the size of a lentil or postage stamp is lifted up using a sharp hook or forceps, or pinched between finger and thumb, cut with a sharp pair of scissors, and 'sown' on to an open wound. Pinch grafts inflict relatively small wounds at the site of their source. French surgeon Louis Ollier (in 1872) and Leipzig surgeon Karl Thiersch (in 1886) claimed to have achieved better results by covering wounds with thin strips of skin – only epidermis, the outer layer. However, Ollier-Thiersch grafts, as they became known, were technically difficult – Thiersch used a sharp razor – and without a blood supply sceptics doubted their capacity to 'take'. In 1893, German surgeon Fedor Krause (1856–1937) introduced the full-thickness skin graft – epidermis and dermis. Positive results were claimed, but the technique creates a large open wound susceptible to 'operative sepsis', a potentially deadly infection, and detriment at the site of its source sometimes as severe as the one it was intended to repair.[7]

The first free skin grafts were autografts, where the skin is the patient's – Reverdin's 'guinea pig' had pinches of skin excised from the undamaged inside of his upper arm. George David Pollock (1817–1897), operating at St George's Hospital London, is reputed to have been the first to perform a homograft (from the 1960s onwards referred to as allograft), where the grafted skin is that of someone other than the recipient. His 'guinea pig' was Anne T., an eight-year-old girl who two years previously had suffered deep burns over both thighs when her dress caught fire, and whose wounds had refused to heal completely. She submitted to an admixture of autograft and homografts – Pollock applied some of his own and that of an unidentified black man. Indeed, skin of a pigment different to that of recipients began to be used in experiments to establish whether or not homografts work. Moreover, in the United States 'Negro-Caucasian cross homotransplantation' was used in pseudo-scientific research into racial difference: what happens when black skin is grafted onto a white person, or white skin onto a black person: would the grafted skin remain the same colour or adopt that of its recipient?[8]

Skin is the place where an individual's identity – age, race and social standing – is assigned and recognized. As art historian James Elkins observed, skin *is* the body, and we ourselves *are* skin.[9] Hence people who relinquish skin to another person are literally and existentially giving something of themselves. Medical historian Susan Lederer identified a variety of skin donors-on-the-hoof: parent, sibling and other family members, friend, and colleague of patient or her parents; medical staff including surgeons, nurses and medical students; strangers who sometimes were paid.[10] Indeed, extensive wounds might be covered with pinches of skin explanted from numerous people – Lederer cites a 1903 case where over 200 people together provided more than 2,400 pinches that were seeded on to the open wounds of a man who had been badly scalded in a train crash in New Jersey.[11]

Lederer's material is drawn from American sources, and records suggest enthusiasm for skin homografting was greater in the United States than in the United Kingdom. Irrespective of possible national differences, it is undoubtedly the case that the thought of 'being skinned alive' makes people squirm. The skin of living people is exquisitely sensitive. It emerges out of the ectoderm, the same embryonic tissue that gives rise to the brain. Spread across it is a vast number of sensory points, which receive sensations ranging from pleasure to pain. Lederer quotes a surgeon who advocated covering the eyes of 'hysterical adults' and 'very nervous children' with a bandage, so they could not 'see the pain' while their skin was cut.[12] Donors-on-the-hoof sometimes sacrificed wages as well as skin, and where they were the recipient's parent, as was frequently the case, it was often tragically common for the child to die before their wounds had healed.

Inflicting open wounds on healthy people in order to repair those of a burn victim who was likely to die was ethically troubling. Alternatives were available: 'surgical waste' and 'refuse of nature', which medical professionals believed they were entitled to withhold and repurpose without knowledge or agreement of their source. For instance, in 1895, C. F. Timmerman picked up a leg that had been amputated at the Amsterdam City Hospital and drove it to the home of his private patient, where, assisted only by the patient's sister, he removed twenty-two pinch grafts and floated them on to the large wound on the patient's chest.[13] Foreskins removed during circumcision were highly prized for their flexibility.[14] 'Refuse of nature' might be suggestive of windfalls, but actually describes the corpse of a stillborn baby. The more audacious surgeons tried zoografts (now known as xenografts) involving mostly the skin of domesticated animals such as pigs, sheep, rabbits, dogs, cats, chickens, but sometimes also rats and frogs.[15]

Burn Shock

By the turn of the nineteenth century, a sizeable mound of evidence had accumulated suggesting that skin homografts 'take' initially but subsequently 'melt' away. However, not every surgeon accepted the inevitability of rejection; their optimism was encouraged by autografts occasionally working, and popular portrayals of skin 'homografters' as heroes in an epic struggle against terrible injury. Sceptics claimed publicity was encouraging a credulous laity to place 'small squares as sacrificial offerings on the altar of self-inflicted martyrdom'.[16] Nonetheless, recipients might have drawn comfort from corporeal representations of sympathy for their plight.

The science of immunology was in the cradle, and incapable of comforting either enthusiast or sceptic of skin homografts. Around the turn of the twentieth century, Austrian pathologist Karl Landsteiner (1868–1943) had discovered that human blood falls into three major groups, which he called A, B and C (later changed to O). Shortly afterwards a student and colleague together identified a fourth rare group known as type AB. It was subsequently established that the four blood groups regard each other as 'foreign' or 'alien', and refuse to mix together in a test tube, an observation that suggested the reason why some blood transfusions were ending badly, sometimes in the recipient's death. Nonetheless it took some years for these findings to be applied. Landsteiner's reputation was established in the United States in 1923 when he was offered and accepted a post at the Rockefeller Institute for Medical Research, New York, and by the award of the Nobel Prize in 1930.

Findings of research on blood transfusion were used in support of the proposal that the same underlying mechanism is responsible for success or failure of all medical procedures involving repurposed body/corpse stuff, including skin. Some surgeons began hoping that survival of skin homografts might be prolonged where sources and recipient shared the same blood group.

Meanwhile, medical attention had been diverted onto the fatal condition known as burn shock, which was responsible for death in the immediate aftermath of injury. Paradoxically, the worse the injury, the less likely victims are to be aware that they are about to die because nerves have been destroyed and they suffer no pain. As pioneering transplant surgeon and Nobel Laureate Joseph Murray (1919–2012) recollected from his wartime service:

> It is a rather eerie experience for the surgeon to see a severely burnt patient, one who had had 50–60 per cent of his total body surface damaged, let us say, walk into the hospital or into the first aid station feeling quite comfortable and asking for a cigarette. The surgeon knows that the individual within a matter of hours will be dead.[17]

Burn shock was attributed to a hypothetical substance called burn toxin. Its existence had first been postulated in the 1870s when it was found that healthy mice died following injections of burned skin. In the mid 1920s, E. C. Davidson of the Henry Ford Hospital, Detroit, began promoting tannic acid as an alternative to antitoxin, which had become standard treatment of burn shock.[18] Painted on to fresh open wounds, tannic acid coagulates with leaking body fluids to form a rigid shell, which, Davidson claimed, 'fixes' burn toxin in the wound and is protective during healing. The technique became known as tanning because tannic acid, which is derived from vegetable matter such as tree bark, has been used for centuries to prevent carcass skin and hide from rotting and transform it into leather. Positive results from tanning began to be reported on both sides of the Atlantic, and although Davidson had initially recommended it for the worst injuries, tanning of less severe burns and scalds became routine.

Shortly after Davidson began promoting tanning, American physiologist Frank Pell Underhill (1877–1932) published a different approach to treating burn shock. Underhill had monitored victims of the Rialto Theater fire in New Haven in 1921, and found that the volume of fluids leaking from open wounds was far greater than hitherto had been credited. He concluded that burn shock is caused by rapid and massive dehydration, and recommended actively and aggressively restoring and maintaining fluid levels in a victim's body.[19] Investigators began exploring how much of which type of fluid was necessary to prevent death. Enthusiasts of tanning began incorporating tubes of tanning jelly in first aid kits on the grounds that immediate application might contain leakage from open wounds.

Treatment incorporating tanning was passive, a matter of waiting for wounds to heal without any other intervention. Its most vocal critics were Vilray Papin Blair (1871–1955) and James Barrett Brown (1899–1971) of the Barnes Hospital, St Louis, Missouri, the teaching hospital of Washington University School of Medicine, an exceptionally well-funded progressive facility. They condemned tanning on the grounds that bacteria flourish under the rigid shell, and the resulting infection exacerbates the injury, sometimes to the extent of necessitating amputation of hand, foot or limb, and even killing the patient.[20] Blair and Brown advocated active treatments that encourage growth of healthy living skin, and keeping the wound surface clean by daily submerging patients in saline baths and frequently changing dressings.

Blair and Brown's opinion was not easily dismissed. When America entered the First World War in 1917, Blair had led a team of American surgeons drafted to repair terrible facial injuries of soldiers fighting in the trenches, and in peacetime was a member of the international cohort of surgeons creating the new medical specialty of plastic reconstructive surgery.[21] Brown was Blair's pupil, then colleague, and finally successor.[22] The pair developed 'split-skin' grafts, a compromise between Ollier-Thiersch thin grafts and Krause's full-thickness ones, where sheets of skin, consisting of epidermis and some dermis – roughly the thickness of sunburned skin – 5 inches (13 cm) in width and 18 inches (46 cm) in length, are removed, typically from back and thigh. Brown claimed to have been inspired by a boyhood recollection of watching men making leather horse harnesses by splitting cow hide into layers. Split-skin grafts 'take' with relative ease, and where the patient is the source, wounds created at the site heal sufficiently quickly to allow successive 'crops'. However, neither Blair nor Brown used cadaver skin. Where the patient had insufficient intact skin, donors-on-the-hoof were recruited, sometimes up to thirty people, each one requiring a physical examination, anaesthesia, and post-operative hospital care.

Blair and Brown cut free hand, using a sharp long thin knife of their own design. But few surgeons possessed the considerable skill required to cut strips of uniform thickness. In 1938, Earl C. Padgett (1897–1946), a surgeon trained by Blair, and George J. Hood, Professor of Mechanical Engineering at the University of Kansas, developed a calibrated dermatome, an instrument that looked like a cross between a potato peeler and a heavy-duty safety razor, which can remove strips of skin of pre-determined thickness.[23] Eleven years later, Harry Brown invented an electric dermatome capable of operating at 8,000 strokes per minute. Its design had been conceived during Brown's imprisonment in a Japanese prisoner of war camp.[24] Some surgeons continue to cut free hand, but automated dermatomes obviate surgical skill, and can be operated by people without medical training. Their invention set in motion the de-medicalization of the extraction of cadaver stuff.

CHAPTER 2

Pioneers of 'Eye Banking'

In September 1942, Academician Derzhavin, President of the Soviet of Antifascist Scientists, and his colleague Academician Pilipchuk, radioed a message from Moscow to the editor of the *Journal of the American Medical Association* expressing their gratitude to the United States of America for its magnificent support of their nation's heroic struggle following Stalin's decision to swap sides and join the Allies' war against the Nazis. As an expression of appreciation, Vladimir Petrovich Filatov (1875–1956), the Soviet Union's leading ophthalmic surgeon, was revealing the location of an almost inexhaustible site for collecting corneas for restoring the sight of men whose cornea had been scarred in battle: the mortuary.[1]

Filatov had been a member of the international cohort of surgeons who innovated surgical techniques and honed their skills while attempting to repair the devastating facial injuries suffered by soldiers in the trenches during the First World War.[2] In 1929, as a public demonstration of his commitment to Stalin's ambition of accelerating the pace of modernization of the Soviet Union, he had declared war on corneal blindness and invited corneal-blind Soviet citizens, numbering, he reckoned, around 100,000, to his hospital in Odessa.[3]

Slightly smaller than a 5-pence coin, the cornea is situated on the eye's exterior, admitting light into the interior. Looking through a damaged cornea is like looking through a rain-lashed or grease-smudged window. Poverty was largely responsible for corneal blindness in the Soviet Union. Trachoma, a painful bacterial infection of the eyes, flourishes in a toxic mixture of filth, flies and poor personal hygiene resulting from inadequate sanitation and restricted access to clean water and soap (and nowadays to cheap antibiotics). Children's corneas were scarred by conjunctivitis of the newborn (*ophthalmic neonatorum*), a consequence of maternal gonorrhoea and chlamydia, or vitamin A deficiency, which is associated with malnutrition. Other culprits included trauma associated with war, civil unrest, domestic violence and absence of eye protection for workers in dirty and dangerous occupations.

Throughout the nineteenth century, surgeons had conducted a fruitless search for a cornea replacement that 'worked'. Corneas excised from the eyes of non-human species or made out of crystal inevitably failed. The introduction of chloroform anaesthetic in the mid-nineteenth century made keratoplasty, the medical term for the surgical procedure, less traumatic for patients, and gave surgeons more time in which to operate. Cutting into the eye carries the risk of infection that complicates the healing process, may necessitate enucleation of the globe, or at worst generalizes and kills the patient. Towards the end of the nineteenth century, risk of infection began to lessen where surgeons adhered to Lister's battle plan against sepsis, which involved copious volumes of carbolic acid.

In December 1905, Eduard Zirm (1863–1944), operating in a hospital in Olomouc (now Moravia, in the Czech Republic), is credited with performing the first successful corneal graft. The source was Karl Brauer, an eleven-year-old boy whose eye had been enucleated when Zirm failed to remove the fragment of iron that had penetrated it. The recipient was Lois Golgar, a day labourer, whose corneas had been scarred by a splash of lime he had been using to whitewash a chicken hut. Golgar regained and retained some sight until his death some two and a half years later.

Filatov was unable to satisfy the growing demand for keratoplasty with surgical waste, and began searching for alternatives. Foetal corneas proved unsuitable when recipients suffered acute astigmatism. In 1931 he enucleated a cadaver eye, and excised and grafted its cornea. It had taken courage to enter the mortuary. As he put it: 'I must confess that I undertook the operation with some trepidation since there were those who warned me of the danger of "cadaver infection", "cadaver toxin", etc.'[4] Fortunately the operation went well, the patient's sight was restored, and it proved to be a turning point in the history of keratoplasty specifically, and the relationship of mortuaries and medicine in general.

Initially, Filatov sought cadaver eyes in mortuaries of hospitals in and around Odessa. But time and place of death remained unpredictable, despite Stalin's brutal suppression of dissent and misconceived policies that created widespread famine and starvation. It was unfeasible to maintain operating theatre, medical staff and potential recipients in a constant state of readiness in case a potential source of corneas died. Filatov was fortunate in gaining admission to the mortuary of the Sklifosovsky Institute, Moscow's central trauma hospital, where death was a permanent presence, and a culture of extracting and repurposing stuff from fresh cadavers had become entrenched. In 1930, surgeon Serge Yudin (1891–1954) had found that transfusions of cadaver blood 'work' when he saved the life of a young engineer, who had slashed his wrists in a suicide attempt, by audaciously transfusing the blood of a man who had died six hours previously.[5] Yudin's methods might conjure gothic images, but he was licensed by the state attorney's office to collect blood from cadavers, and soon ambulances from all over Moscow were delivering corpses of victims of heart attack, stroke, murder or accident; deaths following a tram accident were particularly common. By the mid-1930s at least sixty large centres and more than 500 subsidiary ones were collecting and distributing 'canned' cadaver blood throughout the Soviet Union.

Filatov established that a cornea excised from an eye enucleated from a corpse within twelve hours following confirmation of death remains reusable for around two days if it is immediately immersed in an appropriate solution. He rated cadaver blood best, but paraffin and saline solution were acceptable alternatives – and cooled to 4°C (the standard temperature of domestic refrigerators). He designed a wide-mouthed glass container with a device on its interior base that secures the jelly-like globe, a method of safe storage and transport known as moist chamber or pot, which became the benchmark against which all other methods of storage are still judged.[6] A box marked with a red cross containing a small ice-filled Thermos jug in which full moist chambers had been packed was transported 500 miles by rail to arrive in Odessa the following day.

Filatov was a leading light of Soviet science, awarded the medal of the Hero of Socialist Labour and five Orders of Lenin, elected a member of the Soviet Academy of Sciences and a member of the Presidium of the Ukrainian Supreme Soviet. His final honour was awarded posthumously: in 1962, his portrait appeared on a 42-kopek postage stamp. His most impressive memorial though is the Filatov Experimental Institute of Eye Diseases and Tissue Therapy, opened in 1936 by the Ukrainian Soviet Socialist Republic, as the jewel in the crown of the Soviet Union's scientifically organized fight against blindness and eye diseases. The hospital had 300 beds and modern laboratories, and stood in beautifully landscaped grounds outside the centre of Odessa on a boulevard which runs alongside the Black Sea, and which was lined with former palaces and mansions that after the Revolution had been converted into sanatoria and 'houses of rest', where favoured *apparatchik* spent their holidays or convalesced.

The Ministry of Health sent oculists – mostly women – from every corner of the Soviet Union to the institute to attend a three-month training course in keratoplasty, which Filatov had simplified and standardized by designing instruments such as trephines – similar to a biscuit cutter – that were manufactured in the Krasnogvardeyets Medical Instrument Factory, Leningrad (awarded the Order of Lenin). By August 1949, and despite his hospital's destruction by the Axis powers during their occupation of Odessa between August 1941 and January 1944, Filatov had grafted 1,000 cadaver corneas.[7] By 1955, more than 8,400 Soviet citizens had undergone keratoplasty, 3,500 of whom were patients of Filatov and his colleagues.

Filatov's pioneering work represents another instance of the paradox of the Soviet state under Stalin, supporting measures for improving the quality of its citizens' lives whilst simultaneously subjecting them to wanton and ruthless brutality.[8] Whether or not Filatov had sought and obtained each dead person's surviving kinfolk's agreement to enucleate eyes, or enucleated them without asking, is unknown, and probably unknowable. In an article published in the *American Braille Press* in 1933, Filatov stated that Soviet law required next of kinfolk's agreement, and complained of their refusal to cooperate.

America's First 'Eye Bank'

R. Townley Paton (1901–1984) imported Filatov's scheme into the United States. Paton, a New York ophthalmologist who enjoyed a flourishing and fashionable private practice, had honed his grafting technique during the 1930s using corneas of 'death-row donors': men executed in the electric chair in Sing Sing Prison, around thirty miles north of New York City. A priest sought their agreement, and if they said yes, Paton witnessed their death, enucleated their eyes and, if no operation was scheduled that day, stored them in his kitchen refrigerator. His son recalled how these forays almost led to the loss of the family cook who had been horrified to find a container of human eyeballs on the shelf next to the milk bottle.[9]

Executed prisoners dried up as a source of cadaver corneas following some sensational publicity.[10] Around this time, Thomas J. Watson, President of International Business Machines (IBM), donated $1,000 towards the translation into English of Filatov's publications. Paton read them and decided to adapt Filatov's methods to the landscape of mortuaries, money and medicine in New York City. In 1944, the doors of the Eye-Bank for Sight Restoration, Inc. were opened. Only fifteen American surgeons had any experience of keratoplasty, and the Eye-Bank for Sight Restoration established a programme of scholarships and fellowships to train young ophthalmologists released from military service.[11] Paton reckoned there were 22,000 corneal-blind Americans, although how he arrived at this statistic is unclear; data on the prevalence of blindness and its causes were not routinely collected in the United States. There is evidence that the incidence of trachoma was falling, thanks partly to doctors on Ellis Island who used a buttonhook to lift up the eyelids of would-be immigrants and quarantine anyone suspected of having the disease. 'Teaching material' was found in a weekly free corneal clinic, opened in 1947, to which potential candidates for keratoplasty who were unable to bear the cost of treatment were referred for evaluation.[12]

Paton had decided to include 'bank' in his entity's name in order to capitalize on the public's enthusiastic support of the giant blood bank that the American Red Cross had organized during the Second World War.[13] 'Bank' used in conjunction with body and cadaver stuff is little more than a confusing metaphor, not simply because, unlike body/corpse stuff, coins and bank notes are not putrescible, but also because it provides few clues as to what the variety of entities actually do. Bernard Fantus (1874–1940) is credited as the first physician to describe a scheme for extracting and distributing body/cadaver stuff as a 'bank'. In 1937, he adopted it to convey the inventory method of ensuring fresh blood was reliably available for transfusion in County Cook Hospital, Chicago. As he put it, 'Just as one cannot draw money from a bank unless one has deposited some, so the blood preservation department cannot supply blood unless as much comes in as goes out.'[14] Every hospital department held an account, which had been opened with an initial 'deposit' of fresh blood. Fantus maintained a blood bank account book with columns for 'debits', 'credits' and a running 'balance'. Cash never entered transactions: neither County Cook Hospital, a non-profit public hospital devoted to providing health care to all, nor its patients, could afford the blood of paid professional donors who were the principal source at the time. Recipients' physicians were responsible for ensuring that an equivalent volume of blood was credited to their department's account. The debt could be repaid either by the recipient if and when she or he had recovered, or kinfolk – type of blood deposited need not match that transfused. Where recipients died, next of kin inherited the debt. Fantus refused to accept cadaver blood because, he argued, '[T]here is something revolting to Anglo-Saxon susceptibilities in the proposal', a claim which might be interpreted as either visceral disgust of therapeutic cannibalism/vampirism, or coded opposition to communism under Stalin, or even a toxic mixture of the two. However, in treating everyone's pint of blood as equal, Fantus allowed cross-racial exchange of blood, a controversial practice provoking in racially prejudiced minds fear of miscegenation. When blood banking was adopted elsewhere, as it rapidly was, in some instances blood was identified by race and segregated in separate refrigerators. The practice was prohibited following the adoption of the Civil Rights Act 1954. In 1972, Louisiana became the last state to desegregate blood stored for transfusion.[15]

Ironically, Fantus had appropriated the inventory method of retail financial banking when confidence in their operations had been undermined by many failures during the Great Depression. Perhaps that is why New York state regulators refused to incorporate Paton's eye bank on the grounds that the public might believe its business was money. They acquiesced when Paton agreed to insert a hyphen between 'Eye' and 'Bank' and add 'for Sight Restoration'.[16]

New York Hospital provided the Eye-Bank with a temporary home until its permanent one in the Manhattan Eye, Ear and Throat Hospital was ready (Paton was head of its Eye Department). 'Bank' might have conjured up an image of dollar bills, tellers, safe and vault in regulators' minds, but the Eye-Bank's facilities consisted of a room on the second floor of the nurses' home equipped with a domestic refrigerator and second-hand furniture. Eye-Bank staff were allowed access to the hospital's laboratory to test the health and safety of 'deposits'.[17]

Knights of the Blind

Almost immediately on hearing news of the Eye-Bank's opening, Staten Island Central Lions began fund-raising for it. Yorkville Lions quickly followed suit, and offered to pay the fees of young ophthalmologists eager to attend its training courses.[18] Louisville Kentucky Lions agreed to underwrite the travel expenses of a blind man who wanted to attend its weekly free corneal clinic.[19] Buffalo Host Lions were more ambitious, and set their sights on arranging enucleation and transportation of cadaver eyes within a 150-mile radius of the University of Buffalo Medical School.[20] Their first fund-raising event, held in the autumn of 1945, was a raffle of fifty-five prizes, including a car and a refrigerator. The second was a Turkey Shoot that netted $66.[21]

Filatov, in *My Path in Science*, his autobiography published in English in 1957, attributed his achievements to the application of the principles of dialectical materialism to ophthalmology.[22] When his scheme was imported into the United States of America, it flourished thanks to the enthusiastic support of civil society associations. These icons of the American way of life originated as buttresses against the rugged individuality of the frontier mentality. As Alexis de Tocqueville famously observed in the early nineteenth century:

> Americans make associations to give entertainments, to found seminaries, to build inns, to construct churches, to diffuse books, to send missionaries to the antipodes; in this manner, they found hospitals, prisons, and schools. If it is proposed to inculcate some truth or to foster some feeling by encouragement of a great example, they form a society. Wherever at the head of some new undertaking you see the government in France, or a man of rank in England, in the United States you will be sure to find an association.[23]

Civil society has left its mark on every walk of American life, including the arts, environment and education, but the sector where it has been most active is health care.[24] The Eye-Bank was financed by 'big-money philanthropists', such as Albert G. Milbank, prominent lawyer, financier and philanthropist, whose eponymous foundation awarded an initial grant of $35,000. Paton's wealthy patients (including the Shah of Persia) were another source of funds. Outside New York, its purpose and practices were pioneered and nourished by members of a Lions Club, who were mostly middle- and working-class men – women were finally admitted in 1987 – whom historian Olivier Zunz calls 'mass philanthropists'.[25]

Lions Clubs, launched in 1917 at a convention in Dallas in a blaze of jingoistic rhetoric, was the last of the 'big three' service clubs to be established: Rotary Clubs had begun in 1905, Kiwanis in 1915. Service clubs had enjoyed phenomenal growth during the 1920s, but Lions Clubs rapidly became the largest, at one stage accounting for 90 per cent of all service club members, which meant that one in every 150 American men was a Lion.[26] Its success had been expedited by its founders' decision to allow relatively few people to open a club, so that small towns, some boasting a population of less than 5,000 people, had one, and also permitting more than one Lions Club to be set up in larger cities.[27]

Every service club has a Grand Endeavour. Helen Keller (1880–1968) was responsible for Lions Clubs' choice of sight. Invited to address the annual international convention in 1925 held in Cedar Point, Ohio, she challenged them by saying, 'Will you not constitute yourselves Knights of the Blind in this crusade against darkness?'[28] Keller's speech had a profound impact on her audience, who immediately resolved to adopt sight conservation and support of blind people as Lions' Grand Endeavour. On their return home, Lions began formulating schemes, sometimes in collaboration with the American Foundation for the Blind, aimed either at preventing sight loss or promoting independence in blind people.

Eye banking was an obvious addition to the Lions' Grand Endeavour. Their enthusiasm for it was borne aloft by scientific achievements such as the atomic bomb, which had defeated the Japanese, and medical 'magic bullets' such as penicillin, cortisone and streptomycin, and fed by articles in popular journals such as *Time* magazine and the *Reader's Digest*, and television programmes dramatizing developments in laboratory and operating theatre, and at the hospital bedside.

Lions Clubs' capacity to meet the new challenge was boosted by the pride of Lions multiplying in the immediate aftermath of the Second World War. Growing popularity of associational life during this period has been attributed to Cold War rhetoric. Voluntarism, opponents of communism argued, encourages private initiative untainted by selfish gain, and serves as a bulwark against top-down regimentation of command-and-control Soviet-style politics. Little wonder, then, that civil society organizations were banned behind the Iron Curtain until it was torn down in 1989.

In 1954, the federal government endorsed discretionary philanthropy, where people rather than policymakers and planners decide on which causes and people to support, by adopting tax code 501(c)(3) that, among other concessions, allows individuals to set charitable contributions against tax, and entities relying on these contributions to do the same. The rhetoric of voluntarism, of public responsibility without government compulsion, became a rallying cry of American opponents of socialized medicine, compulsory health insurance and tighter regulation of providers of health care.[29] In contrast, in the United Kingdom discretionary philanthropy was being discouraged, and replaced by a welfare state funded out of general taxation. As a result, almost from the outset, entities extracting cadaver stuff for repurposing in medicine on the opposite sides of the Atlantic had a different ethos and business model.

CHAPTER 3

'Doctor, I see you!': Marketing Corpse Philanthropy

The Eye-Bank appointed Aida d'Acosta Breckenbridge (1884–1962), the first woman to fly solo in an airship, as executive director. Paton's son David described her as 'a formidable presence, a society-based crusader', with a reputation as a formidable fund raiser – she elicited over three million dollars from big-money philanthropists, such as the King of Siam and Mrs Pillsbury of flour fame, towards the new Wilmer Eye Institute's building at Johns Hopkins Hospital, Baltimore.[1]

Breckenbridge persuaded Stanley Resor, president of J. Walter Thompson, the world's first and at the time largest advertising agency, to serve as the Eye-Bank's president, and recruited seventy-five leading figures in politics, science, law, finance, business, advertising and the entertainment industries as members of its Advisory Council, including Eleanor Roosevelt, film star Ethel Barrymore, novelist Booth Tarkington (author of *The Magnificent Ambersons*), presidents of banks and airlines, senators and congressmen.

Breckenbridge's responsibilities included what sociologist Kieran Healy calls 'cultural work', shorthand in this instance for placating opposition to enucleation of cadaver eyes, sidelining rival identities of the modern corpse, convincing policymakers of the rightness of the Eye-Bank's activities, and promoting in the general public 'corpse philanthropy', a term appropriated from medical historian Susan Lederer's 'body philanthropy' to describe a unique expression of posthumous altruism.[2]

'I like to imagine how, after I'm no longer around to enjoy gazing at the pigeons and skyscrapers and pretty girls strolling along E. 42nd St., my eyes may make sight possible for someone else,' is how journalist Eckert Goodman explained what had motivated him to become a corpse philanthropist.[3] The Eye-Bank assisted by providing a card, small enough to fit into purse or wallet, which identified its bearer as someone anticipating admission to an imagined afterlife where they had joined a community of dead people who posthumously had literally given their eyes to the Eye-Bank for Sight Restoration.

The Eye-Bank set in place the convention of calling would-be corpse philanthropists 'donor'. Yet this designation, which was appropriated from living people who have voluntarily had blood extracted out of their outstretched arm at a blood bank, confuses present and past tense. This is the first instance of the category error, where live and dead persons, bodies and corpses, biopolitics and necropolitics are confused. Moreover, in practice, a card bearer's ambition was unlikely to be realized. More accurately, the card confirms that they have pledged their cadaver stuff.

The Feeling Rules of Corpse Philanthropy

Marketing corpse philanthropy required formulating and communicating what sociologist Arlie Hochschild called 'feeling rules'.[4] Rule one is, consider a fresh corpse as just a shell; the person it once contained has now departed.[5] The rule can be difficult to observe around the time of death, when cadaver stuff urgently must be extracted, and when, as Robert Pogue Harrison eloquently puts it, 'In its perfect likeness of the person who has passed away, the corpse withholds a presence at the same time as it renders present an absence. The disquieting character of its presence-at-hand comes precisely from the presence of a void where there was once a person.'[6]

Rule two follows from rule one: a utilitarian calculus is both legitimate and desirable in the immediate aftermath of death: the corpse is of no further use to the person it once contained, but its stuff can improve the lives of others.

Brutal maths was ameliorated by inserting agreement to enucleation into the grieving process. For instance, an article headlined 'Three auto deaths renew six lives' explains how a widow found solace in respecting her husband's wish that his eyes might help others. And the storyline of *The Immortal Eye*, a film produced for television, followed a grief-stricken couple who found meaning by agreeing to enucleation of the eyes of their stillborn baby.[7]

Rule three is keep a check on visceral repugnance. Resistance to extraction of cadaver stuff is typically attributed to the bogeymen of superstition, ignorance and brute force of tradition, and to certain fragments being more strongly associated with a sense of self – eyes, skin, heart and brain as opposed to kidneys and liver. The latter evidentially is the case, but food culture also exercises an inconsistent influence on attitudes towards extraction and repurposing of cadaver stuff, in particular in relation to eyeballs, which are the cadaver part people are the least willing to volunteer, and which are also excluded from western diets.

Enucleated cadaver eyeballs are notable for their absence in *A Gift like the Gifts of God*, the marketing leaflet that staff at J. Walter Thompson helped to produce. The substance of the 'gift' is sight, an abstract capacity carrying a variety of meanings.[8] 'God gave man sight at birth, and man now has sight in his gift to give to another' suggests that corpse philanthropists are comparable to the divine agent of human creation. Readers were assured of their great good fortune to be among the first generation to be offered the opportunity of giving something that would cost them nothing in a monetary sense yet is more precious than gold. Money might be a yardstick of generosity, but the 'gift of sight' is incommensurable in terms of hard cash, and can only be evaluated in the currency of regard.[9]

Breckenbridge provided journalists with 'true and wonderful stories' of ordinary upstanding citizens who had had their sight restored by keratoplasty, which were published under dramatic headlines such as 'It's like being born again!' and 'Doctor, I see you!' The latter article carried a picture of Fred Berg, a bookkeeper in a Brooklyn textile firm, knotting his tie, a simple procedure that he had been unable to perform for more than twenty years, a result the reporter could not resist calling the 'Eye Bank's dividend'.[10] Mrs Ethel McKinnon had been blind for several years, and the first thing she saw when the bandage was taken off her eye was a crucifix on the wall of the ward in St Joseph's Hospital. 'Thank God,' she murmured, 'I can see again.'[11]

The Eye-Bank's marketing messages reached a wide audience. In 1945, the *New York Times* magazine and *Reader's Digest* carried articles about it. Thanks to Eddie Rickenbacker, First World War ace fighter pilot, chairman of the board of Eastern Airlines and member of the Eye-Bank's Council, a copy of the *Reader's Digest* article was inserted into the pocket on the back of every one of its aircraft's seats for a year following publication. Agency stories about the Eye-Bank appeared in hundreds of newspapers throughout the United States, and it was featured on numerous radio programmes. However, in allowing his name to be mentioned in the publicity, Paton stood accused of unethical conduct, of having flouted his profession's prohibition against advertising for patients. As his son David explained, 'having one's name mentioned in news stories was for scoundrels – certainly not for professionals'.[12] Another criticism was that the publicity encouraged false hope; ophthalmologists complained they were being approached by people who had gained the impression that keratoplasty could restore their irremediable blindness. Indeed, an editorial published in the *American Journal of Ophthalmology* claimed that ill-considered publicity 'was making the whole matter ridiculous in the eyes of the profession'.[13]

Home-spun Invocations

Marketing corpse philanthropy allowed Lions to demonstrate their prowess as Knights of the Blind. It was a competitive sport: *The Lion* occasionally carried a report of an impressive tally of eye pledges notched up by a club. For instance, in 1952, Mrs Dan E. Williams became the 20,000th person to pledge their eyes to the Oklahoma Lions Eye Bank.[14] More than 2,200 pledges were accumulated by the Kentucky Eye Foundation thanks to the hard work of 152 Lions and twenty-one Lions' wives who belonged to its Speakers' Bureau and who, during 1960, rehearsed 'the eye bank story' to 379 church, school and social organizations in their state.[15]

County and state fairs offered great opportunities for spreading the message. For instance, in the summer of 1961 the stand of Staunton Lions at Staunton-Augusta Agricultural Fair was emblazoned with the slogan 'Let your eyes live after you!', and Lions took turns to explain to passers-by why their corneas were sorely needed. Some 2,000 visitors took away pledge cards, many of which were signed and returned to the Virginia Eye Bank, then headed by Dr E. G. Gill, a past president of Lions International.[16]

In 1962, Lions in Omaha, Nebraska, kicked off their eye pledge week with television and radio advertisements featuring recipients of cadaver corneas telling their stories. On the final day, marketing was intensified: viewers and listeners were invited to phone a central switchboard manned by Lions who flashed names and addresses to a radio club, which, for its part, transmitted details to radio cars driven by Lions who collected 500 pledges in four hours.[17] When only nine pledge-card bearers were identified in Hannibal, Missouri, a city of 20,000 people, Lions launched a 'Sock-o' campaign. Hannibal Lions were divided into teams named after major football clubs. One campaign poster featured a book of blank pages titled *The Blind Person's Picture Album*; another pictured a seeing-eye dog and was captioned, 'We're putting him out of business'; a third had black pages and represented 'Sunrise, Sunset for the blind'.[18] At the end of the campaign, 194 pledges had been added to the city's credit.

The Lions' efforts might have been amateurish and lacking in sophistication, but their home-spun marketing messages spoke more directly to

Figure 1: A Knight of the Blind marketing corpse philanthropy

what Tocqueville called 'habits of the heart' than the Eye-Bank's hallowed marketing messages that bore the hallmark of Madison Avenue's advertising executives, and situated corpse philanthropy within the daily life of small-town America.

CHAPTER 4

Bioprospecting in Mortuaries

By incorporating the language of testamentary documents in its marketing material, the Eye-Bank encouraged the misconception that eyes are material property, and that their posthumous distribution can be directed in much the same way as, say, silver spoons can be bequeathed to a niece. This is not the case: in law, neither corpse nor cadaver stuff are your property or anyone else's, and so cannot be the subject of a bequest, hence pledges fell into a legal vacuum.

The no-property legal principle applies to both living bodies and corpses albeit for different reasons. Its origin in relation to the living body is the abolition of chattel slavery, in the United Kingdom by Act of Parliament in 1833, and in the United States in 1865 by the thirteenth amendment to the Constitution. The chattel slave's body is the lawful property of her owner who extracts use value out of it through work, including reproductive labour (the slave mother has no legal claim to her children – they are her owner's property), and exchange value by selling it at an auction where their bodies are monetized according to supply and demand, physical condition, skills and such like. Where chattel slavery is prohibited, the living human body belongs to no one, including the person who inhabits it. Indeed, nowadays, voluntary enslavement is prohibited in most legal jurisdictions. Nonetheless, the commonplace that our living body is our own, that it is ours to nurture or mistreat, makes the no-property principle feel counterintuitive. But the moorings of the living body as property that can be bought and sold are different to those of self-propriety and self-ownership, which are founded on human dignity and autonomy, and the right of individuals to determine what happens to their body, secular principles originating with the seventeenth-century English philosopher John Locke. The distinction can be clarified by reference to rape, which constitutes violation of a person's body, and which is prosecuted under a different legal heading to breaking into and entering premises.

Scholars describe the origins of the no-property principle in relation to the corpse as murky and complicated by being understood as different in relation to the buried and unburied corpse. *Haynes' case* is usually cited in relation to the former. It concerned William Haynes who in 1614 was prosecuted for stealing the winding sheets in which three men and a woman had been buried. Judges sitting at the Serjeant's Inn in Fleet Street speculated on who owned the sheets. Clearly, the buried corpses as buried corpses are incapable of owning anything. The answer was found in ecclesiastical law which has it that following divine service and burial in a consecrated grave the spirit departs to the realms of the supernatural, and the material remains become part of the earth: they are *caro data vermibus* – they belong to worms.[1]

Convention has it that *Dr Handyside's case* is the authority for the common law no-property principle in relation to the unburied corpse. Sometime during the 1740s – the exact date is unknown – Handyside, a male midwife, took possession of the corpses of conjoined twins, which he had pickled and displayed in a glass jar. The father asked the court to make Handyside return them to him but he lost the case when the judge decided no one could claim ownership of an unburied corpse. The case might sound like an irrelevant legal curiosity, but it is the precedent, and was decided during a period when scientific interest in human anatomy was growing, and medical museums were emerging in their modern form, stimulating a market in 'interesting' specimens.

Collecting Sites

Irrespective of their dubious legal standing, Paton knew that pledges were incapable of satisfying his demand for cadaver corneas, not simply because pledgers tend to be young, but also because death rarely visits eye hospitals, which typically are without a mortuary. He sought to establish 'collecting sites' in places where fresh corpses rest between death and disposal, but which were originally claimed and occupied for other and specific purposes.

Anthropologist Cori Hayden calls medicine's search for collecting sites 'bioprospecting', a new name for the long-established practice of exploiting local knowledge in the search for naturally occurring stuff that can be repurposed in medicine, which now includes cadaver stuff. However, Hayden draws a crucial distinction between modern and traditional bioprospecting: whereas traditional bioprospectors foraged without any thought of recompense, modern bioprospectors have been placed under some sort of obligation – ethical, financial and social – to their sources.[2] Hayden's fieldwork was conducted in 'biodiversity-rich' Mexico where American drug and biotechnology companies are following in the footsteps of traditional bioprospectors who travelled to regions of the globe that are now called developing nations. In contrast, during the second half of the twentieth century, bioprospecting for collecting sites of cadaver stuff intensified in the globalizing landscape of industrialized and post-industrialized societies.

Representatives of New York City hospitals with busy mortuaries were invited to a meeting where the Eye-Bank's aims and requirements were explained. Those agreeing to cooperate were called Eye-Bank Affiliates, and equipped with the following: moist chambers; labels; instructions on enucleation; criteria of suitability of corpses – all eyes were welcome but those of elderly people were preferable because their corneas 'take' more easily; 'Release' or 'Permission' forms on which to record the agreement of surviving next of kin; a card on which to record the dead person's age, gender, time and cause of death, and any information that might indicate that the corpse harboured infectious agents, such as results of a Wasserman test for syphilis.[3] Collecting sites less than thirty minutes away from the Eye-Bank's headquarters were instructed to pack chambers in a cardboard box; those further away in a Thermos container filled with wet ice.

Eyes had to be enucleated within a few hours after confirmation of death. Someone suitably trained can complete the task in a matter of minutes. Muscles holding the eye in place are divided, and the globe is rotated towards the nose to expose the optic nerve, which is cut. Enucleation might be a relatively straightforward procedure but most affiliates were unproductive, largely because responsibility for the task typically fell on junior medical staff who were loath to ask next of kin to agree to enucleation of the eyes of someone who had just died, were fully occupied dealing with sick patients, and were unwilling to assist an unrelated outside organization. Matters improved when the Eye-Bank offered off-duty junior ophthalmologists a cash 'incentive' of between $10 and $35, depending on time spent and distance travelled.[4]

In 1953, it emerged that between 1952 and 1953, Queens County Assistant Medical Examiner Jacob Werne had clandestinely enucleated 232 eyes of corpses in his possession, which represented almost one in four of those handled by the Eye-Bank during its first five years.[5] Werne was absent on military leave when a Grand Jury examined his working practices, which included claiming to have performed autopsies that had actually been performed by interns or hospital pathologists.[6] Why Werne was so energetic on the Eye-Bank's behalf was unexplained. A careful perusal of his records established that at best 'oral permission' for enucleation had been obtained from next of kin. Werne was accused of 'constantly and flagrantly disregarding rules' and charged with 'constituting himself the ultimate authority'.

The Grand Jury recommended prohibiting medical examiners from cooperating with the Eye-Bank. Yet the premises of coroners/medical examiners are potentially productive collective sites. In law, deaths in accidental, violent, sudden, unattended, unexplained or otherwise suspicious circumstances, or which take place in specific contexts such as prison or mental hospital or under a surgeon's knife, or where the dead person's identity is unknown, must be referred to them.[7] Whether or not the responsibility falls to a coroner, who is an elected official, or to a medical examiner, who is appointed and medically qualified, is a political decision. New York was one of the first states to adopt a medical examiner system.

No one can challenge the coroner/medical examiner's right, either of possession of a corpse, or to withhold from grave or crematory cadaver stuff that might be used as evidence in a prosecution, for instance, a brain extracted out of a battered skull for evidence in a murder trial. However, this latter right was granted in recognition of the public's interest in the accuracy of causes of death, and prosecution of malfeasance; it did not apply to schemes for extracting and withholding cadaver stuff for repurposing in medicine.

The Quasi-property Rule

Two weeks after the Grand Jury's twenty-five-page report was published, New York City faced a flood of actions for damages from next of kin presumably relying on the common law quasi-property ruling, which had originated in the decision in *Pierce v Proprietors of Swan Point Cemetery* (1872), a case concerning a quarrel between a dead man's wife and daughter over where his corpse should be interred. The case provided a precedent for people seeking legal redress for distress caused by a kinfolk's corpse perceived to have been subject to what is sometimes called outrageous conduct, that is, intentionally, recklessly or negligently removed, withheld or mutilated, or not interred or cremated. The decision was a uniquely American acknowledgement of the cultural and emotional significance people invest in the material remains of their dead kinfolk; that granted them, rather than the dead person or their estate, a private personal right to decide on their disposal.

Pierce v Proprietors of Swan Point Cemetery occupied the landscape of modern death. Rapid population growth on both sides of the Atlantic meant not only more people were living but also more were dying. Between 1760 and 1764, around 885,000 people died in England; between 1845 and 1849 that number had more than doubled, to 1.9 million. American demographic data for this period are incomplete but those that are available suggest the same trend several decades later. In 1810 only two cities had populations of over 50,000 (New York and Philadelphia). By 1860 there were sixteen large cities, and by 1920 when half the American population was urban that number had risen to 144.[8]

On both sides of the Atlantic, the Church traditionally was the earthbound guardian of human remains awaiting resurrection, the spirit having departed to the realms of the supernatural. But it was unable to cope with the growing demand for burial. Urban church graveyards in particular were so packed that they sometimes stank and corpse stuff continually surfaced.[9] The great English social and sanitary reformer Edwin Chadwick, in *Supplementary Report on Interment in Towns* (1843), stressed the health risks they posed, as did Mark Twain in *Life on the Mississippi* (1883). Swan Point Cemetery was one of many new secular cemeteries laid out in suburban and rural areas. Some were financed by entrepreneurs; some publicly owned and paid for out of taxation by city authorities. For instance, the City of London Cemetery was opened in 1856, and in 1869 Hart Island, which is now reputed to be the largest publicly funded cemetery in the world, was laid out by New York City authorities, and is where each year several thousand unclaimed corpses of homeless or poor people were buried in trenches dug in its potter's field.

Disputes about burial in secular grounds fell under common law where the principles guiding judges' decisions were based not on principles that regulate the possession and ownership of material property but in recognition of

> considerations arising partly out of the domestic relations, the duties and obligations which spring from family relationship and the ties of blood; partly out of the sentiment so universal among all civilized nations, ancient and modern, that the dead should repose in some spot where they will be secure from profanation; partly out of what is demanded by society for the preservation of the public health, morality and decency, and partly often out of what is required by proper respect for and observance of the wishes of the departed themselves.[10]

Paradoxically, the decision in *Pierce v Proprietors of Swan Point Cemetery* recognized next of kin's sense of a duty of care for a corpse at a time when the likelihood of their viewing and handling it between death and disposal had begun to diminish. Population growth in towns and cities was exercising sanitary reformers who, among other things, began insisting that in overcrowded homes putrefaction of corpses awaiting burial threatened public health. Local government began opening public mortuaries into which 'dangerous' corpses might be sequestered, a measure widely resented because it deprived grieving kinfolk of a material focus of a wake.

Growth in populations and deteriorating living conditions are associated with industrialization and the creation of new markets. Another new understanding of the unburied corpse was as a business opportunity. During the nineteenth century, the funeral industry began professionalizing and marketing services for grief management, including care and disposal of human remains. The American industry's growth was particularly rapid: by the end of the nineteenth century most American communities had undertakers who, for a fee, took care of the assortment of responsibilities such as washing and dressing of corpses which hitherto had been a lay responsibility fulfilled by kinfolk and neighbours.[11]

Sanitary reformers effectively stigmatized the corpse by emphasizing that it was putrescible and potentially a threat to public health. In contrast, funeral directors humanized it by employing the language of identity and personality, and allowing customers to specify clothing choice, hairstyle and cosmetics. Both were responsible for corpses increasingly moving from deathbed to burial plot (or urn) without making a public appearance. If the corpse is viewed, as traditionally happens more frequently in the United States than in Britain, technicians disguise its appearance, making death look like a state of peaceful repose. Enucleators of cadaver eyes cooperate by filling the empty sockets with an eye cap or wadding and closing the eyelid securely.

CHAPTER 5

The Doctrinal Tyranny of Skin

Fire in some form has always been exploited in war. However, during the Second World War it was responsible for frying, roasting and scalding an unprecedented number of people in an unprecedented variety of ways, such as with flame throwers, incendiary bullets and bombs (thermite and phosphorus), and the Bazooka that propels small particles of burning aluminium that inflict multiple deep wounds; flash from burning cordite and other explosives; contact with heated surfaces (exhaust pipes, hot barrels of guns, etc.); super-heated steam from fractured pipes following a direct hit of tank, plane or ship; fractured radiators containing hot glycol; and ignited petrol. Accidental injury away from the battlefield was common, for instance, when uniforms that had been soaked in petrol to kill lice caught fire, or ground crew smoked cigarettes whilst handling high-octane fuel.[1] Moreover, bombing raids ensured civilians did not escape injury. Indeed, as James Barrett Brown observed, fire had become a psychological weapon; as he put it, 'Burns seem so terrible to most people, and even more so with women and children subject to them from military attack, that a new dread is formed, and a point of morale comes up for the military forces as well as the civilian.'[2]

A tremendous burden fell on British hospitals almost immediately following the outbreak of hostilities. In August 1940, the War Office's War Wounds Committee established a subcommittee to review burn care. Its members included Harold Gillies (1882–1960), another member of the international cohort of surgeons that had attempted to repair the devastating facial injuries of combatants in the trench warfare of the First World War, and his protégé Archibald McIndoe (1900–1960), a New Zealand-born plastic surgeon who had established a fashionable private practice, and who in 1938 had been appointed consultant plastic surgeon to the Royal Air Force. Together Gillies and McIndoe succeeded in persuading the military to abandon tanning, particularly of faces and hands, and, following Gillies's visit to specialist centres in the United States, adopt the methods advocated by Vilray Papin Blair and James Barrett Brown. The wisdom of frequent and prolonged soaking of burn victims in saline baths gained further support from observations that airmen shot down over the sea, and hence immersed in seawater, suffered fewer infections.

The Medical Research Council, the British government agency that decides how taxation is spent on biomedical research, identified the Glasgow Royal Infirmary's Burns Unit as a promising place to undertake research into burns. The unit was unique: no other British hospital had a ward dedicated to treating victims of burns and scalds; its staff specialized in burn care whereas elsewhere, victims were cared for – or mostly neglected – by general surgeons, junior doctors and nurses; each year more that 200 patients were admitted, whereas a general hospital typically admitted fewer than ten.

In 1942, bacteriologist Leonard Colebrook (1883–1967) was placed in charge of the Burns Unit.[3] Colebrook, or Coli as he was affectionately known (a pun on *Escherichia coli,* a bacteria which plays several different roles in what follows), was one of two distinguished pupils of Almroth Wright (1861–1947) (the other was Alexander Fleming (1881–1955) of penicillin fame). Wright, an outspoken opponent of women's suffrage, was a great believer in the capacity of vaccines and antitoxins to combat infections, and oversaw their preparation and large-scale manufacture in laboratories of St Mary's Hospital, London. Colebrook's interest in burns had been sparked during the First World War when a pilot and his gunner had been brought into St Mary's 'absolutely charred'.[4] Both he and Fleming tried every available treatment, but the young men died. The experience led them to publish a paper in *The Lancet* advocating early and extensive skin grafting.[5]

Colebrook's reputation as a bacteriologist had been secured during the 1930s when he oversaw a remarkable reduction in the number of women dying of puerperal (childbed) sepsis in Queen Charlotte's Maternity Hospital, London. Britain's high rates of maternal mortality and morbidity were a particular concern of feminists and policymakers during the interwar years. In 1928, a decision was taken to transfer the hospital to larger premises with an isolation block for women with puerperal fever. The new building facilitated research into puerperal infection by providing investigators with easy access to 'guinea pigs'; it also had a suite of laboratories that had been funded by the Medical Research Council, the

Rockefeller Foundation and the Bernhard Baron Memorial Fund (Bernhard Baron (1850–1929) had been a wealthy cigarette manufacturer, and his company Carreras is now part of Rothmans International).

Colebrook seized the opportunity to ensure the building's design incorporated features that might reduce the risk of contact infection – the route Ignaz Semmelweiss (1818–1865) had identified as responsible for puerperal sepsis almost a century previously – such as ensuring doors were without handles, which notoriously are contaminated with bacteria. He repeatedly reminded staff that 'socially clean' hands are not 'safe hands', that bacteria in nose or mouth can infect women, and encouraged them to wear gloves (easier to sterilize than hands), gown and face mask when attending women in labour. However, a majority of staff, especially those providing domiciliary care, refused to heed his advice.

Colebrook was an early adopter of promising antiseptics. For instance, he recommended Dettol®, the trade name of a disinfectant effective against staphylococci, which has a low toxicity for human skin, which Reckitt & Son, a manufacturer of household products, launched on the British market in 1932.[6] Fewer women died of puerperal fever largely because of Colebrook's experiments with Prontosil, the brand name of the sulphonamide, the first effective antibacterial drug, which he began prescribing shortly after its market launch. In the five years prior to its adoption, 112 of the 495 women admitted to Queen Charlotte's Hospital suffering from puerperal fever died (an average of 23 per cent); between 1936 and 1937, twelve of the 219 women admitted died (an average of 5.5 per cent).[7] Colebrook's role in reducing deaths from puerperal fever was recognized in 1944 by the Royal College of Obstetricians and Gynaecologists, which elected him as a member in the first year following a change in its rules allowing recognition of people outside its own medical specialty who had contributed to women's health.

The open wound in a newly delivered woman's uterus and that of a burn victim are both susceptible to potentially fatal bacterial infection. In 1939, Colebrook was appointed colonel in the Royal Army Medical Corps, bacteriological consultant to the British Expeditionary Force, and a member of the War Office's War Wounds Committee. His task was to supervise research on sulphonamides in the prevention of wound infection. Its urgency was impressed upon him during a tour of casualty stations in France. On his return to England he began experimenting in Rooksdown House, a lunatic asylum near Basingstoke, Hampshire, which had been requisitioned by the War Office and converted into a plastic surgery hospital overseen by Gillies.

It was generally held that the source of infection was lurking on the patient's person when they were admitted to hospital. However, from his experience with puerperal fever, Colebrook was well aware that hospitals are both the site on which medicine triumphs and also the place where grievous harm frequently is done. In the Glasgow Burns Unit he found bacteria flourishing in bedclothes and dust. The cross-infection rate more than halved following his insistence that woollen blankets were replaced with linen sheets, ward floors were regularly oiled, and patients' wounds dressed using the 'no touch' technique.

Medawar and Homograft Rejection

Colebrook continued to cherish an ambition of advancing skin homografts, and in his uniform as a full staff colonel went to Oxford to enlist biologist Peter Medawar (1915–1987).[8] Medawar, in *Memoirs of a Thinking Radish*, his autobiography, describes the drama that had turned skin homografts into a lifetime obsession, and which in many respects was similar to the one that Colebrook had confronted in the previous world war. In the summer of 1940, during the Battle of Britain, he and his wife Jean were relaxing in the garden of their home in Oxford when they heard an aeroplane flying low, followed by a loud explosion. It was a British bomber and the pilot survived but was badly burned. Medawar phoned the Radcliffe Infirmary, the hospital where the young man was being treated, to offer his help. The doctors asked him if he knew how to prevent skin homografts from sloughing off. The question proved a turning point in his life. As he put it:

> This conjunction of events had first made me aware of the body's exquisite powers of discrimination also fixed my career as a scientist. I was henceforward to devote the greater part of my time, thought, and creative energy to discovering how the body discriminates between its own and other living cells.[9]

Medawar was working in the Oxford laboratory of Howard W. Florey (1896–1968), who in 1945 shared the Nobel Prize for Medicine with Ernest Chain (1906–1979) and Alexander Fleming for their work on penicillin. In his first (unsuccessful) investigations into why skin homografts fail and whether this reaction might be circumvented, a 'living soup' of human skin was sprayed on to open wounds. He laid some of the blame for failure on the complicated and lengthy supply chain connecting his laboratory bench in Oxford, hospitals where burned patients were being treated, and

Rooksdown House where he collected surgical waste from operations conducted by Gillies, and where Colebrook had briefly worked.

The Glasgow Burns Unit offered a promising place in which to pursue his research. Thanks to Colebrook its environment was relatively clean, and hence the risk of infection being responsible for homograft failure somewhat reduced. The collecting site was 'in house', and he had an enthusiastic and experienced clinical collaborator in surgeon Thomas Gibson (1915–1993). Their first experiment involved documenting and comparing the natural history of skin auto- and homografts. Their 'guinea pig' was Mrs McK., a young woman whose back had been badly burned when she fell onto a gas fire whilst experiencing an epileptic seizure. Pinches of her and her brother's skin were grafted onto her wounds, and at regular intervals a pinch of each was removed and examined under a microscope. At first there was no evident difference between autografts and homografts but after a few days the homografts took on a different appearance to the autografts; they looked inflamed.

From his clinical experience Gibson had gained a shrewd suspicion that a second set of skin homografts fares worse than the first. To test this out a second set of homografts was applied to Mrs McK.'s wounds. Whereas the first set had experienced a period of grace before rejection, the second set was attacked almost immediately and provoked a violent inflammatory reaction that Medawar and Gibson called the 'second set response'. In the report of their findings published in 1943 in the *Journal of Anatomy*, they speculated that the process is similar to that which daily leads to the elimination of bacteria or viruses foreign to the body, an idea that may owe an intellectual debt to Colebrook, and which implicated white blood cells, or leucocytes, which contain the antigens that sensitize the body to foreign proteins such as bacteria and viruses.[10] Medawar reported to the Medical Research Council that, at present, permanent survival of skin homografts was impossible and suggested that their only clinical value was as a temporary biological wound dressing.

Homostatic Stuff

Medawar's conclusion echoed that of Brown, who the previous year had proposed treating skin homografts as temporary biological wound dressing (in the same publication he revealed his first homograft of cadaver skin).[11] Brown acknowledged that, as things stood, science had nothing to offer to prevent skin homografts from inevitably melting away but, he averred, whilst they remain intact they alleviate pain, seal in body fluids, protect wounds from infection and afford patients respite until they have recovered sufficiently for autografts safely to be undertaken.

Brown was then serving as chief of plastic surgery at Valley Forge General Hospital, a massive military hospital in Phoenixville, Pennsylvania. Before the war, he had investigated the science of skin rejection, influenced perhaps by his colleague Leo Loeb (1869–1959), a genetically orientated physiologist, one of the few people undertaking systematic research into skin homografts, including determining strength and timing of rejection.[12] In 1937, Brown had joined the growing band of surgeons confirming that skin exchanged between genetically identical (monozygotic) twins is permanently accepted, a finding taken to suggest that genetic affinity somehow is crucial in determining the success of homografts, and that matching blood group of sources and recipient is insufficient to prevent rejection. However, their experiments had limited clinical value other than as a test of monozygocity.

In peacetime, Medawar continued to pursue his ambition of discovering how one person could, as he put it, walk around in the skin of another. Moreover, he believed that if a method was found of making skin homografts 'take' permanently, then transplants of every other type of stuff would succeed, an assertion known as 'the doctrinal tyranny of skin homografts'.

Almost immediately the doctrinal tyranny was challenged by evidence drawn from clinical practice that homografts of some cadaver stuff, in particular corneas, are capable of permanent survival, where a homograft of skin of the same source is rejected; in other words, stuff is not equally provocative in arousing an antagonistic response from the recipient's body; so-called 'privileged' stuff repurposed in a homograft is exempt from or withstands rejection by its recipient. Medawar disagreed; he claimed some sites on the body afford foreign stuff sanctuary from the recipient's defence mechanisms. In order to ascertain their location, his team inserted slivers of skin into different parts of the bodies of experimental animals, and found skin homografts surviving indefinitely in the brain, a finding he claimed explained the success of keratoplasty.[13]

The case for privileged stuff was elaborated at an international scientific symposium held in London in 1954. Fragments of the human body/corpse, it was proposed, are either 'homostatic' or 'homovital'.[14] Homostatic stuff – sometimes referred to as 'avascular', or 'non-vital' or 'non-viable' or even 'dead' – is bloodless, performs mostly mechanical services, and for reasons then poorly understood, despite being 'foreign', is tolerated by a recipient's body. Homovital stuff – skin, kidneys, hearts, lungs and so on – depends on blood for survival, and is fated to be attacked and rejected by recipients.

Paradoxically, although skin is the exemplar of homovitality, treated as a biological wound dressing, without the expectation of a permanent 'take', it can be handled as if it is homostatic. In effect, skin homografts now had two careers: a potential one as permanent replacement, and an actual one as biological wound dressing.

CHAPTER 6

Growth Hormone Soup

In 1921, Herbert McLean Evans (1882–1971) announced that immature rats that had undergone a course of daily injections of a soup of anterior lobes had grown three times heavier than their untreated littermates. The rats were not fat: large bones accounted for most of their extra weight.[1] Evans, a doctor by training, and professor of anatomy at the University of California at Berkeley, called the substance responsible for promoting bone growth 'pea soup', because carcass pituitary glands are pea sized. It was more potage than consommé, and its recipe was simple: carefully separate the pituitary gland's anterior and posterior lobes; soak the anterior lobes in alcohol; pound the mixture in a mortar for fifteen minutes; centrifuge the resulting sludge for half an hour.[2]

That the pituitary gland was somehow involved in bone growth had been realized in the late nineteenth century, following the discovery of pituitary tumours in the corpses of very tall people and people who had acromegaly, a condition where bones and soft tissue start growing imperceptibly, eventually resulting in a great lantern jaw and disproportionately large hands and feet. Gigantism occurs in young people, whose long bones are still capable of growth, whereas acromegaly affects older people.

Evans was one of several scientists attempting to identify the substance in the pituitary gland responsible for these perversions of growth. A majority of his contemporaries had been experimenting by feeding immature laboratory animals a diet of carcass pituitary glands. The rats ate the glands with gusto but Evans suspected – correctly – that stomach acid destroys the activity of the putative growth-promoting substance, and had injected the pituitary gland soup into the rats' blood stream. Another reason for his success was that scientists in his laboratory developed a method for performing a hypophysectomy – the surgical removal of the pituitary gland – without killing the experimental animal or sending it into a coma or leaving fragments behind. Complete hypophysectomy is essential in order to render an experimental animal a blank canvas on which the activities of the various substances produced in the pituitary gland's anterior lobe may be revealed. It is a tricky procedure: the pituitary gland is heavily defended within the skull, resting in a bony cradle under the brain and behind the eyes. Evans's team succeeded by fashioning new instruments with which to drill a hole in the rat's skull close to the eye, and sucking out the exposed pituitary gland through a pipette.

Growth hormone was Evans's third claim to fame: he had formulated Evans' Blue, a diazo dye, sometimes called T1824, which is still sometimes used in laboratories to measure blood volume. Together with his long-time colleague Katherine J. Scott Bishop (1889–1976), he identified the 'anti-sterility' vitamin, which he called Alpha-Tocopherol (tocopherol is Greek for pregnancy), and which is now known as vitamin E.[3]

In 1923, the American Medical Association awarded Evans its gold medal. Recognition of his achievements coincided with the Nobel Prize for Medicine being conferred on Frederick Banting (1891–1941) and Charles Best (1899–1978) for isolating in a soup of carcass pancreases a substance named insulin, which was the first effective treatment of type-1 diabetes. Results of their experiment had been announced in 1921. Mass production of insulin began two years later through an unprecedented collaboration of the University of Toronto (Banting and Best's employer), Eli Lilly and Co., a pharmaceutical company, and Chicago meatpackers who provided vast numbers of carcass pancreases.[4] Eli Lilly obtained the rights to market insulin in the United States, Central America and South America, and its profits soared in 1923, with about half deriving from insulin, even though its commercial distribution did not begin until October of that year.[5] Information about how much money the Chicago meatpackers made out of the deal is unavailable.

The 'endocrinological gold rush' began. The term was coined by British scientist Alan S. Parkes (1900–1990) to capture the feverish excitement surrounding research into substances, now known as hormones, that might be extracted from carcass fragments, and also how some of his colleagues bent the rules of professional probity in their efforts to gain international fame and fortune.[6] Within a relatively short space of time, scientists succeeded in isolating substances now known as sex hormones: oestrone (1929), progesterone (1934) and androsterone (1934). However, a reproductive endocrinologist described bagging these hormones as a wild turkey shoot in comparison to the challenge posed by the elusive growth-promoting substance.⁷

Hindsight makes the enormity of the challenge easy to gauge. The pituitary gland's anterior lobe produces (at least) six distinctive hormones responsible for major physiological effects: growth hormone (GH); adrenotrophic hormone (ACTH), which stimulates the adrenal gland; follicle stimulating hormone (FSH) and luteinizing hormone (LH), responsible for ovarian and testicular function; thyroid-stimulating hormone (TSH); and prolactin, which stimulates milk production in breast tissue. These are protein hormones with an enormously complex molecular structure, but scientists were equipped with rudimentary mapping techniques. Progress everywhere was painfully slow, frustrated by the refusal of rival scientists to agree on which experimental animal to use – some preferred rats, others mice, rabbits or dogs – and on soup recipe. Moreover, the instant a scientist published a finding a rival would challenge it. Nonetheless, every competitor agreed that in addition to a growth-promoting substance, a remarkable number of substances is produced within the pituitary gland's anterior lobe. The pea-sized bit of gristle that hitherto was believed to be the source of nasal phlegm, an abject disgusting substance, began to be spoken of with awe. As one investigator put it, 'That this small gland, which in man averages less than 0.5 gm in weight, secretes this number of hormones as separate entities throughout the entire secretory processes taxes the imagination.'⁸ Not only was it exceptionally productive, its substances determined the status of virtually every bodily function, including some essential for life. It began to be referred to as the master gland, or the general headquarters of the endocrine system, or the conductor of the endocrine orchestra. This mixture of metaphors is an indication of a lack of agreement over whether its secretions work independently or collaboratively.

Biotrash

Evans was rarely to be seen working at the laboratory bench. He concentrated on provisioning his laboratory. Evidence of the lengths to which he was prepared to go first emerged when, as a medical student seeking human testicles in the freshest possible condition, he had gained admission to executions at San Quentin prison. Standing beneath the gallows, and without bothering to take the dead man's trousers down or waiting for the attending doctor to hold up his hand to indicate that the heart had stopped beating, he used scissors and knife to cut off the scrotum. Towards the end of his career, Evans defended himself by claiming he removed the testicles only of corpses unclaimed by kinfolk.⁹

The scientific ambition of identifying and isolating a substance capable of promoting bone growth whetted a gargantuan appetite for carcass pituitary glands that only America's meatpacking industry was capable of satisfying. The industry had emerged shortly after the Civil War when the meat barons, as its founders came to be known, seized the opportunity created by the widening rail network and began transporting livestock from farms and ranches, mostly in the west, to the railway hub in Chicago. Here they set up feedlots, slaughterhouses and processing facilities organized around the disassembly line, the industry's defining technology, created in the 1880s by Philip Danforth Armour (1832–1901), founder of the largest of the Chicago meatpackers. The disassembly line, which still operates, industrialized the butcher's craft; some early ones had the capacity to kill up to 2,500 animals bred for meat each day, the annual output of several farms or a whole ranch. Carcasses are attached upside down to a hook on a conveyer belt and taken past a succession of workstations, each one staffed by a man who performs the same task over and over again, until they are completely disassembled into various fragments.[10]

Innovations in refrigeration extended the shelf life of fresh meat and facilitated its dispatch by rail to feed the growing appetites of human carnivores, mostly in the urban east. But table meat accounts for less than half of a carcass. For centuries carcass skin has been repurposed into leather goods, bone into cutlery and combs and fat into soap and candles, but little use had been found for the remainder, which is why wherever possible livestock were slaughtered near to flowing water into which it could be thrown. However, meat barons famously declared war on waste. One of their favourite maxims was 'you either sell it or smell it'.[11] In their operations, a single creature might be transformed into hundreds of different things: gut became tennis racket strings; bristles hairbrushes; hooves glue; fat soap; bones fertilizer or knife handles; head and offal were rendered into animal feedstuffs; and scraps of flesh were stripped of bone and other extraneous stuff and incorporated into foods such as sausage meat, tinned pork and beans and edible fats.

The English language has a host of different terms for waste, such as garbage, trash, rubbish, detritus and dirt, which are often used interchangeably.[12] The neologism 'biotrash' refers to waste that is organic and putrescible; it has vegetable, animal and human antecedents.[13] Every society produces waste.[14] But, as sociologist Martin O'Brien observed, the creation and magical transmogrification of rubbish drove the Industrial Revolution.[15]

Profits from repurposing the vast volume of biotrash created on their disassembly lines were the key to the meatpackers' phenomenal growth and profitability; it allowed them to undercut the prices at which traditional butchers sold table meat, and drive them out of business.[16] Working conditions on the disassembly line were unpleasant and often dangerous. In *The Jungle* (1906), Upton Sinclair (1878–1968) famously exposed the physical and economic hardships endured by the workforce who lived in Packington, the insalubrious residential area surrounding the Chicago meatpackers' plants. But what shocked the public most was Sinclair's revelation of convenience foods contaminated by dead rats, rat droppings and sawdust. The book had been aimed at the public's heart but, as Sinclair complained, it had hit its stomach. Public outrage forced the federal government rapidly to adopt the Pure Foods and Drug Act 1906, which created the Food and Drug Administration (FDA), the agency responsible for ensuring the safety of foods and medicines.

By 1920, meatpacking was the largest industry in the United States, and responsible for one-tenth of its gross national product. However, behind this impressive statistic was an industry that had fallen into the doldrums. A pre-war investigation by the Federal Trade Commission had found that the five largest meatpackers – Armour, Swift, Morrisey, Wilson and Cudahy – had captured more than half of the nation's meat market from their competitors through a secret corporate alliance, called a trust, where prices are fixed and markets carved up at the expense of suppliers, customers and competitors. The trust was 'busted' but punishment was postponed during the First World War when, paradoxically, the industry flourished thanks to government support. However, shortly after hostilities ceased, the meat market plunged into recession, and in 1920 meatpackers were forced to divest themselves of downstream activities such as butchers' shops, cold storage warehouses, market newspapers and journals, but allowed to retain their hold on slaughter and meatpacking.[17] Faced with few opportunities for developing their interests in meat, the meatpackers looked favourably on any novel method of repurposing abattoir biotrash.

Cycles of Credit

Extirpating carcass pituitary glands out of their heavily defended position in the skull is labour-intensive. Carcass heads were cut off at the fifth station of the disassembly line, following knock, shackle, stick and bleed, and thrown down a chute to the by-products floor below, where women sorted, cleaned and routed stuff destined for the manufacture of by-products.[18] Carcass skull bones are thick and only a man, who was paid significantly more than women, had the physical strength to operate the splitting wheel that was used to crack them open.[19] Dissecting out each pituitary gland from its inaccessible and heavily fortified position was women's work. It was difficult: as endocrinologist George Corner (1889–1981) succinctly put it, 'these small objects buried in the floor of the brain cavity are damnably

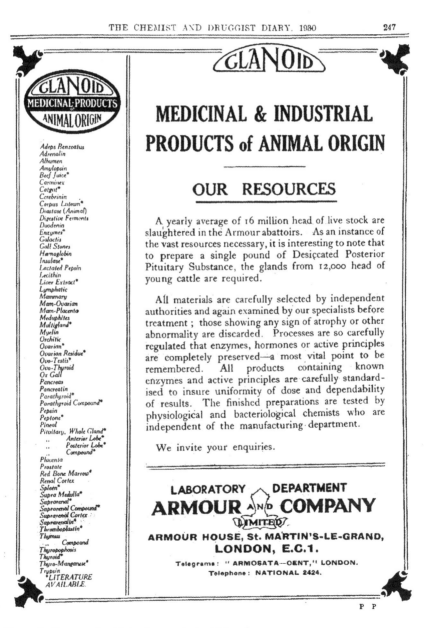

Figure 2: Repurposing biotrash on an industrial scale

hard to reach and dig out of the surrounding bone'.[20] At the most, one woman could retrieve sixty carcass pituitary glands each day.

In 1923, Evans was awarded a grant of $1,700 by the University of California to pay a small local slaughterhouse $804 to extirpate 100,000 carcass pituitary glands, and spend $887 on hiring a car that delivered around 400 glands daily to his laboratory, which meant each gland cost around 20 cents.[21] The following year the slaughterhouse demanded half as much again, thereby increasing the cost of each gland to around 26 cents.

Evans had insufficient funds to meet his investigators' demand for carcass pituitary glands and asked Chicago meatpackers if they would provide them in exchange for first sight of research findings. Armour agreed to Evans's terms. Insulin had cast scientists as alchemists who could transform abattoir waste into gold. Armour was already extirpating vast numbers of carcass pituitary glands, using the anterior lobe in a putative growth-promoting preparation, and the posterior lobe in pituitrin, a safer alternative to ergot, derived from fungi, and a traditional but sometimes dangerous method of accelerating or inducing labour.[22] Pituitrin was very expensive, as only minute traces of it are present in each posterior lobe.

The agreement with Armour aligned Evans's laboratory bench with the disassembly line. In 1930, the result of the first test on a patient of growth hormone isolated in his laboratory in a soup of carcass pituitary glands that Armour had supplied was announced. The research subject was Miss J. M., aged nine and a half, who was 36 inches (91 cm) tall, 16 inches (41 cm) less than the average height of girls of her age. William Engelbach, a New York doctor and leading endocrinologist, administered a course of daily injections of Evans's growth-promoting substance. Engelbach declared its potency superior to that of other, expensive, purportedly growth-promoting substances available on the market.[23] However, Miss J. M. was the first and only person to receive it. The experiment had cast Evans as both Armour's beneficiary and competitor. Fearful of Armour withdrawing support, Evans tactfully claimed scientists in his laboratory were so completely occupied with basic research that they had no time to prepare growth hormone for patients.[24] When asked he would recommend Armour's products.

CHAPTER 7

The American Market for a Growth-promoting Substance

Stimulating demand in the United States for a growth-promoting substance was easy: Americans expect to be big; abundance of size and quantity is highly prized in most aspects of American life. Moreover, during the twentieth century the range of normal height has shifted upwards: the average American man currently stands around 70 inches (1.78 m) tall compared to just over 66 inches (1.67 m) in 1906, and the average American woman now stands at around 64 inches (1.63 m).[1] Above-average height is associated with attractiveness, athletic prowess and upward social mobility. Studies have repeatedly found that taller people on average earn more money and are more likely to be promoted than their shorter colleagues. Another much-quoted statistic is that the taller candidate tends to win the presidential election.

Short stature is associated with a variety of disadvantages. The physical environment caters to people of average height and people at the short end of the normal curve may be unable to reach ordinary things, such as drinking fountains, letterboxes, door handles and lift buttons. Short stature cannot be hidden; people tend to stare at people with dwarfism, and subject them to callousness beyond that shown to almost any other disabled group.[2] Short stature is often mistaken for immaturity and powerlessness – little people are often infantilized and patronized. In the interwar period, 'midgets' – the term was drawn from the midge, an annoying small insect, and is now considered deeply offensive – were still being exhibited in freak shows, a popular source of entertainment originating in the nineteenth century, sometimes individually but more often were presented in a group as if they belonged to a kind of subspecies. The Century of Progress Exposition in Chicago in 1933 featured a Midget Village, complete with elected mayor, occupying a stage set. It was billed as a reproduction, reduced to midget scale, of an ancient Bavarian city, with forty-five buildings, its own police and fire departments, a school, a church and its own souvenir store. Dwarfism carries cultural baggage of special even magical status, a commonplace exploited in *Snow White and the Seven Dwarfs*, a film made by Walt Disney in 1937.

Historian Robert Bogdan claims that exploiting people with visible physical anomalies as entertainment diminished in popularity in response to the medicalization of disability, through which 'freaks' became 'patients'.[3] He locates the turning point in the interwar years, when genes began to be incorporated into explanations of physical difference. A genetic inheritance is immutable but around the same time the idea that the human body is malleable and perfectible was being popularized. This was the moment when medicine entered the market for cosmetic enhancement. Surgeons began applying techniques developed to repair facial injuries of combatants in the First World War trenches to help civilians become more content with their appearance.[4] Similarly, meatpackers' marketing of growth-promoting substances suggested short stature was a disease that medicine could now ameliorate. Put another way, relatively short stature in children was pharmaceuticalized.

Judging from the letters Evans received, some people invested great hope in carcass growth hormone.[5] A young man complained: 'I am only four foot 10 inches and it is very disgusting to me, I am just a back number, at times I think it's going to hurt my marriage, I think I would even be a guinea pig. I'd like to be at least 4 inches taller.' The father of an eighteen-year-old woman, who stood at 4 feet 4 inches (1.32 m), claimed her life was not worth living: 'What chances has she got of getting a job? She tried last summer but with no success. What chance has she of marriage? Only to one like herself which will multiply our burden. What chance have we for peace of mind?' A seventeen-year-old man, who was 5 feet tall (1.52 m), believed growth hormone would make bullies leave him in peace. As he put it:

> My friends always call me names concerning my height. I can't go out of our own house without being teased. I feel so neglected and self-conscious. Sometimes I get to the point of fighting it out with taller boys whenever they would look down on me. I have no friends who are my height level. All of them are tall. I could not study my lessons very well.

Short stature was even seen as un-American: 'I am a Chinese American girl. Now my problem is that I am too short, only five foot two inches. I hate it.

I wish I were five foot 11 inches or more. You look lousy and unimpressive when tiny but not when you are statuesque.'

Rise and Fall of Carcass Growth Hormone

The American Medical Association's Council on Pharmacy and Chemistry, which sat in judgment on the scientific claims of novel medicines, heaped opprobrium on carcass growth hormone preparations and organotherapies.[6] Indeed, during the interwar years, sceptics called physicians who prescribed them 'endocriminologists'.[7]

In March 1944, the journal *Science* published an article announcing the isolation of a comparatively pure bovine growth hormone.[8] The authors were Evans and Choh Hao Li, a biochemist whom Evans had appointed in 1938 to clarify pituitary gland soups produced in his laboratory.[9] Li had accepted the position because few people were prepared to hire a Chinese biochemist (for the same reason, he had been unable to find a landlord prepared to rent a room to him and his wife).[10] Evans proved an excellent judge of scientific potential: Li was an outstanding scientist – twice nominated to receive the Nobel Prize – who devoted his long career to the chemistry of the hormones of the anterior lobe of the pituitary gland. But Evans paid him a miserly salary and relegated him to a cupboard-sized, overheated basement laboratory through which the university's main steam lines sizzled and rats ran. Li was often obliged to leave his bed in the middle of the night to open its windows in order to prevent excess heat spoiling a delicate experiment.[11]

A 'pure' preparation of growth hormone was one eliciting only bone growth in hypophysectomized, immature laboratory rats; it was incapable of producing any of the actions attributed to other secretions of the pituitary gland's anterior lobe. A 'pure' preparation could be used in experiments that might convince the scientifically minded that carcass growth hormone could make children grow taller. The popular press heralded Evans and Li's achievement as an important contribution to the war effort: underneath a photograph published in *Time* magazine of Evans and Li staring at a glass flask containing a clear fluid is a caption reading, 'Enough of this might turn Japs into giants', the implication being that an inferiority complex about their short stature was responsible for Japanese aggression and hence the war in the Pacific.[12] Another journalist wrote to Evans confirming the existence of a large market for a growth-promoting substance: 'The world is waiting, especially we here in the East who are tired of wearing elevator shoes and submitting to other forms of artificial height.'[13]

It had taken two pounds of carcass pituitary glands to produce one thousandth of an ounce of growth hormone. Obtaining such a large quantity had proved a considerable challenge.[14] In 1942, the federal government had limited the number of animals being slaughtered, imposed meat rationing, issued coupons to civilians, but held down the retail price of meat. As a result of these measures, Americans who together before the war had been eating 400 million lb (180 million kg) of red meat each week now had to suffer the ignominy of a diet of fowl and herrings. It was relatively easy for authorities to monitor the activities of the large meatpackers, and the number of animals processed on their disassembly lines fell. But smaller abattoirs mostly escaped scrutiny and a thriving under-the-counter trade developed which was reputed to be shifting as many animals as the law-abiding meatpackers. 'Bootlegger' meat sellers suspected scientists seeking carcass pituitary glands were federal government agents and refused to help out. To make matters worse, able-bodied men were being enlisted into the armed forces, and meatpackers lacked the manpower to operate the splitting wheel. As a favour to the University of California, a slaughterhouse in Oakland agreed to collect some but the most it could supply met just under one-tenth of Li's monthly consumption of 70 lb (32 kg). Only by dint of begging, cajoling and calling in favours did Evans managed to obtain sufficient glands to keep Li occupied.

Preparations for a clinical trial of 'pure' carcass growth hormone began. The State Virus Laboratory agreed to sterilize the soup, which was likely to be contaminated as its main ingredient had been extracted on a meatpacker's disassembly line. Investigators carried out safety tests on themselves by injecting a small amount into a muscle. Unfortunately, another crisis in the meat industry forced the trial to be postponed. In the autumn of 1946, the number of animals slaughtered fell to 80 per cent of 1945 levels, and in Chicago 5,000 packinghouse workers were laid off.[15] Armour had no carcass pituitary glands to spare; it was hundreds of pounds behind in the production of its own products and needed at least twelve months to meet outstanding orders.[16] Evans was compelled to import carcass pituitary glands from New Zealand and South America at exorbitant prices. Between 1947 and 1948 he spent more than $16,000 on carcass pituitary glands, compared to $5,000 in the previous year.[17]

The 'guinea pig' was F. H., a young woman aged sixteen years and eight months who was 50 inches (1.27 m) tall and who in the previous year had had the pituitary glands of two freshly slaughtered calves grafted into her stomach in an attempt to make her grow taller. In the spring of 1947, F. H.

was admitted to a tiny private room in the hospital of the University of California Medical School, San Francisco, placed on an exacting regime of daily injections of bovine growth hormone and fed a diet of milkshake fortified with eggs, sugar and cream. At the end of three months she had grown ⅜ inch (0.95 cm). On her release from hospital she continued submitting to injections for several months but her height remained unchanged. The last her doctors heard of her was when she wrote to them from Paris, France, asking if it was ok for her to get married.[18] Carcass growth hormone might make the bones of hypophysectomized, immature laboratory rats grow but despite many claims to the contrary it failed to 'work' in children. In 1949, when Armour stopped marketing it for humans in the United States, any remaining commercial optimism in carcass growth hormone was abandoned.

Into the Mortuary

In August 1958, Maurice Raben (1915–1977), an endocrinologist at Tufts University, Boston, announced that the height of a patient, a teenager who had stopped growing when he was just 4 feet 2 inches (1.27 m) tall, had increased by 2 inches (5 cm) following a course of injections of human growth hormone (hGH). When treatment ended in 1962 the young man had reached the height of 5 feet 4½ inches (1.64 m).[19] Raben's news was shortly followed by an announcement that a tiny eleven-year-old girl had grown 3 inches (8 cm) in height following a six-month course of hGH prepared in Li's new laboratory on the campus of the University of California, San Francisco.[20]

Raben might have beaten Li in the publicity stakes, but it was Li who had established why growth hormone isolated in carcass pituitary glands was incapable of making children grow taller: carcass growth hormone 'works' in cattle, rats and fish, but not in guinea pigs, monkeys or humans; fish growth hormone works in fish but not rats; human growth hormone works in monkeys, and monkey growth hormone works in humans.[21]

That only primate growth hormone works in humans was surprising, even mysterious. Insulin isolated in carcass pancreases is an effective treatment of type-1 diabetes. Moreover, in 1949, injections of carcass adrenocorticotrophic hormone (ACTH), another of the hormones produced in the pituitary gland's anterior lobe, had enabled people long crippled with arthritis to walk, or even dance.[22] ACTH stimulates production of another hormone in the adrenal glands, a synthetic version of which is known as cortisone, which is now ranked alongside penicillin and

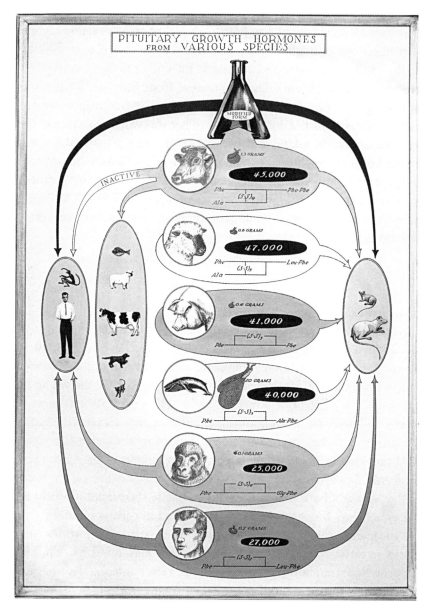

Figure 3: Which growth hormone 'works' where

streptomycin as one of the pharmaceutical 'magic bullets' that shortly after the Second World War dramatically strengthened modern medicine's armoury.

The research required pituitary glands of many different species. Li was an effective bioprospector; he contacted zoos, farms and fisheries, in and outside the United States. Eli Lilly and Co. provided monkey pituitary glands. The pharmaceutical company was producing polio vaccine, and was breeding and sacrificing thousands of monkeys to prepare a kidney soup that the poliovirus relishes. By the end of his career Li could brandish an inch-thick book listing names of investigators in around fifty countries who had visited his laboratory, and whom he asked in return to send him the pituitary glands of animals that inhabit deserts, savannahs and forests.[23] Rolf Luft, a Swedish doctor working at the Karolinska Hospital, Stockholm, had provided freeze-dried human pituitary glands. Luft was convinced that the pituitary gland is somehow involved in the development of cancer, a disease characterized by the unchecked growth of rogue cells, and in the early 1950s, had began hypophysectomizing patients with advanced breast or prostate gland cancer in the belief that the radical operation would extend their lives.[24] During its brief popularity, the growth hormone theory of cancer produced few human pituitary glands. Instead, scientists and clinicians were confronted with the challenge of establishing in the mortuary collecting sites capable of satisfying an appetite that had been whetted on the disassembly line.

CHAPTER 8

Civilian Burns: Prevention and Treatment

In the aftermath of war, and the infancy of the British welfare state, attention turned to civilian burns. However, prevention, not treatment, was placed centre stage. An editorial published in the *Lancet* in December 1946 under the headline, 'Death in the Fireplace', challenged official British policy that placed responsibility for a burned or scalded child on its parents. The cause was not parental carelessness or neglect.

> In the broadest sense, bad housing and overcrowding are the principal ministers of death and injury. There is no room for children to play, except around the grate. Kettles and saucepans are often upset, because the grates are ill designed or out of repair. Overcrowding of both persons and equipment makes it difficult to keep dangerous things in safe places, and to prevent children from interfering with them. Fireguards for coal, gas, and electric fires are absent or wholly inadequate. The garments of women and children are commonly made of material that burns like a torch.[1]

The Children and Young Persons Act (1908) had made compulsory the use of fireguards in homes where children under seven years of age were living; it had also empowered a judge to impose a fine of up to £10 on parents of a child who suffered a serious burn from an unguarded coal fire. But a child's death or terrible injury was considered sufficient punishment, and the act was seldom enforced. The clause concerning unguarded fires had been inserted in response to several articles published in medical journals expressing concern over the number of children dying following a drastic burn, and coroners calling attention to the dangers of unguarded coal fires and inflammable clothing, particularly flannelette (brushed cotton or winceyette) nightwear that was cheap and warm, and hence very popular. In the same year as the act was placed on the statute book, the Home Office appointed a committee to consider flameproofing flannelette, which published its findings in the *Second Report of the Departmental Committee on the Law Relating to Coroners* (1910).

Colebrook was chief agitator; he was confronted daily by children suffering the horrible consequences of a burn or scald. In 1943, he had left Glasgow to head the new Burns Unit at the Birmingham Accident Hospital, which two years previously the Medical Research Council had established as a centre of experimental trauma care.[2] Colebrook had been invited to take up the position, and the Bernhard Baron Trust contributed £5,000 to equip his bacteriological laboratory.[3]

Birmingham was a centre of heavy engineering industry – it was known as the workshop of Britain. Many of its numerous factories had been requisitioned for the war effort, and were manufacturing ammunition and military equipment, including the Spitfire fighter aircraft. Little wonder then that the German Luftwaffe heavily and repeatedly bombed it. Nonetheless, the first patients admitted to the Burns Unit were soldiers wounded during the Normandy Campaign. Civilian casualties began to be admitted in August 1944.

Vera Colebrook, Colebrook's second wife, was instrumental in encouraging him to venture out of the laboratory and build a public health platform on which to campaign for the introduction of measures that would afford children greater protection from burns and scalds. She was a war widow working as a freelance broadcaster with the BBC, whom Colebrook, a widower, began courting in 1946. Whilst waiting for him to complete some work in the laboratory, Colebrook asked her to chat to some of the patients on the ward in order to boost their morale. Expecting to see war casualties, she was shocked to find that the majority were children. Something had to be done to prevent such great suffering.[4]

Deaths at all ages from burns or scalds had fallen from around 2,500 annually at the turn of the twentieth century to around 1,200 before the Second World War, an improvement generally attributed to electric lighting replacing naked gas flames and paraffin lamps, and the secular trend towards smaller families reducing overcrowding in homes. Whether or not this was the case was impossible to prove: details of the circumstances surrounding fatal accidents were absent, incomplete or inconsistently recorded on death certificates. Much more was known about the rate and type of workplace accidents than domestic ones: factories were inspected, and accidents reported; the home environment was not monitored.

More reliable data on frequency and type of injury were required not only to identify effective methods of prevention but also to draw public and policymakers' attention to the enormity of pain, misery, deformities and disability experienced by survivors, and the burden their injuries were imposing on hospital resources and staff.

Colebrook was uniquely well placed to collect data. Most victims of drastic burn or scald were admitted to general hospitals that treated relatively few, whereas each year his specialist unit in Birmingham admitted around 400 cases, and treated many more who did not require hospitalization. However, in generalizing his data to the British population, Colebrook faced the challenge that Birmingham was unrepresentative. Much of the city's housing had been destroyed in heavy bombing and much of the accommodation left standing was overcrowded slums. Nonetheless, Colebrook insisted that Birmingham's tally of domestic burns and scalds was fairly typical of British cities.[5]

Colebrook established that the home environment is more dangerous than the workplace, and that industrial accidents tended to be less serious than domestic ones. Around half of the victims of domestic burns and scalds admitted to the Burns Unit were children under five years of age. Boys were more likely than girls to tip over themselves a saucepan or kettle filled with boiling water. Thirty per cent of Birmingham homes lacked a bath, and 20 per cent were making do with a cold tap in the kitchen. In these homes both boys and girls were sometimes scalded when they accidentally fell into a galvanized tin bath full of boiling water while their carer went to fetch cold water and soap. In many of Birmingham's houses the only source of heating was an unguarded open coal, gas or electric fire. Colebrook found that clothes catching fire were responsible for more than one-third of admissions and four out of five deaths, that girls were at greatest risk of this particular accident, suffered the worst injuries and were most likely to die, although the same fate was suffered by a significant number of frail elderly women. Sudden unconsciousness, for instance, in people who suffered a stroke, or who were inebriated, or experiencing an epileptic seizure, and who fell on to an unguarded fire, was responsible for some of the deepest and most disfiguring injuries (Medawar's Glasgow 'guinea pig' fell into this group).[6]

'Prevention is the best treatment' is the mantra of first responders and medical staff treating victims of burns and scalds. Colebrook's crowning achievement was the Heating Appliances (Fireguards) Act (1952), which required all gas and electric fires to be guarded when sold. Gillies, McIndoe and Medawar's names feature in a list of influential people who supported the proposals set out in a pamphlet containing statistics and shocking photographs such as the one below, which the Colebrooks sent to every Member of Parliament.

12

C.G. Aged 3. Nightdress touched fire shown below.
36 per cent of her body burned.
Stay in hospital: 174 days so far.

Figure 4: Cause and effect: the case for fire safety legislation

When neither Labour nor Conservative governments were willing to support Colebrook's model act, MP Denys Bullard (Conservative) agreed to sponsor it. In March 1952, it was adopted unopposed by the House of Commons.

The act, which came into effect on 1 October 1954, dealt only with new appliances (unfortunately the British Standards Institute's original design failed to prevent small hands from grabbing hold of the electric element or touching gas flames). But it was estimated that between four and five million appliances bought before 1954 were still in use. The national press took up the Colebrooks' campaign, and in 1957 the government instructed nationalized gas and electricity industries and local government to encourage people to fix guards on older appliances.[7] Their price varied from between £1 and £4. A different approach was required to encourage people to place a fireguard in front of domestic coal fires. Some local authorities offered to lend or subsidize the cost of fireguards to needy people, but take up was disappointing. In 1958 the government sponsored the Royal Society for the Prevention of Accidents' campaign 'Guard that fire'. Its message, in small print, was 'fire guards are cheaper than your child's life'.

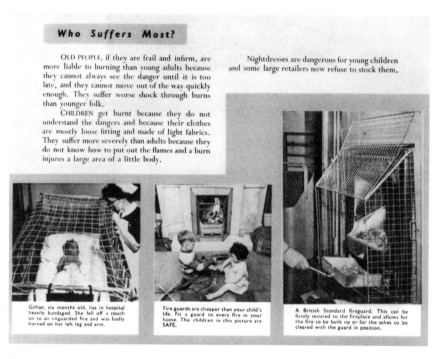

Figure 5: Preventing great suffering

The risks associated with unguarded coal fires diminished as they were replaced by alternative sources of heating such as central heating and paraffin heaters. The dangers presented by poorly designed unstable paraffin heaters were addressed in the Oil Burners Standards Act 1960 and regulations under the Consumer Protection Act 1962.

Flameproofed garments are more expensive. Colebrook suspected that lobbying by the cotton trade was behind the decision to rely on labelling when Parliament adopted the Fabrics (Misdescription) Act 1913. Had the statute required all children's nightwear to be flameproof, he estimated that each year the lives of 100 British children would have been saved, and countless others would have been spared pain and long-term disability.

In 1959, the British Standards Institute issued BSI 3121, which described how to make children's clothing inflammable. Adoption was voluntary, but was pioneered by the manufacturer of the Ladybird range of children's clothing, a popular brand, and major chain store retailers such as Woolworths, British Home Stores, Littlewoods, Marks & Spencer and the Lewis's group. But others dragged their feet. In 1960, the Consumer Advisory Council of the British Standards Institute tested ten nightdresses by holding them against an open fire, and found that only one did not go up in flames in under twenty seconds – the cheaper the product, the more quickly it caught fire. The *Daily Mail* newspaper seized the gauntlet, and in October 1964 the sale of children's nightdresses made out of flammable material was banned under the Children's Nightdresses Regulation 1964.

Despite these various measures, the annual death toll exacted by burns or scalds continued to hover at just under 1,000. And because chances of survival were gradually improving, it looked as if the number of accidents might actually be increasing. The landscape of risk undergoes regular upheaval, necessitating regular redrawing of epidemiological maps and reconsideration of policy on prevention. For instance, fast food outlets using hot fat have overtaken heavy engineering plants using molten metals in the most dangerous workplace league. And despite improvements in housing and living standards, the home remains a dangerous place. In particular, flammable liquids and self-inflicted burns (the latter reflecting, perhaps, a change in method of suicide following the replacement of coal gas by non-poisonous gas in the late 1960s) were responsible for many more admissions to the Birmingham Burns Unit.[8]

A National Plan for Burn Care

The campaign to reduce the incidence of burns or scalds did not divert Colebrook from his ambition of defeating the scourge of infection that exacerbated burn victims' pain, fever, wasting and misery consequent on the stench of sepsis, which frequently killed them. The challenge had become more urgent following the discovery in 1940 by D.A.K. Black, working at the Nuffield Institute, Oxford, that between three and four litres of half-strength plasma, which had just become available, were effective in preventing death from burn shock.[9] More survivors focused attention on treatment at later stages, when infection is the principal threat.

Some of the routes taken by pathogens responsible for wound infection had been mapped. In the 1930s, bacteriologist Robert Cruickshank (1899–1974), who in 1955, following the death of Alexander Fleming, was appointed principal of the Wright-Fleming Institute of Microbiology, St Mary's Hospital, London, had discovered an exceptional burden of bacteria in the air of the Glasgow Burns Unit. Colebrook identified their source in the Birmingham Burns Unit: pus-filled dressings changed on an open ward – in other words, patients posed a danger to one another.[10] Improving air hygiene in the Birmingham Accident Hospital was a challenge; it had been established in 1941 in the Victorian building vacated by the Queen's Hospital, which had been transferred to new buildings on the Birmingham University campus. Isolating patients from one another was impossible: the wards were of Nightingale's design, without cubicles or curtains; the hospital was overcrowded, with patients' beds overflowing into corridors. Nonetheless, Colebrook reduced the risk of airborne infection by establishing a 'dressing station', an air-conditioned room separated from the ward by an airlock, where wound dressings were changed using the no-touch technique. Death from sepsis fell from ten in every hundred patients to just over four.[11]

In *A New Approach to the Treatment of Burns and Scalds* (1950), Colebrook set out an ambitious national plan of burns centres.[12] Ideally, every large town and city should have a burns unit, preferably a purpose-built bungalow where patients are isolated from one another, and wounds dressed in a space entered through an airlock. Colebrook reckoned if the 25,000 or so victims admitted each year to a hospital in England and Wales received treatment in a unit modelled on his prototype, fewer would die, wound infection would be contained, skin grafts would be more likely to take, and the average length of stay of around 40 days (a few patients stayed more than a year) might be reduced by up to ten days, which, according to his back-of-the-envelope calculation would in turn save British taxpayers around £1.3 million each year.[13]

Burn treatment had fallen off central government's agenda. The British military was strapped for cash, and British servicemen wounded in the Korean War were treated in American, Commonwealth or United Nations mobile army surgical hospitals (MASH), and the seriously injured were admitted to American military hospitals in Japan and National Health Service (NHS) hospitals in the United Kingdom.[14] The home front was also neglected: a consensus had emerged that in the event of an attack involving atomic weapons, the devastation would be so stupendous that any attempt to plan medical relief of civilian casualties was futile.[15]

The few NHS facilities dedicated to burn care were makeshift adaptations of wards or Nissan huts in hospitals where people wounded in the Second World War had been treated. Unfortunately, caring for patients in close proximity to each other greatly increases the likelihood of cross infection, which occasionally reached catastrophic proportions and resulted in wards being closed. Elsewhere, burn victims were admitted either to the septic wing or 'dirty' ward of a general hospital, and put in the charge of inexperienced junior doctors. Nurses were responsible for changing stinking dressings full of pus. More importantly, though, patients – often children who were scared to death – experienced terrible and prolonged suffering.

In June 1952, the Ministry of Health circulated *Treatment of Burns* to the Regional Hospital Boards, central government's agents exercising control over hospitals within specific geographical areas; fourteen in England and Wales (with London and the Homes Counties split into four); Wales constituting one; five in Scotland. The circular drew attention to Colebrook's national plan, and advised each region to establish one regional burns unit, preferably close to where burns are likely to occur, such as places of heavy industry.[16] Central government might issue advice on planning, but the general emphasis was on devolution and autonomy. Each board had inherited different financial, human and material resources. As a result, a postcode lottery determined quality of burn care.

CHAPTER 9

Extending Shelf Life After Death

Fifteen in every thousand Second World War casualties had suffered a drastic burn or scald.[1] But unfortunately, despite hot war ending, the prospect of far greater numbers loomed. Relations between erstwhile allies the United States of America, the United Kingdom and the Soviet Union were deteriorating, and in 1947 the American government's optimism that its monopoly of thermo-nuclear weapons would afford its own citizens and those of its allies immunity against the threat of attack was dashed by information about the Soviets' heavy investment in their development. Shortly after the Soviets' first test explosion in the summer of 1949 the United States government was presented with a forecast that by 1956 – the so-called 'year of maximum danger' – the Kremlin's arsenal would include atomic bombs and bombers capable of reaching it from bases in northern Russia. Thermo-nuclear weapons concentrate the destructive power of heat. It was estimated that two out of every three survivors of a nuclear attack would have suffered a thermal injury, although not all of these would be serious and warrant medical treatment. This forecast was based on surveys of Hiroshima, where around 70,000 people had been killed and 70,000 injured. These statistics probably underestimated the number of casualties – counting the dead had proved far easier than counting survivors – and exclude burn victims who had sought pain relief in Hiroshima's numerous rivers and whose corpses were washed out to sea, as well as people who died sometime after the explosion from the havoc wreaked on their body by gamma radiation.[2] Hospitals had been flattened and medical staff killed or injured; it was evident that medical care would be unavailable for days, or even weeks, and many injured survivors would die. Chaos would ensue. The Joint Panel on Medical Aspects of Atomic Warfare, a collaboration of the United States Army and Navy, advised anticipating how armed forces might function in conditions of thermo-nuclear war. Investigations into the effects of exposure to radiation were instituted, which are now notorious for including thirty-one secret medical experiments conducted without the knowledge or agreement of 700 civilian 'guinea pigs', some of whom were prisoners.[3]

The United States Navy's preparations included investigating how to treat injuries inflicted by conventional weapons in hot wars fought in countries where communist China and the Soviet Union were extending their influence. In 1949, Rear Admiral C. A. Swanson, navy surgeon general, announced that a human bone bank had been opened in the Naval Hospital, the flagship hospital of the United States Navy, in Bethesda, just outside Washington DC and facing the campus of the federal government's National Institutes of Health. This was not the first occasion on which bone had been conscripted into military medicine. During the First World War, it had been repurposed in innovative procedures intended to save shattered limbs from amputation and reconstruct jawbones that had been destroyed by shrapnel. In peacetime, it was repurposed as replacement for a section of bone that had been removed because it held a life-threatening malignancy or had been damaged beyond repair, or to stabilize a spine twisted by scoliosis. Bone returned to military service during the Second World War.

Initially the repurposed bone had been that of the patient, who underwent separate incisions on different sites of their body: one where suitable bone is removed and another where the bone is repurposed. Operations were lengthy, and patients experienced twice as much pain and an extended recovery. Moreover, some people, particularly children, have little bone to spare.

In 1945, surgeon William H. Von Lackam, at the New York Orthopedic Hospital, had come up with the idea of a 'bone bank', which was actually a domestic refrigerator stocked with surgical waste. However, the shelf life of a refrigerator's contents is inconveniently brief. Domestic markets for freezers and frozen foods had just been launched. As meat bones remain edible following freezing and thawing, Lackam's New York colleagues explored whether the technology could be exploited in medicine. Experiments on laboratory animals found grafts of frozen bone both safe and effective.[4]

These bone banks were 'stashes', industry insiders' shorthand for a stockpile of surgical waste a surgeon accumulates for repurposing in his practice. Stashes proliferated, and became commonplace in the backstage of hospitals, where the only indication that body/cadaver stuff was being stored in a particular domestic refrigerator or freezer standing in a corridor, office or staff room was a notice stuck on the door warning, 'No food to be kept in here'. This was not simply to prevent contamination by whatever staff had brought in for lunch, but whenever the door of refrigerator or freezer is opened the temperature rises and the shelf life of its contents is shortened.

The Naval Hospital's bone bank was not a stash but a research facility created by orthopaedic surgeon and serving naval officer George W. Hyatt (1920–1993) for investigations into how bone might be repurposed to meet the demands of navy medicine, which is delivered in onshore hospitals in the United States and foreign ports, in combat zones and on board hospital ships.

Hyatt's bone bank consisted of a freezer with a capacity of around four cubic feet. He originally intended stocking it with surgical waste collected in the Naval Hospital's operating theatres, mostly ribs removed during thoracotomies, and long bones of amputated limbs. A progress report submitted to Rear Admiral Swanson includes a photograph of a surgeon standing next to an operating table, using the 'no-touch' technique to transfer bone that he had just removed from a patient to a bone bank technician. The patient on the operating table is swathed in drapes, and everyone else is wearing gown, gloves and mask.[5]

When it became clear that few Naval Hospital patients had bone removed during surgery, Hyatt decided to follow his orthopaedic colleagues who had ventured into the mortuary. However, whereas their ambition was more easily to replenish their stash, Hyatt's grew at the mortuary door. As he explained: 'It was at this point that the concept of a "tissue bank" was developed, for in the course of getting the bone it was obvious that one must go through the skin, fascia, and other structures useful as allografts.'[6]

Hyatt's proposal to extract and experiment on a variety of homostatic cadaver stuff was supported financially by the Navy's Bureau of Medicine and approved by its legal counsel. In May 1951, the Navy Tissue Bank moved into a suite of rooms in the basement of a wing of Building One (the main building), which included an operating theatre for its exclusive use, scrub room, a workroom for preparing sterile supplies and equipment, and a processing room holding the deep freeze and freeze drier.[7] Why Hyatt pinned the slogan *Ex morte vita* (from death comes life) on the operating theatre door is unclear. Perhaps Mary Shelley's *Frankenstein* was his inspiration. But a mission statement in Latin printed in gothic typeface is more suggestive of medieval necromancy than an ambition to repair injuries sustained defending the free world.

Figure 6: Navy Tissue Bank, 1958: disassembly and inventory

Techniques of Preservation

In 1912, Alexis Carrel (1873–1944), at the Rockefeller Institute of Medicine, New York, grafted dog skin that had been suspended in liquid paraffin and held in an icebox, and gained the reputation as the first scientist to succeed in extending the shelf life of skin. However, no modification of his method proved capable of extending shelf life of skin beyond a month or so. Optimism was renewed shortly after the Second World War when British scientists Chris Polge, Audrey Smith and Alan Parkes, working in the Medical Research Council's London laboratory, serendipitously found that glycerol protects soft stuff from damage during freezing and defrosting.[8] Their research material was fowl semen, and almost immediately livestock and dairy industries exploited their finding to create banks of frozen bull semen for artificial insemination of cows, transforming practices and productivity of dairy and livestock industries. Cryopreservation of body/corpse stuff became a 'hot' research topic, with findings shared at international symposia, notably one organized by the CIBA Foundation, and held in London in March 1953, which was attended by Medawar and his team, and Hyatt and other members of Navy Tissue Bank staff.[9]

Freezing extends shelf life indefinitely but it was highly unlikely that a freezer would be available in a combat zone. In January 1950, Hyatt submitted a request to the Naval Medical Research Institute for financial support for a programme of tests aimed at establishing whether freeze-dried cadaver stuff had the capacity to withstand combat-zone conditions.[10]

In freeze-drying, sometimes called lyophilization, deep-frozen stuff is placed in a vacuum chamber where the fluid is sublimated; in other words, it is dried. In the early 1930s, biochemist Earl Flosdorf, then attached to the University of Pennsylvania, found its first medical application when he succeeded in freeze-drying human blood plasma and sera. But it remained a laboratory curiosity until a method was found for freeze-drying on a larger scale. The food industry quickly recognized its commercial potential, and in 1938 instant coffee was launched on the market. During the Second World War, the American Red Cross dispatched bottles of freeze-dried American blood plasma to Britain.[11] After the war, Flosdorf adapted equipment to freeze-dry penicillin. The idea of experimenting with freeze-drying had come to Hyatt while working on an alarm system to warn against accidental defrosting of the freezer's contents.[12] He began collaborating with Flosdorf, now director of research and development at J. Stokes Machine Company, a Philadelphia manufacturer of machinery for the mass production of pharmaceuticals, which had added freeze-driers to its product range.

Freeze-dried bone recovered from and grafted into dogs worked just as well as fresh and frozen bone. Subsequently, fourteen patients at the Naval Hospital underwent successful grafts of freeze-dried cadaver bone.

Establishing a Collecting Site

It was 3 o'clock one morning last week when a car carrying four bluejackets [enlisted sailors] plunged over an embankment and hit a tree in Arlington, Va. Two of the sailors were scarcely hurt, but two died with broken necks. At 7 a.m. word of the deaths was passed to the 'decedent affairs desk' at the US Naval Hospital at Bethesda, Md., which in turn called Commander George W. Hyatt, director of the hospital's tissue bank. Dr Hyatt, an orthopaedic surgeon, seized the chance to turn a loss of life into a lifesaving procedure. He arranged for the bodies to be moved 20 miles to the hospital's morgue, then turned to 'the toughest part of my job': telephoning the two families to notify them of the deaths. Dr Hyatt waited an hour or so for the first shock to wear off, then called back: Would the families consent to having parts of the sailors' bodies taken for the hospital's tissue bank? Both agreed.[13]

This is how in May 1956 the Navy Tissue Bank was introduced to readers of *Time* magazine. A two-column article, complementing the issue's cover feature on the Navy's role in the atomic age, reveals that Hyatt had solved the knotty perennial problem of how to ensure speedy notification of the whereabouts of fresh corpses by embedding the Navy Tissue Bank in the routines of the hospital's decedent affairs office, which is responsible for notifying next of kin of a patient's death, seeking agreement for post-mortems, and arranging disposal of unclaimed corpses.

Kinfolk were seldom present at the deathbed. The Naval Hospital's patients were drawn from the United States Navy's enormous contingent of personnel and their dependants from every corner of the United States. A member of the bank's team equipped with a prepared script would seek agreement by telephone.[14] At the time, 'telephonic consent' was novel; not every American home had a telephone. Indeed, a former Tissue Bank technician admitted that with hindsight its methods of obtaining agreement were 'loosey goosey'.[15]

The two young men referred to in *Time* magazine were typical of those whose corpses were disassembled in the Tissue Bank's operating theatre in its first two decades: a majority had died following an accident or suicide; most were relatively young (the average age was forty-seven); very few were over seventy-five; men and women were almost equal in number.[16]

Stocking the Bank

'Full-tissue recovery' is labour-intensive. It took three physicians, a nurse and six technicians – some experienced, some trainee – between eight and twelve hours to 'tear a corpse apart'.[17]

Kinfolk were assured that during disassembly the proprieties that ensure dignity in death were fully observed: the corpse's face was covered, speech forbidden and hand signals used as communication. An elaborate (and expensive) aseptic protocol was observed. The corpse was scrubbed clean, the skin of its trunk and lower extremities shaved, and it was covered in sterile drapes and linen. The team wore gown, hat, mask, gloves and muslin

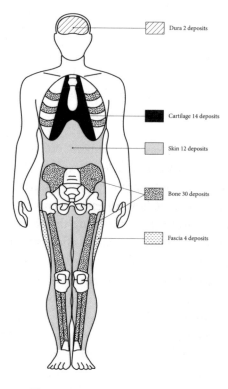

Figure 7: How to disassemble a corpse, 1963

boots, which were changed whenever a new site was tackled. Altogether, five hampers of linen, seventy-five pairs of gloves and glove wrappers, fifty gowns, 400 towels, 200 sheets and a host of miscellaneous materials, including 500 wrappers, were used each time.[18]

When nothing eligible for repurposing remained on the corpse, it was reassembled: wooden dowels – like broom poles – replaced long bones; empty eye sockets were filled (cadaver eyes were delivered to the Lions Eye Bank of Washington DC); plastic tubes were attached to what remained of arteries to facilitate embalming.[19]

Cottage industry

Hyatt referred to the Bank as a 'cottage industry' to emphasize how every task required hands-on skills.[20] His staff attended 'bone bank school' where they followed a six-month syllabus covering anatomy and physiology and the bank's methods of extracting, processing, packaging and storing cadaver stuff.[21] For instance, they were taught how to strip cadaver bone of extraneous matter, and turn it on a lathe, or split it into 'matchsticks', or grind it into a powder, or consolidate it into a 'bone burger', and how to cut cadaver skin into convenient sizes before wrapping it in Cellophane.

Shelf life was extended by refrigeration, freezing or freeze-drying. It took fourteen days for all the frozen fluid in bone to sublimate; softer cadaver stuff required far less time. Stored in vacuum-sealed glass bottles at room temperature, freeze-dried cadaver stuff, as one observer commented, might easily be mistaken for specimens in a medical museum.[22] Each item's biography was recorded on a register that had been created with the help of Remington Rand. A quick glance was sufficient to ascertain the source's identity, results of various safety tests, when it had been made, where and how it had been repurposed, and if it had 'worked'.

A neologism for these items used in what follows is 'cadaver artefact', a clumsy name that combines a reminder of the source, and emphasizes that each item differed because its final form depended on the source's shape

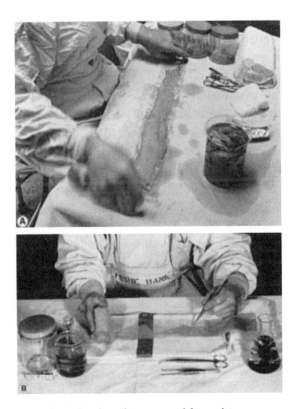

Figure 8: Preparing cadaver skin for refrigerator and freeze drier

and size and technician's technique, in much the same way as meat loafs vary according to type and quality of meat, recipe and the cook's skill.

Testing Cadaver Artefacts in Combat Zones

The first opportunity for establishing whether freeze-dried cadaver artefacts 'work' arrived on 25 June 1950, when Kim Il Sung, communist leader of North Korea, assisted by the Soviet Union and Maoist China, invaded South Korea. President Harry Truman responded by offering all possible support to South Korea. The United Nations Security Council approved Truman's intervention, and combat and medical support units from nineteen countries began fighting alongside the South Korean army.

The casualty rate was high despite the military superiority of the United States and its allies. At the end of two years of heavy fighting, around 54,000 Americans had died and 103,000 had been injured – the latter number excludes the many thousands of men who were slightly wounded, treated, and returned to their units without their injury being recorded.[23] Front lines, with bunkers, barbed wire and sandbagged trenches, were relatively static, resembling those of First World War battlefields. Indeed, many of the injuries were similar save for burns, which increased in number following innovations in conventional weaponry that exploited heat: one in every fifty casualties admitted to hospital had suffered a serious burn, a significantly higher rate than that in the Second World War.[24]

Although the Navy Tissue Bank's stock of freeze-dried cadaver artefacts was small, they proved their worth. Indeed, demand for freeze-dried veins and arteries that might prevent amputation of a limb shattered by a high-velocity bullet was particularly high, and the bank was forced to raid civilian surgeons' stashes.

In July 1953, following newly elected President Eisenhower's threat to deploy atomic weapons, North Korea and its allies agreed a ceasefire. By 1957, the bank's shelves held a large stock including 180,000 square centimetres of skin, various bone artefacts, dura mater, cartilage and arteries.[25] Peace offered few opportunities for testing cadaver artefacts. Civilian surgeons were recruited as collaborators if they accepted liability where a procedure resulted in injury, and agreed to provide extensive data on outcomes over a specified period, usually four years. But meeting the second requirement was difficult. For whereas the Navy Tissue Bank easily kept track of military recipients of its artefacts through their pension records, civilian surgeons were constrained by patient mobility and their hospital's limited financial and administrative resources. The Vietnam War obviated the need for civilian collaborators. Its dates are usually given as 1959 to 1975, although no one day can be singled out as its start, and the United States began regular bombing of the Democratic Republic of Vietnam (North Vietnam) in 1964, around the same time as increasing numbers of American troops began entering the country.

Fighting in the Vietnam War was a matter of days of dangerous patrolling or of manning isolated outposts under constant harassment, punctuated by occasional clashes on a larger scale. The casualty rate was high: around 47,000 Americans were killed in action, 75,000 were permanently disabled, 80,000 seriously wounded and 150,000 lightly wounded (although these data pale into insignificance when placed next to the one million Vietnamese combatant and the two million civilian deaths).[26] The American death rate was lower than in previous wars because helicopter ambulance evacuation to hospital ships, *Repose* and *Sanctuary*, and the Navy Station Hospital in Da Nang, South Vietnam, allowed severely wounded men who previously would probably have died to survive.[27] However, improved rates of immediate survival meant that many more bodies required treatment and repair.

The bank's stock was rapidly exhausted. In 1969, a second collecting site and processing facility was opened at the Naval Regional Medical Center, San Diego, California (closed in 1999). Productivity increased: the cumulative total of corpses disassembled during its first seventeen years of operation to 1967 stood at 809;[28] by 1977 it had reached 1,500.[29]

When hostilities in the Vietnam War ceased President Richard Nixon, aided by Secretary of State Henry Kissinger, presided over a dramatic contraction in American military power. Congress agreed drastic cuts in the defence budget; the Draft ended, and the military became an all-volunteer force. The Navy Tissue Bank was downsized, and its staff turned their attention away from research on war wounds, and began focusing on transplantation of homovital stuff, specifically bone marrow and kidneys.

CHAPTER 10

Cadaver Eyes, Death Denial and the National Health Service

'Never in the field of human conflict has so much been owed by so many to so few', is how Winston Churchill famously praised the airmen who in 1940 defeated the German Luftwaffe in the Battle of Britain. Sometimes called 'Britain's finest', the majority of the young men had been drawn from university flying clubs and similar elite associations. Their Spitfire and Hurricane planes were undoubtedly exceptional fighting machines but when hit by the enemy, which often happened, their design made it difficult for crew to escape from the cockpit. For some the price of survival was 'airman's burn', a pattern of damage caused by fuel tanks situated in the front of the plane catching fire and effectively acting like a blowtorch on exposed face and hands.

Archibald McIndoe orchestrated the airmen's physical and social reconstruction.[1] Consultant plastic surgeon to the Royal Air Force when hostilities began, the War Office put him to work in the Queen Victoria Hospital, in East Grinstead, a town in the southeast corner of England.[2] The small cottage hospital had been allocated a role as a centre for treating burned combatants.

McIndoe arranged the transfer of 649 men who had sustained airman's burn to the Queen Victoria Hospital, where over several years they underwent numerous surgical procedures to rebuild their face and hands. Some had not worn protective goggles on the grounds that they interfered with their ability to spot enemy aircraft, and their eyelids had been burned. McIndoe prioritized their repair, but the procedure sometimes failed to work in time to prevent exposed corneas drying out, ulcerating, scarring and corneal blindness ensuing. In 1947, the ten-bedded Corneo-Plastic Unit was opened at the hospital, and Benjamin Rycroft (1902–1967), who before the war had gained some experience of keratoplasty, was appointed ophthalmic surgeon. His caseload was heavy: in addition to blinded airmen it included combatants whose corneas had been damaged when their tank caught fire or in a mine explosion.

The Welsh pioneer of keratoplasty was James William Tudor Thomas (1893–1976), who in 1935 published an account of his experience of fifty-six corneal grafts in order to convince his colleagues that the procedure was not a hopeless gamble.[3] His 'guinea pigs' were typical civilian victims of corneal blindness; their corneas had been damaged either by infection (Tudor Thomas's cohort included several people whose corneas had been scarred by neonatal conjunctivitis) or a work-related accident – a splash of caustic soda had ruined the cornea of L. H's left eye, and F.J.F.W's right eye had been damaged by a limewash burn.

The corneas repurposed by Tudor Thomas had either been found in surgical waste or volunteered by irremediably blind donors-on-the-hoof. Sources' identities were withheld from recipients. Secrecy facilitates the transformation of the 'windows of a person's soul' into anonymous replacement parts; it also frustrates recipients' capacity personally to compensate sources in the currency of regard. One of Tudor Thomas's patients complained to readers of the *Daily Mail* of her inability to express her gratitude personally: 'It is wonderful, and yet I do not know to this day the names of the man and woman whose sacrifice has lifted me out of my prison.'[4]

Rycroft repurposed corneas excised from eyes enucleated from corpses resting in the Queen Victoria Hospital's mortuary. At the time, the corneas of elderly people were considered best for grafting purposes because they 'take' more easily than younger ones. But death was an infrequent visitor to the Queen Victoria Hospital – a majority of local people died in their own beds or in a private nursing home. Rycroft reckoned that without access to busier mortuaries it would take around three years to clear his waiting list.[5]

McIndoe championed his patients' cause both in and out of the hospital.[6] His various schemes for keeping their spirits up during lengthy and arduous treatment included establishing a facility for manufacturing aircraft instruments set up in the hospital grounds that offered paid work; cajoling local residents to welcome them into their homes, often with faces covered in tramlines of stitches or swathed in bandages; and persuading local police to ignore drunken behaviour, an important concession because the young men had formed the Guinea Pig Club, a drinking fellowship which continues to meet, albeit in gatherings of a more sober nature.

In June 1949, McIndoe announced the opening of what he called an eye bank, but which in practice was a scheme for encouraging pledges aimed primarily at people living in the vicinity of the Queen Victoria Hospital. A pledge form was issued on which was printed the telephone number of the Corneo-Plastic Unit, and an instruction to next of kin to get in touch as soon as possible after the pledger's death had been confirmed. An accompanying letter advised the pledger to keep the pledge form separate from their will, which typically is read after the corpse has been buried or cremated and certainly too late for the eyes to be enucleated for repurposing.

Death Denial in the NHS

British ophthalmologists had been unimpressed by the Eye-Bank for Sight Restoration when it opened in New York in 1944. One sceptic had opined, 'There is no such thing as an "eye bank" because leaving eyes to the hospital in a will was hardly practicable.'[7] A lawyer enquiring how his client might bequeath her eyes was advised by a Faculty of Ophthalmologists' spokesperson to tell her 'to see that someone is at hand immediately she dies to remove her eyes and put them in the fridge. If they were not wanted within a reasonable time it would be just too bad.'[8]

Five years later, the eminent ophthalmologist Stewart Duke-Elder (1898–1978) marvelled at how the Eye-Bank's supply chain delivered enucleated cadaver eyes to an operating theatre before, as he put it, the soul of their source had reached heaven.[9] But his colleagues remained unenthusiastic about eye banking. Whereas Rycroft's ambition was to restore the sight of a finite number of young heroic war-wounded men, their patients included people of every age and walk of life, complaining of a variety of eye problems, from relatively trivial to potentially fatal. Moreover, their caseloads had begun growing inexorably on 5 July 1948, the Appointed Day, when the NHS began offering health care free at the point of delivery to every British subject. Indeed, the waiting list for Moorfields Hospital, London, Britain's leading eye hospital, was so long that its medical staff feared easier availability of cadaver eyes would only serve to lengthen it.[10]

McIndoe hated the very idea of the NHS. According to Elaine Blond (1902–1985), he envisaged it run by 'pen pushers' demanding medical staff fill out forms at the expense of patient care.[11] She was his close friend and chief benefactor, a British big-money philanthropist, whose wealth had been inherited from her father Michael Marks, founder of the Marks & Spencer retail empire, and whose second husband Neville Blond's family business manufactured underwear for its stores.

McIndoe's antipathy was partly borne out of experience. In 1943, he had asked the Peanut Club to support a new children's burns ward at the Queen

Victoria Hospital. The *Kent and Sussex Courier* had created the club in 1931 as a fund-raising vehicle for local hospitals; it was primarily for children – its motto was 'Happy smiles and helpful deeds' – but adults and pet animals were allowed to join. By the 1940s it boasted over 600,000 members, each of whom had paid 12 pennies (5 new pence) for an enamel peanut badge. The club's members raised £24,000 towards the children's ward by staging events such as an old-time music ball at the Tunbridge Wells Assembly Hall where McIndoe was guest of honour.

The Peanut Club was one of many civil society organizations supporting British hospitals in cash and kind that closed in anticipation of nationalization.[12] The South East Metropolitan Regional Hospital Board, which now had oversight of the Queen Victoria Hospital, refused to allow the children's ward to be built and, adding insult to injury, claimed the cash the Peanut Club had raised towards the children's ward was now Ministry of Health property, and demanded the trustees hand it over. Mrs Gordon Clemetson, a journalist on the *Kent and Sussex Courier*, better known as Aunt Agatha, the Peanut Club's public face, was outraged; she declared she would rather go to prison than submit to this piece of 'red tapism'.[13] She escaped having to make this sacrifice when a Peanut Club member appealed directly to Labour Prime Minister Clement Attlee, who arranged for the children's ward to be built. Queen Elizabeth the Queen Mother opened it in 1955.[14]

Ministry of Health staff fuelled McIndoe's antipathy by responding with dismay to the publicity surrounding his eye bank. They disapproved of New York's Madison Avenue style of marketing of corpse philanthropy. Moreover, allusions to death, however indirect, were an embarrassment to bureaucrats responsible for putting practical flesh on the NHS's legislative skeleton, and concerned to remove fear of pain – even death – that was holding back some people, particularly the elderly, from seeking medical care. One of their number explained his objections as follows:

> It is bad enough that a person who has gone into hospital and thinks he has only to have an x-ray should be met at the door and asked to sign a form giving his blessing to the surgeon's doing their worst. Do not leave him to think he will come out blind as well as cut up. And even more, do not let him think that his condition is so bad that all that remains is to dispose of odd bits and pieces.[15]

In 1955, British social anthropologist Geoffrey Gorer became the first scholar to observe how during the first half of the twentieth century, throughout industrialized, urbanized and technologically sophisticated parts of the world, death was increasingly ignored and denied. French historian Philippe Ariès took up the death-denial theme in *The Hour of Our Death* (1977), a monumental monograph, where he observes that by the twentieth century, 'Everything in town goes on as if nobody died anymore.'[16] Death, which hitherto had been viewed as inevitable and normal, accepted with equanimity and assimilated by society, had been medicalized, by which Ariès meant it had been sequestered out of private homes and surrounding communities into hospitals, where it was treated as a professional and technical responsibility. Ariès implicated families complicit for reasons of social and economic convenience. There is some merit in his accusation. But he might have observed that sequestration gathered pace during the 1950s when medicine began to offer heroic medical innovations such as aggressive cancer chemotherapies and daring cardiac surgery, procedures performed on people in poor health with relatively high failure rates.

Death is unwelcome wherever it stands for medical impotence, which is how it is seen in modern acute hospitals. But this identity had a specific political salience within the young NHS, which was dedicated to improving the health of the British population by providing access to people hitherto denied treatment for reasons of poverty. This laudable ambition was rapidly achieved: in 1949, the first full year of operation, 2,788,000 English and Welsh people completed a course of hospital treatment; in 1958 that number had increased to 3,783,000, and in 1969 it had reached 4,968,000.[17] However – and perhaps deliberately – published data fail to distinguish between people discharged through the hospital's front door and those whose departure was organized by funeral directors through the mortuary's back door.

Affording Eye Pledges a Legal Footing

Shortly after McIndoe opened his eye bank he was disabused of his assumption that the Anatomy Act 1832 applied. The statute had been adopted by Parliament to provide medical students with a reliable supply of cadavers for dissection, and at the same time to quell public outrage about grave robbers, sometimes called resurrectionists because of their rude intrusion into the afterlife, and murderers such as the infamous serial killers Burke and Hare who between 1827 and 1828 sold the corpses of seventeen victims to Dr Robert Knox, a private anatomy tutor in Edinburgh. The illicit traffic in corpses had begun in the late eighteenth century when the gallows, the only legitimate source of cadavers for dissection, proved incapable of satisfying the demand that was growing rapidly following expansion in numbers of students attending medical schools where human anatomy and dissection were central to the curriculum.

The Anatomy Act introduced corpse philanthropy into the United Kingdom. One section allowed people during their life to pledge their corpse, either in writing or verbally, for dissection in a medical school. This provision pandered to men such as Jeremy Bentham (1748–1832), the English utilitarian philosopher who demonstrated his faith in anatomy's capacity to contribute to the public good by directing that his corpse be dissected during a public lecture. However, men like Bentham were exceptional in being accustomed to exercising political and social authority – in 1831, a mere 4,500 men out of a population of more than 2.6 million people, were entitled to vote – and by being likely to have drawn up a testamentary document setting out how their property should be distributed following their death. Death and disposal of corpses, and distribution of property, held other meanings for the overwhelming majority of people who actually had little, if any, choice in the political and social arrangements into which they were born, and in which their lives were lived, and whose material possessions were few. Moreover, dissection was considered shameful because it had been integral to the punishment of capital crime, which included execution and denial of what was popularly understood as a decent burial.[18]

McIndoe instituted 'Sight for the Blind', a campaign to afford eye pledges a legal footing. It was supported by the community in and around East Grinstead, an area stitched into the 'stockbroker belt', so-called because its easy commute into London's financial centre made it popular with people working in London's financial industry. Politically this was solid Conservative Party territory. Its residents in general opposed the ethos of the post-war settlement, which, among other things, was intent on replacing big-money and mass and discretionary philanthropy, which had been responsible for much of the landscape of welfare in the United Kingdom, with services paid for out of general taxation and planned and delivered by central and local government.

Mrs Clemetson was one of McIndoe's staunchest allies. Bolstered by the victory over the Peanut Club children's ward, in December 1951, two months after a general election had returned the Conservative Party to government, the *Kent and Sussex Courier* opened a register in which people could record their willingness to pledge their eyes to the Queen Victoria Hospital.[19] McIndoe acknowledged their 'offer', a subtle indication that their pledge might be legally and practically unrealizable.

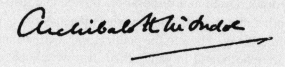

Figure 9: Acknowledging a pledge

The public meeting campaigners held at East Grinstead Town Hall was reported in the national press. The BBC televised a documentary about keratoplasty. Two MPs, Horace King (Labour) and Gerald Williams (Conservative), took up the gauntlet, and in May 1952, in the presence of Prime Minister Winston Churchill, and with McIndoe and Rycroft looking down from the Stranger's Gallery, the Corneal Grafting Bill was placed before the House of Commons under the new ten-minute rule, which allowed back-bench MPs to introduce emergency legislation and speak on it for ten minutes, when day-to-day business had been completed. The bill met no opposition and the Corneal Grafting Act was placed on the statute book in a matter of weeks.

CHAPTER 11

Whose Corpse Is It?

One section of the Corneal Grafting Act 1952 allowed people during their lifetime to pledge their eyes either in writing at any time or orally during their last illness in the presence of two or more witnesses. An editorial in the *Kent and Sussex Courier* crowed, 'Any fears that people would recoil from the idea of having their eyes thus used were obviously groundless as every week brought fresh evidence of the depths of public sympathy and the keen desire to help.'[1]

McIndoe and Rycroft had relied on the British public extending their enthusiastic wartime support of blood banking to pledge their eyes to young men whose sight had been compromised defending the nation. While in every war bodies are damaged or destroyed, in modern warfare it is deemed insufficient for young men to kill and be killed, to be injured or to injure; survivors are required actively and visibly to participate in social, moral and physical reconstruction. This point is forcefully made by cultural critic Elaine Scarry in *The Body in Pain*, a meditation on the vulnerability of the human body.[2] She might have observed that, in war war, civilians enthusiastically, figuratively and literally are obliged to give something of themselves.

By the time the Corneal Grafting Act was placed on the statute book, a majority of candidates for keratoplasty was civilian. New feeling rules were required. Sacralized marketing messages work less effectively on the heartstrings of the British public than on American ones, but sympathy for children's plight is universal. 'Some blind boy or girl may see again – through the eyes of Sir Winston Churchill,' the *Daily Sketch,* a tabloid newspaper, told its readers in May 1956. However, it was Lady Clementine Churchill (1885–1977), not her husband, who had been moved by the 'Eye Will Crusade' led by Francis Stanislaw, a theatre producer.[3] Stanislaw had been inspired by *The Gift*, a play written by Mary Lumsden, a personal friend of Rycroft, which was set in the Harley Street consulting room of the fictitious ophthalmologist Sir David Crossley, and which explored Lady Elizabeth Crossley's decision to pledge her eyes to a blind relative. All the actors in the Savoy Repertory Players, the company of amateur thespians that had staged the play, had been moved to pledge their eyes to the Royal Eye Hospital, south London. Esmond Knight (1906–1987), star of West End stage and Hollywood actor, who had been blinded in action against the German battleship *Bismarck*, became the Crusade's president. Its roll of glamorous vice presidents included the actresses Nora Swinbourne and Heather Sears, ballet dancer Anton Dolin and composer Sir Arthur Bliss.

The Royal National Institute for the Blind (RNIB, since 2002 called the Royal National Institute of Blind People), Britain's leading association for blind and partially sighted people, had refused to support Stanislaw's Eye Will Campaign. Indeed, it considered him a dangerous nuisance, meddling in a medical matter that was the proper concern of ophthalmologists. Whereas the Lions had enthusiastically added keratoplasty to the various services they undertook as Knights of the Blind, the RNIB's work began when an ophthalmologist admitted that medicine had nothing further to offer a patient. Moreover, in conveying the impression that blindness is curable, marketing of eye pledges was damaging its objective of demonstrating that blindness or impaired sight is not a barrier to a full and useful life. Evidence was accumulating that the marketing of pledges was misleading; it was fostering hope of sight restoration where none existed and, in some instances, persuading people recently registered as blind to refuse rehabilitation and trek from one doctor to another in search of a cure. A home teacher of blind people contacted the organization with a case in point:

> We have only this week had to re-classify a blind man who was formerly a hairdresser from 'employable' to 'not capable' for the reason that while he has the potential for work he will not accept rehabilitation and training and placing because either the ophthalmic surgeon or the man thinks the ophthalmic surgeon told him he will get his sight back, and then the man thinks he will be able to carry on hairdressing.

The man had been registered as blind for more than three years.[4]

Nonetheless, the RNIB received a growing number of enquiries about pledging. Many expressed a wish to restore the sight of a child resident in one of its Sunshine Homes for Blind Babies, sometimes specifying that the preferred recipient was a little girl. The RNIB replied that it would be delighted if a child could regain sight through such generosity, and despite its reservations, enclosed a copy of *A Note on the Corneal Grafting Act*, which included a pledge form and answers to frequently asked questions (FAQs) that included: 'I am in the services. Can I bequeath my eyes?' (yes); 'I am very elderly. Will my eyes be of any use?' (certainly); 'Do you require the eyes of dogs' (no).[5] This last FAQ suggests that some people confused the RNIB with the charity Guide Dogs for the Blind.

In February 1961, Stanislaw's mother Ada became the first 'crusader' to have her pledge realized. 'I want my eyes to go to a child,' she had said. 'I want them to be of the greatest possible use.' Her corneas restored a child's sight. 'My mother wanted a chance to go on being useful beyond the grave,' Stanislaw explained to readers of the *Reader's Digest*. Its editor commented approvingly, 'If being useful adds meaning to life, there are few better ways than to pass on the gift of sight.'[6]

Unpledged and Pledged Cadaver Eyes

Keratoplasty was considered a relatively minor surgical procedure – 'not as bad as having a tooth out,' one of Rycroft's patients remarked.[7] It was performed under local anaesthetic to minimize the risk of vomiting, a frequent consequence of general anaesthesia, which would threaten the graft's security. After the operation, patients remained in bed for days with their head immobilized by sand bags. A typical hospital stay was between three and four weeks, considerably longer than for a majority of other ophthalmological procedures, creating difficulties in securing a hospital bed.

Death is unpredictable; enucleated cadaver eyes in a moist chamber have a brief shelf life; availability of hospital bed, operating theatre and theatre staff could not be guaranteed; and potential recipients might be unwilling to undergo an operation and lengthy hospital stay at short notice. Where availability of theatre, theatre staff, hospital bed and patient coincided, a section of the Corneal Grafting Act allowed the *person lawfully in possession* of a corpse to agree to enucleation unless they had *reason to believe* that either the dead person during their life or following their death their surviving spouse or next of kin would object (emphases added). What constituted a *reason to believe* was not defined, and no explanation was provided as to how an objection should be either recorded or ascertained.

The *person lawfully in possession* of a corpse describes a legal role. Who can fill it is decided by where and under what circumstances death occurs. Where someone dies of 'natural' causes in a private home, the surviving spouse, or executor, or the householder may fill the role. Where someone dies in hospital the Management Committee or Board of Governors or such like nominates the role's occupant, whose rights and responsibilities in relation to the corpse are relinquished the moment the corpse is claimed by surviving spouse, next of kin or executor. Coroners take lawful possession of a corpse where circumstances surrounding death require investigation, for instance, where malfeasance is suspected, or where the death was unexpected, accidental, on an operating table, or in an institution including psychiatric hospital and prison, or where its cause is uncertain. No one can trump a coroner's right of possession. However, a source of confusion, which was compounded as more and more cadaver stuff was sought, was the difference between legal and physical possession, which arose where coroners were in legal but not physical possession of a corpse – some people were under the impression that physical possession constituted legal possession.

Beyond the medico-legal sphere, the main responsibility of the person in possession is safe disposal of a corpse. The cost of disposal of an unclaimed corpse is met out of the public purse. However, a section of the Anatomy Act 1832 allowed unclaimed corpses to be sent to a medical school, effectively providing workhouse wardens with a means of sparing the public purse from paying the cost of burial, albeit a cheap one in a potter's field. The Anatomy Act had been followed by the Poor Law Amendment Act 1834, sometimes referred to as the 'Whig Starvation and Infanticide Act' because of its austere and punitive dealings with anyone facing destitution or frailty, which had greatly increased intake and mortality in workhouses. Little wonder then that historian Ruth Richardson found that during the nineteenth century a majority of corpses dissected by medical students had been transferred from a workhouse.[8]

Paradoxically, the Corneal Grafting Act was much friendlier towards enucleators than its nineteenth-century predecessor had been to dissectors. For whereas anyone caught flouting the Anatomy Act faced punishment of either three months, imprisonment or a fine not exceeding £50, Parliament provided no penalty for infringing the Corneal Grafting Act: there was no punishment for enucleating cadaver eyes in the absence of pledge or next of kin's agreement. Moreover, common law failed to provide aggrieved British next of kin with a means of redress comparable to that provided by the American quasi-property ruling.

Law, policy and practice on arrangements for extracting and repurposing cadaver stuff struggle for coherence. On paper the Corneal Grafting Act was more permissive than the Anatomy Act, but the political landscape of Britain in the 1950s was wholly different to that of the 1830s. British subjects were enjoying mass democracy and a welfare state intent on creating a more socially just and more materially equal society. In a matter of days after the Corneal Grafting Act had received royal assent, the Ministry of Health circulated a memo to every NHS hospital counselling restraint: hospitals without an ophthalmologist but where death was a frequent visitor should not become collecting sites.[9]

Exercise of restraint was wholly voluntary. Contrary to McIndoe's fear of finding himself overwhelmed by the demands of petty bureaucrats, Ministry of Health staff did not direct medical practice, including that incorporating repurposed cadaver stuff. The ministry drew up the financial framework and issued some general advice about desirable outcomes, but never imposed a national plan. Instead, it left details to be worked out at the periphery. Moreover, following a machinery-of-government reshuffle in January 1951, its remit was greatly reduced when its responsibility for housing and local government was transferred to the new Ministry of Town and Country Planning, and its minister denied a seat in Cabinet, developments which, as one commentator put it, transformed it from one of the most impressive mansions in Whitehall into a semi-detached villa.[10] The Ministry of Health ceased to offer attractive career opportunities, and its staff were unlikely to be made up of high flyers. Its work was confined to prompting and encouraging the evolution of medical practice, as one of those attempting so to do put it: 'You help professional opinion to form itself spontaneously.'[11]

At the same time, the authority of the medical profession was growing. In the heated negotiations leading up to the creation of the NHS, its representatives had won important concessions that gave senior doctors considerable say both in how the service was run and on their own practice. Little wonder then that a consultant ophthalmologist at the Wolverhampton Eye Hospital reported that no one had ever interfered with his work, and that he was completely free to decide what he should or should not do.[12] What this means is that both realization of pledges and enucleation were determined by professional idiosyncrasy and happenstance.

CHAPTER 12

Collecting British Cadaver Pituitary Glands

Rumours of Raben and Li's success at making children grow taller began circulating inside and outside the United States before their reports were published in scientific journals. Frank George Young (1908–1988), the Sir William Dunn Professor of Biochemistry at Cambridge University, recognized and seized the opportunity that they had created.[1]

Hitherto, British scientists had been unable to gain hands-on experience of isolating hormones in carcass pituitary glands because the British meat industry had been incapable of meeting their demand for research material. The industry was dominated by farmers/butchers, sometimes pillars of their local community and conservative in outlook, who each week slaughtered one or two animals, sufficient to meet local demand for the Sunday roast. An economic incentive to industrialize had been absent: home-produced meat commanded a premium over the price of imports of beef from Ireland, Argentina and the United States, lamb from Australia and New Zealand, and bacon from Denmark, not because its quality was higher but because British people were patriotic.

Unfortunately, the carcasses in which the 'roast beef of olde England' was found were likely to be contaminated by foot and mouth disease, anthrax, murrain or tuberculosis. In the late nineteenth century, local government, charged with responsibility for safeguarding public health, began appointing meat inspectors who had the authority to conduct the 'poke and sniff' test to identify and exclude diseased carcasses and rotten meat.[2] But they were few in number and, in order to rationalize their work, abattoirs began to be built where for a small fee farmers, butchers and private individuals could take their animals to be killed. Modernization of livestock slaughter proceeded at a snail's pace. In 1937, central government decided to intervene, and established the Livestock Commission. War disrupted its plans but also allowed the Ministry of Food to commandeer the most up-to-date slaughterhouses in the British Isles in order to reduce the country's dependence on expensive imports of carcass glands, chiefly pancreases for insulin.[3] The scheme was initially restricted to abattoirs that had refrigeration, and which killed relatively large numbers of animals, such as the Greenock Abattoir where each year over 45,000 animals were slaughtered (although this is nothing compared to the daily throughput of a typical Chicago meatpacking disassembly line of 2,500 animals). It was extended to smaller abattoirs when civil servants hit on the idea of commandeering refrigerated ice-cream vans that were standing idle because of the milk shortage.[4]

Despite wartime efforts, Britain failed to approach self-sufficiency in carcass glands of any kind. Peace did little to improve the situation. The war had placed Britain in substantial debt to the United States; sterling was weak, the dollar strong, and the cost of American imports escalated. In 1947, a short course of American carcass growth hormone cost £60, or $240;[5] the Treasury maintained tight control of its dollar account and permission had to be obtained for imports from the United States of both raw material and finished goods, including medicines, until the late 1950s.

Setting up the British Collection of Cadaver Pituitary Glands

Young persuaded the Medical Research Council's Clinical Endocrinology Committee to support a British test of species specificity of growth hormone.[6] Glaxo, a British manufacturer of polio vaccine, agreed to extract the pituitary glands of rhesus monkeys that it was breeding and killing in order to prepare a kidney soup that the polio virus relishes. Around 600 monkeys were killed each month but their pituitary glands yield thirty-three times less growth hormone than human ones. Pathologists known to be sympathetic to research were asked to extract cadaver pituitary glands and send them by post in batches of fifty, packed in a polythene container, to the Protein Hut in Tennis Court Road, Cambridge, which was an annexe of Young's department housing a research bench and laboratory animals, and which had been erected as a temporary measure. Periodically, the University of Cambridge had considered pulling it down.

British confirmation of the species-specificity of growth hormone was found.[7] The Medical Research Council wasted no time in appointing pathologists and biochemists to a committee that would be responsible for creating and maintaining a high-volume national collection of cadaver pituitary glands, and organizing production of hGH for a clinical trial. Paediatric endocrinologist James Mourilyan Tanner (1910–2010), head of the Growth Disorder Clinic at Great Ormond Street Hospital for children in London, was appointed chair of a committee of clinicians who would supervise the clinical trial.

Tanner is renowned for drawing up charts of typical growth in the height of children. But why some children grew taller and others did not was mostly a mystery. Tanner's charts were used to identify 'presumed growth-hormone-lacking-dwarfs'. By 1977, when the clinical trial was completed, 642 children who were considerably shorter than their peers had been recruited as research subjects, just over half of whom had been monitored at one of the fourteen growth centres in British teaching hospitals, and the remainder at Great Ormond Street Hospital.[8]

Children admitted to the clinical trial would receive thrice-weekly injections of human growth hormone (hGH) over several years. Roughly speaking, one cadaver pituitary gland yields hGH sufficient for one injection, which meant that the annual ratio of child to corpses was 1:150. The scale of demand for cadaver pituitary glands was daunting, unprecedented and, hindsight confirms, has never been equalled. In 1959, a letter was circulated to pathologists throughout the country, whose responsibilities include the mortuary, drawing their attention to American reports of hGH 'working' and inviting their support.[9]

Neglected Premises

There are two types of mortuary in the United Kingdom: hospital and public. Both are closed to the public. The typical hospital mortuary is at the end of the corridor leading to laundry, heating plant and such like; the typical public mortuary is an anonymous, official-looking building. Both have a sign posted on their entrance door warning 'No unauthorized entry'.

There was a good chance that the Medical Research Council's letter was delivered to a poorly equipped mortuary in dilapidated premises. The NHS had inherited a stock of hospitals ranging from elite metropolitan teaching hospitals with an average of over 500 beds, financed by voluntary donations, to tiny rural cottage hospitals, some of which had been built by civil society organizations, and some originating as Poor Law institutions and paid for by local taxpayers.[10] Some had suffered bomb damage; others were outdated or neglected. However, in the austere economic climate of the war's aftermath there was a dearth of capital, and replacing buildings was a lower priority than buying modern equipment such as X-ray machines.

Public mortuaries originated in the Public Health Act 1875, which required local government to provide somewhere for corpses to rest between death and disposal. The measure was directed at overcrowded slums where the 'smell of death' threatened rapidly to become offensive, and where a putrefying corpse might become a health hazard. A corpse's removal was often resisted, not simply because it deprived people of the focus of a wake, but because it was suspected that the measure was subterfuge to furnish anatomy rooms of medical schools with 'teaching material'.

Wealthy municipal governments had complied with the statute by erecting new public mortuaries – by the twentieth century London boasted around 100 – and less affluent, mostly rural ones by commandeering space in a workhouse mortuary. By the 1950s, the stock of public mortuaries ranged from busy purpose-built ones to what were described as historical relics used as 'an improvised left-luggage office for a passing corpse' half a dozen times a year at the most.[11] Those falling into the latter category lacked refrigeration, scales and even hot water. Modernization was a low priority for local governments struggling to fulfil other responsibilities, such as building much-needed decent houses.

Culture of Entitlement

The Corneal Grafting Act was on the statute book, and the British public was being encouraged to pledge their eyes. But encouraging the British public to pledge their cadaver pituitary glands was never considered. Instead, the Medical Research Council assumed that the Anatomy Act 1832 had conferred on professionals an entitlement to extract and withhold cadaver stuff for repurposing without the knowledge or agreement of dead people during their lifetime, or, following their death, that of their next of kin. In effect, it believed the act had instituted in mortuaries a 'no-place', a conceptual space where corpses legitimately ceased to be the material remains of dead people and became a source of cadaver stuff, in much the same way as, within an abattoir, as anthropologist Noëlie Vialles puts it, there is a conceptual space where an animal becomes edible, and its material remains become a carcass that can be disassembled and repurposed into meat, meat products and so-called by-products.[12]

Shortly after the letter to pathologists was circulated, Her Majesty's Inspector of Anatomy advised the council that it was mistaken: the Anatomy Act was silent on extraction of cadaver stuff for repurposing in medicine.[13] Civil servants responsible for the preparatory work on the Corneal Grafting Bill had known this was the case but had decided in the interests of what they considered good taste to exclude other cadaver stuff from the statute. As someone working in the Cabinet Office put it: 'A certain romanticism surrounds the idea of giving eyes so that someone else may see. To extend the scheme to other tissue is to reduce it to the level of a butcher's shop.'[14]

Subsequent attempts to have the law clarified had been resisted on various grounds: the applications for which it was being sought were experimental; legislation might prejudice 'local arrangements', that is, informal agreements between professionals and mortuary staff; it was probably best not to enquire into what was going on until some difficulty arose.

Following the publication of a flurry of critical articles in both the medical and popular press, senior clinicians, fearful of a scandal damaging their professional reputations, joined forces and demanded official action.[15] Faced with the possibility of a legal challenge or being forced into an embarrassing cover-up, the Ministry of Health caved in to their demand. During the drafting stage the bill was called the Human Tissue Grafting Bill to signal its continuity with the Corneal Grafting Act that it would repeal and replace. But by the time the paperwork reached the House of Commons, 'grafting' had been dropped. Moreover, unlike the Corneal Grafting Act, which was a private member's bill, the Human Tissue Bill was a Conservative government measure – the Cabinet had decided to take the lead on an issue that had become a major consideration of public interest, and on which religious and sentimental objection might be encountered.[16] Yet the Second Reading of the Bill – the first opportunity that MPs would have to discuss it – took place a few minutes before 11 p.m. on 20 December 1960. Many MPs would either have begun or were eager to begin their Christmas holiday, and journalists (if any were present) would have been loath to report on a gruesome topic for fear of dampening their readers' seasonal cheer.

Edith Pitt, parliamentary secretary to the Ministry of Health, introduced the bill, and explained that the government had decided to act because it was now possible to graft other cadaver fragments, referring specifically to bone, arteries and skin. The latter, she explained, was particularly valuable in treating people with a serious burn. Moreover, the wording of the bill was deliberately vague, referring to 'any part of the body' in order to allow for the possibility that in the future other cadaver fragments might be sought.[17]

Like its predecessor the Corneal Grafting Act, the Human Tissue Act 1961 empowered the person lawfully in possession of the corpse to authorize removal of cadaver stuff where she or he had no reason to believe that either the dead person during their life, or their next of kin following their death, objected.[18] However, what was new was a stipulation that this belief must be corroborated by evidence derived from what was called 'reasonable enquiry as may be practicable'. But *reasonable* and *practicable* were undefined. The vagueness was deliberate, intended to allow health-care professionals considerable latitude whilst preserving the impression that human decencies towards corpses and grieving kinfolk had been respected.[19]

Again, like its predecessor the Corneal Grafting Act 1952, the Human Tissue Act 1961 provided no penalties for infringement. Absence was justified on the ground that none was applicable because the law does not recognize a right of property in the human body – in other words, taking without agreement cannot in this case constitute theft.[20] In effect, in the event of protest the Human Tissue Act provided a legal shelter for anyone interfering with a corpse in a manner of which the dead person during their lifetime and their kinfolk following their death would disapprove. In a letter to Enoch Powell, Minister for Health, a contemporary critic of the act described it as a law made mainly for the protection of the law breaker, and claimed the spirit in which it had been drafted would advocate raising the speed limit in the Royal Parks to suit the convenience of road hogs who consistently exceeded the present one. The writer went on to predict that the act would bring both law and medicine into disrepute.[21]

Culture of Tipping

On 17 June 2001, the front page of the *Sunday Herald*, a Scottish tabloid, was emblazoned with the headline 'Thousands of Scots babies' bones removed'. The article exposed a collection of over 2,100 dead children's thighbones, that had been begun in 1959 and completed in 1972, and which had been organized by the Medical Research Council as part of an investigation into the effect of hydrogen bomb tests then being carried out all over the world. Fall-out dust from atomic bomb explosions contains Strontium-90, which is absorbed through soil and roots into grass eaten by cows and excreted in their milk. Babies are at a high risk of exposure to Strontium-90: they drink relatively large quantities of milk – equivalent to fifteen litres each day by an adult – and the worry was it might cause incurable childhood cancers. The investigation involved measuring the concentration of Strontium-90 in dead children's thighbones, and its findings contributed to the success of campaigns that led to the signing of a partial test-ban treaty in 1963.

The Scottish Executive instituted a public inquiry. It concluded that the Human Tissue Act 1961 had effectively legitimated the mortuary no-place, and commented critically on the paternalistic attitude of medical professionals who had failed clearly to inform parents, who had agreed to the post-mortem examination of their child, of their intentions. In its evidence, the Medical Research Council said that at the time it did not see itself as having any duty to look over the shoulders of independent professionals in terms of how they implemented legal and ethical guidelines.[22]

The informal network of pathologists in and around Glasgow responsible for the collection may have regarded themselves as privy to official secrets, and as contributing to the Cold War effort.[23] However, the cadaver pituitary gland collection was organized around a different consideration: the Medical Research Council offered a shilling (5 new pence) for each one.[24] In official papers, the shilling is referred to variously as 'unofficial honorarium', 'reimbursement of incidental expenses', 'token of appreciation', 'nominal fee', 'handling charge', 'incentive' and 'perq' – short for perquisites, but, in practice, it was a tip.

Mortuary staff wages were significantly below the national average, and tips were relied on to achieve a living wage. These were given by a variety of people: pathologists for assisting at a coroner's autopsy (the 'going rate' was around a tenth of the fee); funeral directors for measuring and weighing a corpse so that a coffin of the correct size could be built; police for cleaning, laying out and generally making presentable a corpse that needed identification – the more advanced its decomposition the more generous the tip.[25]

The Medical Research Council's cheque was made out to the pathologist responsible for the mortuary, but that was not for whom the cash was intended. A pathologist would have been insulted by a sum that was relatively trivial in relation to their salary and, as sociologist Viviana Zelizer points out in *The Social Meaning of Money* (1994), a tip marks a distinction in social status of tipper and tipped, and hence would demean their professional status.[26] Tipping mortuary staff who occupied the lowest rung of the medical hierarchy was appropriate. Nonetheless, pathologists accept fees for performing coroners' post-mortem examinations, which arguably are the same as a tip: both are payments for services beyond a contracted job description; both are neither charity nor gift.

Most tips were received for performing relatively humdrum tasks, but each shilling would have been earned. As the meatpackers had found, it is not easy to extirpate the pituitary gland out of its heavily defended position within the skull. First, the skull bone is exposed by slicing from ear to ear its skin cover, pulling a flap away from the bone, turning it inside out over the face and hooking it under the chin. Next, a large oval-shaped hole – like a skullcap – is cut out of the exposed skull bone using a hand-held circular saw (like a cordless kitchen mixer with a rotary blade), and the bone removed with the aid of hammer and chisel. The brain is released by cutting the optic nerves that tether it in place, and lifted out to expose the pituitary gland in the *sella turca* (Turkish saddle), its protective bony cradle, which is smashed with a chisel so that the pituitary gland can be freed using scalpel or scissors.

The instructions circulated by the Medical Research Council stated that immediately after extirpation, cadaver pituitary glands should be dropped into a stoppered glass bottle containing a large excess of acetone; the bottle should be kept in the post-mortem room or preferably in a refrigerator; and glands were to be sent through the mail to the Department of Biochemistry at the University of Cambridge, together with a note of expenses incurred which would be reimbursed by the Medical Research Council.[27]

CHAPTER 13

Lionizing American Eye Banks

A survey carried out in 1955 by the Eye Bank Committee of the American Academy of Ophthalmology and Otolaryngology found that a patchwork of eye banks had been created in the United States during the decade following the opening of the Eye-Bank for Sight Restoration. Thirteen of the twenty eye banks that had responded were situated within a hospital and under the control of ophthalmologists; the remainder were freestanding lay organizations. A majority of the latter were thought to be moneymaking schemes, possibly health hazards, and more likely than not infuriating local ophthalmologists by refusing to submit to their professional authority.[1]

Demand for cadaver eyes was growing. The Eye Bank Committee proposed creating a national entity that would redistribute among its members cadaver eyes surplus to immediate local requirement, promote rigorous scientific and ethical standards, approve publicity and organize lobbying of legislatures. In October 1961, at the Palmer House Hotel, Chicago, the Eye Bank Association of America (EBAA) was formed with a board of directors, half of whom were ophthalmologists and half laymen.[2] Its code of ethics included a requirement that eye banks' boards of directors include a qualified ophthalmologist, and that members had to be non-profit organizations.[3] Membership was voluntary: thirty-five eye banks were accepted as charter members.

Lionizing Eye Banks

The decision to allow lay representatives to sit on the Eye Bank Association of America's board of directors acknowledged Lions' crucial support of the

American eye banking industry. But it was haphazard; each Lions Club was free to decide whether or not and how to support eye banking.

Lions' contribution was most visible in capital projects achieved largely thanks to the federal government's Hill–Burton Program that matched funds raised by civil society organizations. The programme had been introduced under the Hospital Survey and Construction Act 1946, and was named after its sponsors, Senators Lister Hill and Harold H. Burton, who envisaged it modernizing the nation's hospital stock and moving patients' beds closer to the laboratory bench. Neither senator had any time for the hospital nationalization that was being proposed in the United Kingdom; their ambition was to protect heterogeneity in hospital ownership and foster competition. The programme encouraged a building boom: between 1946 and 1960 just over 700 new hospitals were built, only one of which was for profit, the remainder being owned either by non-profit organizations or local government. In effect, the programme supported the planning of America's health-care infrastructure by lay people, and was responsible for creating a large stock of small hospitals, each stressing its importance to the local community.[4]

Raising sufficient funds for capital projects required the cooperation of several Lions Clubs. On 13 April 1975, members of Missouri Lions Clubs gathered in Columbia at the ceremony dedicating the opening of the Missouri Lions Eye Clinic and Missouri Lions Eye Bank, both part of the Eye Research Foundation of Missouri. Each Lion present recalled how their club had contributed to the $351,366 raised towards the building's cost: Cape Girardeau Lions had organized a pancake day and shooting matches; Jefferson City (Evening) Lions had sponsored a basketball game between Kansas City Chiefs and the St Louis Cardinals; the Wright City Club had put on a donkey basketball game for its members; Brunswick Lions had held a light bulb and broom sale; other clubs had organized tractor and pony pulls, run concession stands at fairs, sold candy and raffled beef.[5]

'The new Lions Eye Institute gives us two and a half times as much space as we had before and we can probably double the number of patients we see each day,' explained Ray Records, the first full-time head of the Department of Ophthalmology at the University of Nebraska, Omaha, at its dedication ceremony in 1975.[6] It had taken 190 Nebraskan Lions Clubs five years to raise $250,000, their share of the building costs, which had been matched by the Hill–Burton Program.

One or several smallish Lions Clubs could raise dollars sufficient to equip an eye bank laboratory. Theodore Lawill, head of ophthalmology at the University of Kansas Medical Center, said the value of these donations far exceeded their cost in dollars because modern technology attracted investigators and ophthalmic surgeons.[7]

Another small-scale project was equipping collecting sites with enucleating kits consisting of sterile instruments, sterile clothes (gown, gloves and cap), wooden replacement eyes and shipping containers. For instance, in 1954 Iowa Lions agreed to support the work of Lion A. E. Braley, Chief of the Department of Ophthalmology, University of Iowa Medical Center. On their return home, they ordered large quantities of brooms, light bulbs and fruitcakes from suppliers who advertised regularly in *The Lion*, and went door to door in order to sell them to friends and neighbours. Profits were used to buy enucleating kits, which were placed strategically throughout the state.[8] In 1960, Lions of Bridgetown, New Jersey, presented their local hospital with a kit, and a Thermos shipping container that had been designed by Joseph A. Albrecht, Lion and Buffalo Eye Bank's first president.[9] Lions also supported eye banks in kind, by staffing telephones and providing administrative support.

Lionizing Enucleation

The value of the infrastructure Lions helped to build could be realized only if corpses were identified. Unfortunately, as Haywood Snipes, a Poplar Bluff Lion and Missouri State Sight chairman put it, 'people don't always die at convenient hours'.[10] The law exacerbated the difficulties created by death's unpredictability by insisting that cadaver eyes must be enucleated by someone holding a medical licence. But it was often impossible for a physician to reach the corpse within the time frame allowed, especially where the deathbed was not in a hospital. A solution to this perennial difficulty was identified in funeral directors, many of whom were Lions.[11] Funeral directors are experts in 'grief therapy', experienced in talking to newly bereaved kinfolk facing difficult decisions, and unlikely to feel uncomfortable about raising the delicate issue of enucleation. Moreover, only a minimum of training was necessary: a majority had acquired rudimentary knowledge of human anatomy and microbiology and the basic surgical skills required to become a licensed embalmer.

For their part, funeral directors were keen to find ways of dispelling their reputation as 'hucksters' who exploit people at their most vulnerable, an accusation which during the 1960s was being levelled in official inquiries and by consumer champions, of whom Jessica Mitford (1917–1996) is undoubtedly the best known. In *The American Way of Death* (1963), which zoomed to the top of the *New York Times* best-seller list shortly after it was published, Mitford used a mixture of social criticism, witty satire and investigative journalism to indict funeral directors as heartless profiteers, confidence men and social pariahs. Greed, not grief, was behind their speedy arrival at the deathbed, because once a body was in their possession the customer was most unlikely to remove it and take their custom elsewhere. 'Body grabbing' was said to be especially rife in the mortuaries of large metropolitan hospitals, where mortuary attendants were reputed to accept bribes in return for notifying a funeral director whenever a corpse was admitted. Funeral directors were also accused of inflating their charges by insisting – falsely – that embalming was a legal requirement.

Lions Clubs exercised their considerable lobbying skills to persuade state politicians to adopt 'morticians' statutes', which permitted licensed embalmers to enucleate cadaver eyes. In 1969, six states had adopted a 'morticians' statute'.[12] Twenty years later, thirty-eight states had one.[13] Lions organized training. Theodore Genest, second vice president of the Alfred, Florida, Lions Club, was one of forty-six funeral directors who in 1973 had been trained at the Bascom Palmer Eye Institute, Miami.[14] In 1982, Clyde Damron, Wichita Lion, mortician and president of the Kansas Eye Enucleation Association reported that 194 morticians had been trained to enucleate cadaver eyes and his association intended ensuring a trained enucleator was available in each of the state's 105 counties.[15]

Lionizing Supply Chains

At 4.15 p.m. on a quiet Sunday afternoon in December 1955, Highway Patrolman Richard Reddick received a call on the Iowa state radio network ordering him to take a special eye container to the hospital at Moline, Illinois. A patient had just made a deathbed request that one of his eyes be donated to the Iowa Eye Bank. At 8.45 p.m., Patrolman Reddick delivered it, packed in ice, to the University of Iowa Medical Center where, at 10 a.m. the following morning, a corneal graft was performed. This was the first occasion on which Iowa Lions and the state governor transported cadaver eyes under an arrangement with the Highway Patrol. Lions also volunteered as cadaver eye couriers. For instance, between 1954 and 1969, a network of Buffalo Lions collected and delivered over 5,000 cadaver eyes, 3,000 of which were grafted and the remainder used in research.[16]

Charity begins at home was the rule: local surgeons typically had first call on a pair of enucleated cadaver eyes.[17] In 1962, the Eyeball Network of Lions radio hams began arranging distribution of cadaver eyes surplus to local demand, or meeting urgent requests.[18] Twice every day, early morning and evening, for around fifteen minutes, the channel would be opened. 'What happened, for example, is Oklahoma City Lion Travis Harris, a blind radio ham, reported that a patient in a local hospital would lose their sight unless a graft was quickly arranged but Oklahoma Sight Conservation Inc., had none available. Hams listening all across the country were instructed to contact their local bank.' When a surplus cadaver eye was found, shipment was arranged, often with the cooperation of airlines and state highway police.[19] By 1964, the network had grown to 130 radio hams in sixty-eight cities and thirty-five states and had placed 674 cadaver eyes sometimes where collecting site and operating theatre were separated by a great distance. By August 1977, it had arranged the redistribution of 8,964 cadaver eyes.[20]

International Lions

'The eye surgeon is in an enviable position in being able to advance world peace by taking his skills abroad,' observed John Harry King Jr (1911–1986), an ophthalmologist with a flourishing private practice, Medical Director of the International Eye-Bank, and a Washington DC Lion.[21] In Cold War battles against communism at home and overseas, ophthalmology became a tool of American diplomacy.

The Lion periodically published a photograph of one or two Lions in an airport car park or about to board a plane holding aloft a large box emblazoned with 'HUMAN EYES – FRAGILE'. Underneath, a caption explains that the box was being taken to a hospital somewhere in Asia, Africa or Latin America. Packed in the box were several moist chambers and a powerful message: 'no other expression of international aid can compare to this gift of eyes'. This is how in 1965 Dr Francis K. Pan, chairman of the Hong Kong Eye Bank and Research Foundation (established by Hong Kong Lions of District 303) framed his thanks to the Iowa Lions Eye Bank. Their gift of two cadaver eyes had been used to restore the sight of two young Chinese women, one of whom had lost her sight as a result of a childhood accident, the other had been blinded as an adult and had been forced to leave her work in a textile factory.[22]

This particular shipment of eyes had been organized by the Eyeball Network and radio hams in the Far East following 'Operation Vision: Hong Kong', a mission led in 1961 by King, ostensibly to assist blind refugees regarded as 'burdens by the Red Chinese'. Among other things, members of the Hong Kong Ophthalmological Society were taught how to perform keratoplasty, and a local source of cadaver eyes was arranged, the latter requiring a revision of the law by the British colonial authorities. The 'missionaries' – ophthalmologists and nurses – had volunteered their services; surgical instruments had been provided at cost; funds for transportation and living expenses had been raised by members of the Pharmaceutical Manufacturers Association in Washington DC, which also contributed medicines.[23]

The International Eye-Bank, which had opened the doors of its headquarters at the Washington Hospital Center in Washington DC in February 1961, was a project of the Medical International Co-operation Organization (MEDICO), which had been founded in 1958 by Thomas A. Dooley (1927–1961),[24] physician and CIA informer.[25] In 1962, following Dooley's premature death, MEDICO merged with the Committee for American Relief Everywhere (CARE), which had been incorporated in 1945 to provide humanitarian assistance to needy peoples in foreign countries (CARE-MEDICO was disbanded in 1979). American and Canadian service organizations, including Lions Clubs, supported their missions; for instance, in 1967 Missouri Lions purchased surgical instruments for two ophthalmologists serving in the US Army Medical Corps who wanted to restore the sight of corneal-blind Vietnamese civilians.[26]

Lionism was exported and embedded in civil society in many parts of the world, save those under communist governments that considered a flourishing civil society an ideological threat. Whereas during the 1920s and '30s overseas Lions Clubs had been set up by American expats in China, Cuba, Honolulu and Hawaii, in the post-war era local citizens established them. International expansion of the organization was rapid: in 1950 there were 8,055 clubs in twenty-eight nations; by 1970 there were 24,391 clubs in 126 nations whose members were styling themselves Knights of the Blind.[27]

CHAPTER 14

A Gland Lost is a Gland Wasted

Following the announcements of Raben and Li's discovery that only growth hormone isolated in primate pituitary glands 'works' in humans, parents of a child failing to grow in height, or physicians acting on their behalf, began asking mortuary staff to extirpate cadaver pituitary glands. Demand grew and competition intensified. Transactions were conducted without paperwork at the mortuary's back door, and so the price a gland commanded is unrecorded. But the sum was sufficient to attract 'pituitary pirates', freelance traders with ruthless tactics, whose nickname conjures up the image of reddish pea-sized cadaver pituitary glands piled high in a treasure chest.[1]

A course of treatment for one child required several hundred cadaver pituitary glands, but biochemist Choh Hao Li needed many thousands to fulfil his ambition of completing a map of hGH's giant complex molecular structure. This would confirm his international reputation in the field of the biochemistry of hormones of the anterior lobe of the pituitary gland, and also indicate how hGH might be synthesized. Synthetic human growth hormone would open an escape route out of the mortuary and, if capable of mass production, be financially rewarding.

Preparing human growth hormone (hGH) for injecting into a child is relatively easy. Early recipes involved picking cadaver pituitary glands free of gristle and bone, grinding them in an electric blender or mincer, repeatedly washing and filtering the mash or mince, and spreading it out on a tray to dry into a lumpy powder. Next the powder is dissolved to create a soup in which hGH can be isolated using chemical precipitation.

Inexpert hands extract a tiny amount of crude hGH out of the soup, whereas hGH prepared in Li's Hormone Research Laboratory was guaranteed to be pure and potent. Parents and physicians asked Li to isolate it in

Figure 10: Preparing cadaver pituitary gland soup

cadaver pituitary glands that had been collected to treat specific children. His terms were as follows: he kept half the glands for his research in exchange for hGH isolated in the remainder.

Roberto F. Escamilla was one of several clinicians who agreed to Li's terms. He was a physician working at the University of California's San Francisco Hospital, who had collaborated with Li and Evans in the experiment with carcass growth hormone in which F. H. had failed to grow. In July 1960, two years after the British scheme had been established, Escamilla launched the San Francisco Pituitary Bank.[2] Supported financially by philanthropists and the Medical School, it was modelled on eye banks, and provided with a motto, 'A gland lost is a gland wasted', which faintly echoes the meatpackers' maxim 'You either sell it or smell it'.

In 1961, Escamilla approached the College of American Pathologists (CAP), which had been established in 1946 to promote the professional and economic interests of the medical specialty, to establish whether or not and how many of its 3,000 members were prepared to cooperate with the San Francisco Pituitary Bank. Three hundred and two agreed, and together promised to provide around 10,000 cadaver pituitary glands annually.[3] Significantly, 419 pathologists refused because they were committed to support others, perhaps one of the two other high-volume processors: Raben, who was handling annually around 15,000 cadaver pituitary glands, and Alfred E. Wilhelmi (1910–1994), biochemist and pituitary hormone enthusiast, at Emory University, Atlanta, Georgia, whose share was around 3,500.[4] The College of American Pathologists urged its members not to interpret Escamilla's enquiry as an attempt to trespass on other people's schemes.[5] Nonetheless, Escamilla retreated and the San Francisco Pituitary Bank's collecting sites were restricted to mortuaries in and around the Bay Area.

Li had other sources. Entries in his laboratory workbooks situate his Hormone Research Laboratory at the hub of a network of mortuaries in the United States and overseas. For example, in 1961, he received cadaver pituitary glands extirpated in mortuaries in California – mostly in the Napa Valley and Los Angeles – and in Italy, Brazil, Puerto Rico, Argentina and Chile. The latter was a particularly fruitful source because Hector Croxatto, in the Laboratory of Physiology, Universidad Catolica, who had collaborated with Li on research, had secured the services of a freelance collector who toured local mortuaries tipping staff. Over a decade beginning in 1959, around 17,000 cadaver pituitary glands were flown from Chile to San Francisco.[6]

Organizing a National Collection

Scientists and clinicians complained that competition was distracting them from their work, and moreover cadaver pituitary glands were being wasted by being used inappropriately on children whose short stature was the result of conditions other than a malfunctioning pituitary gland. Morris Graff, executive secretary of the Endocrinology Study Section of the National Institutes of Health, decided to stack the odds in favour of research by investing federal taxes in a national scheme for extirpating and distributing either cadaver pituitary glands or hGH.

In January 1962, Graff invited representatives of the College of American Pathologists, scientists including Li, physicians and parents of children failing to grow in height to a meeting in Miami, Florida, where it was agreed to convene an ad hoc committee that would draw up a proposal for government financial support.[7] In April 1963, five years after the British collection had been established, the National Pituitary Agency opened for business under contract with the National Institute of Arthritis and Metabolic Diseases, one of the institutes of the National Institutes of Health.

The challenge the agency faced was daunting. It was reckoned that of the 1.8 million people who died each year in the United States, around 325,000 underwent a post-mortem, and that in two out of every three the skull was opened, allowing access to the pituitary gland. If every one of these were extracted then each year the agency would receive around 200,000 cadaver pituitary glands.[8] It was decided to offer mortuary staff $2 for each cadaver pituitary gland, and also to reimburse shipping costs.[9] Two dollars was far more generous than the shilling (then worth 14 cents) offered by the Medical Research Council. Perhaps it met or even exceeded the current rate in American mortuaries. It was intended as an incentive but was described as a 'payment for service' in order to prevent attracting 'meddlesome attention'.[10] The College of American Pathologists assured its members that the cash in no way impugned their idealism or dedication to medical science.

Instructions were circulated to pathologists: all cadaver pituitary glands were useful save those that were grossly infected; age and type of person were immaterial. Additionally, pituitary glands of embalmed corpses were also acceptable, despite doubts that active hGH could be found in them.[11] The pituitary glands preferably should be extirpated within seven but up to forty-eight hours after death. Preservation was either by soaking in acetone or freezing. Acetone-preserved glands were to be stored in a polythene bottle that could be sent through the mail to the National Pituitary Agency's headquarters in Baltimore. Frozen glands were wrapped individually until the mid-1970s, when extirpators began objecting on the grounds that it was too time consuming. From then onwards, they were frozen and stored in bulk. A dry-ice container was developed for shipping.

In its first eight months the National Pituitary Agency handled 11,057 cadaver pituitary glands. Two years later, it received 70,188 cadaver pituitary glands, and from then onwards its annual throughput varied between 70,000 and 80,000. The peak of 83,091 was achieved in 1975, when around one in every twenty-one corpses placed in a grave or crematory in the United States was without a pituitary gland.

Surrendered collecting sites contributed to the early growth in volume of the agency's collection. For instance, in 1964, the University of California

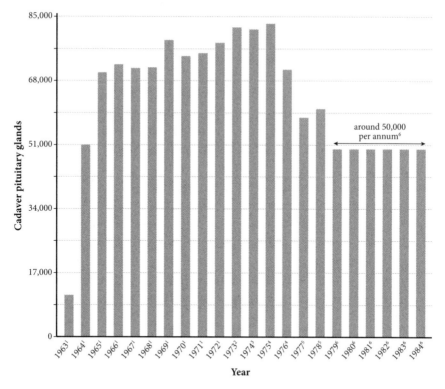

Figure 11: The National Pituitary Agency's collection of American cadaver pituitary glands

Los Angeles Medical School handed its over.[12] Two years later the Veterans Administration ceded its Human Pituitary Bank to the agency. It was a valuable acquisition: Veterans Administration hospitals experience relatively high death rates – veterans typically are economically and socially disadvantaged and suffer physical and mental health problems – and were performing around 28,000 post-mortems each year.

Physician William E. Latimer was employed to stake out new collecting sites. It was clearly impossible for him to visit every mortuary. Instead, as a kind of simulacrum he distributed a washable wall chart of the bronchial-pulmonary segments bearing the legend, 'Every pituitary is needed. Have you been saving them?'[13]

Another strategy was to recruit the parents of children admitted to a clinical trial of the agency's hGH to tour mortuaries in their neighbourhood. Some had hands-on involvement in their child's treatment, receiving in their home nuggets of hGH out of which to prepare and deliver thrice-weekly injections. In December 1965, Robert Blizzard, the agency's first director, established Human Growth Inc., a non-profit organization based in the agency's office. The foundation's growth was rapid: in five years it had recruited 600 members, mostly kinfolk of research subjects, organized in fourteen chapters scattered throughout the United States and Canada. Raising funds for research was one of the foundation's original aims. But some members also volunteered to act as couriers. For instance, during his stopovers in various cities, a commercial airline pilot and father of two short children would plead his case at various mortuary doors and fly off with cadaver pituitary glands in his cockpit.[14] In 1968 he persuaded his employer, Trans World Airlines, to allow other staff members to collect and transport frozen cadaver pituitary glands.[15] Unfortunately, flights were often delayed and glands would arrive defrosted and smelling. Commercial couriers proved more reliable but were expensive.

The College of American Pathologists fulfilled its responsibility as co-sponsor of the National Pituitary Agency by regularly exhorting its members in its newsletter. For instance, in 1972, President William J. Reals pointed out:

> The autopsy can serve not only to advance our knowledge of the causes of disease but also in the case of the pituitary, to cure an otherwise hopeless condition – hypopituitary dwarfism. The tragedy of this disease is that it afflicts children who are doomed to remain doll-like and tiny unless this precious gift of human growth hormone can be given to them in therapeutic amounts.[16]

'Doll-like' suggests that without hGH, short-statured children would never develop into competent adults. In 1973, college members who had not yet agreed to support the National Pituitary Agency were exhorted: 'Next time you shave, look yourself in the eye and ask, "Am I participating in this vital program?" For bearded pathologists, ask yourselves, NOW!'[17] Following his appointment in 1970 as the National Pituitary Agency's director, Australian endocrinologist Salvatore Raiti, who had been working in Great Ormond Street children's hospital in London, the principal site of the Medical Research Council's clinical research into hGH, published an annual letter of thanks. Another tactic was an appeal from an hGH recipient. In 1976, a fourteen-year-old girl wrote: '[W]ithout growth hormone therapy, I would barely grow an inch per year. I have known sadness, and now know hope. You have made this change possible. Thank you.'[18]

The cash 'incentive' was responsible for the San Francisco Pituitary Bank's closure. Reassured by the National Pituitary Agency that it would not trespass in his collecting sites, Escamilla had decided to remain independent, relying on the persuasive power of a physician employed part-time to enlist and sustain the cooperation of pathologists in the Bay Area by phoning them once or twice each week, paying an annual visit, and inviting them to the bank's annual 'pituitary party' where champagne was served, favours handed out to the ladies and framed certificates of appreciation awarded to that year's most productive collecting sites. The Golden Gateway Center was popular, but the most successful party venue was the Fleet Admiral Nimitz Officers Club on Treasure Island in San Francisco Bay, where in 1972 and 1973 the Bohemian Club's Strollers entertained pathologists. Mortuary staff were not invited, but were presented with a bottle of whisky every Christmas. When Bay Area mortuaries switched allegiance to the National Pituitary Agency, Escamilla turned his sights overseas. But inevitably the San Francisco Pituitary Bank was unsustainable, and in November 1975 it closed. Over the fifteen or so years of its existence it had handled 46,154 cadaver pituitary glands, seven out of ten of which had been extirpated in American mortuaries, the remainder overseas.[19]

Marketing Corpse Philanthropy

Whereas the Medical Research Council's collection was clandestine, both the San Francisco Pituitary Bank and the National Pituitary Agency initially courted public approval. Indeed, the agency's founders believed that an informed public might insist that a post-mortem examination is performed on their kinfolk's corpse during which the skull is opened in order to facilitate the pituitary gland's extirpation.[20]

In 1957, California had adopted a law allowing people to pledge cadaver stuff to a licensed agency such as a teaching hospital or a tissue bank. It was exploited by the San Francisco Pituitary Bank in marketing leaflets that explained that carcass growth hormone was now known to be ineffective, and a vast number of cadaver pituitary glands was needed in order to help children whose height was predicted to fall well below the accepted norm. On receipt of a signed and witnessed form the bank issued a pledge card that identified the bearer as a 'Pituitary Bank donor', and urged pledgers to inform their physician and kinfolk of their commitment.[21]

For his part Li promoted pledging whenever invited to explain his latest achievement to the general public. For instance, he mentioned pledges in January 1964 during an interview broadcast on the University Explorer radio series by CBS-KCBS, and again in March 1965 when he appeared on Arthur Godfrey's radio show.[22] The bank forged links with service clubs and parent-teacher associations in the Bay Area. It was listed in a brochure providing 'Medical donor's information' distributed to the 14,000 members of the Bay Area Funeral Society, and advertised on posters hung in state prisons and correction facilities.

The National Pituitary Agency's marketing of pledging had national ambitions. The popular press was provided with 'true and wonderful' stories of how it was rescuing from a 'doll-like existence' children who were normal in every respect save for an inability to grow in height. For instance, the August 1965 edition of *Parade*, which claimed to be the most widely read magazine in the United States, included a three-page article under the banner headline, 'We can end dwarfism!'[23] The short stature of a seventeen-year-old young man is emphasized in a photograph in which two much taller friends bracket him. Sporting jeans and a checked shirt, he is clearly not a 'freak' but a typical healthy and intelligent American high school student. Before treatment with the National Pituitary Agency's hGH, he was four feet (1.22 m) tall, and at school had heard himself called Shorty, Shrimp and Pee-Wee. With treatment he had grown five inches (13 cm), and if sufficient cadaver pituitary glands became available might reach five feet (1.52 m), the Agency's target height.

The *Parade* article closed with an appeal to readers on behalf of potential recipients of hGH: 'If you can, won't you please help, especially before it's too late, and the bones of some undersized boy or girl have fused, forcing the poor child into a life of hellish dwarfism.' A pledge form is printed underneath: 'This is to certify that I, _____, wish to contribute my pituitary gland for the promotion of medical science and clinical research.' Members of the Human Growth Foundation organized 'pituitary drives', campaigns to persuade people to pledge their pituitary glands to the National Pituitary Agency.[24] However, judging from the San Francisco Pituitary Bank's records, few hearts and minds were captured: in 1975, when the bank closed, just over 1,000 pledges had been recorded, and fewer than fifty had been realized.[25]

Mapping the Molecule

'Phe-Pro-The-Ile-Pro-Lev-Ser . . . .' is the code that controls growth' is how a journalist announced Li's achievement in identifying the name and position of the 188 amino acids that he claimed made up the hGH molecule. 'It is a chain with two loops, one containing six sub-units, the other 93 sub-units,' Li explained. 'It's like a long highway with two looping side roads off the main route.'[26] In the extensive media coverage his achievement was likened to finding the solution to an enormous three-dimensional jigsaw puzzle. Exceptionally large hGH molecules had been broken into fragments typically of five to ten amino acids. Each fragment was carefully scrutinized, sketches of possible amino sequences drawn, which might be pieced together so that a complete map might emerge. The long and exacting research had consumed 5,000 cadaver pituitary glands supplied by the San Francisco Pituitary Bank, the National Pituitary Agency and probably some of Li's undeclared sources.[27]

Four years later, Li announced he had synthesized hGH in the laboratory. The synthesis was based on his map of the hGH molecule:

> We had to put 188 amino acids together one block at a time, and then we had to build two bridges – a small one, like the Golden Gate Bridge, and a large one, like the San Francisco to Oakland Bay Bridge. We built the highway first block by block. That took two years. Then we spent two years on the bridges. The bridges are chemical bonds connecting a pair of specific amino acids on the long 'highway'. Without them, the compound doesn't function.

The whole process had cost one million dollars.[28]

Television, radio, magazines and newspapers around the world celebrated Li's achievement, calling it one of the biochemical landmarks of the twentieth century, and deserving of the award of a Nobel Prize. A synthetic growth hormone would facilitate research that hitherto had been stalled by the limited supply of cadaver-derived hGH. Many people stood to gain from its success: the University of California's press release claimed that not only can hGH stimulate growth, but it is also implicated in the aetiology of various diseases, including cancer.[29]

On 27 April 1971 Li filed US Patent, Serial No. 17,811, Synthetic Growth Hormone and Method of Producing It. He then assigned the patent to the Hormone Research Foundation of which he was president. Yet he acknowledged that his synthetic hGH was 90 per cent less potent than hGH isolated in cadaver pituitary glands, a paradox he attributed to impurities creeping into the raw material. However, almost immediately a different convincing explanation emerged: Li's 1966 map was incorrect. In a letter published in *Nature New Biology*, Hugh D. Diall, assistant professor of medicine at the Endocrine Unit of Massachusetts General Hospital, submitted evidence that the hGH molecule has 191 amino acids whereas Li's had only 188.[30] The absence of three amino acids accounted for its relative weakness. The escape route out of the mortuary mapped by Li had led to a dead end.

CHAPTER 15

Who's in the Mortuary?

A one-litre brown glass bottle with the capacity to hold up to fifty cadaver pituitary glands suspended in acetone became a fixture in British mortuaries. It was frequently full. Between 1959 and 1985, mortuary staff extirpated roughly one million cadaver pituitary glands. At the zenith of their endeavours, which was reached in 1976, one in eight British corpses placed in a grave or crematory was without a pituitary gland.

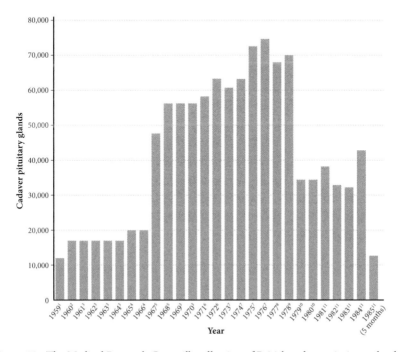

Figure 12: The Medical Research Council's collection of British cadaver pituitary glands

Many more people dying away from home, making it more likely that between death and disposal their corpse rested in either hospital or public mortuary, were providing mortuary staff with numerous opportunities for claiming the Medical Research Council's shilling tip.

	Lunatic asylum %	Hospital %	Workhouse %	Elsewhere, mostly at home %	Total deaths number
1897	1.3	4.4	7.6	86.7	541,487
1907	2.0	6.7	10.0	81.3	524,221
1957	2.9	43.8	2.9	50.4	514,870
1967	3.1	51.5	3.5	41.9	542,516

Figure 13: Where English and Welsh people died

This secular trend continued, and by 2007/9, around one in five (20 per cent) of English people died at home (Welsh people are excluded from these data because in 1999 health policy was devolved to the Welsh government).[1]

Death was exiled from private homes on both sides of the Atlantic, but corpses were more completely removed from the visible surface of the British landscape than that of the United States. Historian Philippe Ariès attributed this to the American funeral industry's aggressive marketing that repeatedly reminded the public of death's inevitability.[2] This undoubtedly was the case, but he might also have observed that the number of British people dying in hospital rose sharply shortly after the NHS began extending health care free at the point of delivery to everyone, a benefit refused to a significant proportion of the American population. The frequency of death's visits to British hospitals increased following admission of many more people in poor health who were offered heroic treatments – open heart surgery, aggressive cancer therapies, and such like – which sometimes fail. In short, the modern death of British people has been more thoroughly medicalized than that of Americans.

Busy Coroners

Deaths away from home are more likely to be referred to a coroner for investigation and to result in corpses being held in a mortuary until the coroner agrees to their release.

Foul play was not responsible for the increase in the weight of coroners' workload. This was the conclusion of the Committee on Death Certification and Coroners (the Broderick Committee, named after its chairman Norman Broderick), which had been appointed in March 1965 by the Home Office

	Registered deaths (thousands)	Deaths reported to coroners – number (thousands)	Deaths reported to a coroner – as a % of registered deaths
1920	466.1	53.7	11.5
1950	510.3	83.6	16.4
1960	526.3	101.1	19.2
1970	575.2	133.4	23.2
1980	581.4	170.2	29.3

Figure 14: Deaths reported to English and Welsh coroners

to investigate the British Medical Association's claim in *Deaths in the Community* (1964) that deficiencies in the process of death certification, and mistakes on the part of coroners, were allowing many murderers to escape detection.[3]

The Broderick Committee blamed the Births and Deaths Registration Act 1926, which had introduced the requirement of recording the main cause of death and other contributory factors on the death certificate, and stated that these facts must be supplied by a licensed medical practitioner who had either cared for the dead person during their final illness or had seen the corpse to confirm death. However, several different doctors care for patients admitted to modern hospitals, and a junior one typically confirms death. This combination of circumstances frequently created uncertainty about who should or was competent to provide the necessary information, which a coroner was called upon to sort out. Successive governments failed to heed the Broderick Committee's recommendation that the statute should be reviewed, and in 1999 it emerged that Harold Shipman, a general practitioner working in Hyde, Cheshire, had murdered at least 215 of his patients, and for twenty-four years had succesfully hidden his criminality behind weaknesses in the system of death certification.[4]

A post-mortem examination is not always necessary to decide what should be put on the death certificate, but almost invariably the coroner ordered one, which is why a growing majority were performed under a coroner's order.

No one can protest when a coroner orders a post-mortem examination. Around eight out of ten were performed on elderly people. In his evidence

	Total deaths number	Total post-mortem examinations number	Coroners' post-mortem examinations		Hospital post-mortem examinations	
			number	%	number	%
1962	557,636	132,915	86,270	64.9	46,645	35.1
1984	566,883	158,104	138,071	87.3	20,033	12.7

Figure 15: Likelihood of post-mortem examination in England and Wales

to the Broderick Committee, Dr Powell, a pathologist working at Bridgend General Hospital, described these post-mortem examinations as 'peculiarly onerous and time consuming and it is a workload which I would be pleased to shed'.[5] But a fee compensated him for each one, and they were quickly done: mortuary staff would prepare the corpse.

Bioavailability

Anthropologist Lawrence Cohen appropriated 'bioavailability' from pharmacology to capture the social and economic circumstances of people who are kidney donors-on-the-hoof. Cohen was investigating the illicit explant/transplant industry in India that emerged during the 1980s; unsurprisingly, he found that desperately poor people largely provisioned it.[6] However, poverty is a necessary but insufficient condition for bioavailability; it is qualified by other social factors. Cohen found rural kidney sellers are primarily men, often small-scale farmers, faced with high levels of indebtedness, whereas urban kidney sellers are predominantly women, often forced to sell a kidney because of their husband's unemployment, migration or, in some cases, alcoholism.

In allowing workhouse wardens to avoid the cost of burial of unclaimed corpses by sending them to a medical school, the Anatomy Act 1832 had effectively established poverty, and mental and physical frailty, as criteria of posthumous bioavailability of British corpses albeit for the purposes of dissection.[7] Historian Ruth Richardson protested that unclaimed did not necessarily mean unloved: next of kin might have failed to claim a corpse because they were unable to bear the cost of its disposal. However, by the 1950s, few corpses were unclaimed for reasons of poverty. The Anatomy Act had encouraged a variety of popular saving schemes through which the poorest sections of the population during their life could accumulate the cost of disposal of their corpse. Subsequently the welfare state recognized the salience of popular understandings of 'a decent funeral' by including a death grant in the National Insurance Scheme.

Richardson was challenging the implication that corpses unclaimed from the workhouse were those of people who had experienced social death prior to physical death. Socially dead people have been excluded from, or are incapable of participating in, everyday social life, and their ties to kinship and friendship have been disentangled.[8] The modern category describes three often-overlapping groups.[9] The first includes people in the final stages of a chronic and lengthy mental or physical illness. The very old form the second group. Survival into old age, even extreme old age, is increasingly common, but in societies where great value is placed on people's contribution to the economy, institutional and structural forces devalue, stereotype and exclude older people. Elderly people may also suffer a contraction of their social network due to immobility or death of family and friends. Many are likely to die alone, at home or in a nursing home or institution. In many respects the third group encapsulates the first two: it comprises people whose personhood has been lost, typically as a result of dementia.

The world does not suffer much of a disruption when a socially dead person experiences physical death. It is often unobserved by kinfolk who have already experienced 'anticipatory grief', resulting in emotional and/or physical withdrawal. Ariès cited a survey conducted in 1963 that had found three in every four bereaved respondents had been absent at their kinfolk's deathbed.[10] Paradoxically, socially dead people are much more likely to experience physical death outside a private home, and in circumstances that are highly likely to be referred to a coroner for investigation.

There is a good chance that before their biological death the elderly person out of whose corpse the pituitary gland was extirpated had been socially dead. Instructions attached to the original letter the Medical Research Council sent to pathologists stated: 'All or a fraction of any post-mortem pituitaries which are not the seat of primary disease will be useful.' Primary disease of the pituitary gland is rare. The instructions described what might be called the medical criteria of bioavailability, observation of which is one of the measures employed to minimize harming potential recipients of body and cadaver stuff.

Infection is a hazard of mortuary work, and mortuary staff are acutely aware of the pathogens lurking in corpses. Mr Brine, whose thirty-year career in mortuaries had begun in 1955, recalled using common sense and the poke and sniff test to decide against extirpating pituitary glands of decomposing corpses and those of people who had died as a result of taking an overdose or poison.[11] Mr Biddle did much the same thing, and from 1970 extirpated the pituitary gland from every corpse he considered suitable in his mortuary and that of four neighbouring hospitals.[12] Nonetheless, in *Checkout*, a documentary televised on Channel 4 in 1992, it was alleged that other technicians were less conscientious, and had forwarded pituitaries indiscriminately.[13]

When the Medical Research Council launched its collection of cadaver pituitary glands, absence of yellow jaundice was a medical criterion of bioavailability of blood donors-on-the-hoof, which was established by taking a medical history, a method known to be unreliable: symptoms could have been mild and people unaware that they were ill, or they might not recollect having had the illness. But dead people cannot be interrogated about their medical history. Nor could their kinfolk as the pituitary glands were being extirpated without their knowledge or agreement.

The Medical Research Council relied on 'terminal sterilization', that is, procedures that hopefully eliminate infectious pathogens – for instance, milk is made safer to drink by heating, a process known as pasteurization. The method used to eliminate the risk of transmitting yellow jaundice was soaking in acetone. The council had sought the opinion of Wilson Smith (1897–1965), a virologist working in the Department of Bacteriology at University College London Hospital. He advised:

> One cannot, of course, state that it is absolutely impossible for some hypothetical agent to survive. This applies also to the virus of infective hepatitis because satisfactory investigations of its resistance to physical and chemical agents have not yet been possible. In my opinion, however, the chances of virus survival under the conditions outlined are so slight as to be negligible, and I personally would not hesitate to proceed with the trials.[14]

Reassurance about the effectiveness of terminal sterilization meant there was almost nothing to prevent any dead person's pituitary gland from being extirpated.

CHAPTER 16

Representational Dilemmas in Marketing Eye Pledges

The Human Tissue Act 1961 allowed British people to pledge ('offer' is the official term) fragments of their corpse. The pledge had either to be written or during a final illness declared in front of two witnesses. Its authority was stronger than that allowed by its predecessor, the Corneal Grafting Act 1952: pledgers had an 'overriding right' to dispose of their corpse and its fragments as they saw fit, and next of kin were denied a veto of a dead person's stated wishes. Nonetheless, Minister for Health Enoch Powell warned against ignoring kinfolk's objections on the grounds that 'far more harm can be done to the cause by a single case in which a strongly held scruple is overridden than perhaps a temporary loss of opportunity'.[1]

During the House of Commons debate on the Human Tissue Bill held shortly before Christmas 1960, Powell announced that the RNIB had accepted responsibility for marketing cadaver eye pledges.[2] As his parliamentary secretary Edith Pitt explained, the reason why encouraging pledges of other cadaver stuff had been rejected was that

> [T]here is already public interest, and great sympathy with the thought that by donating one's eyes after death one can help the living. There has been a change in the public outlook or we should not be talking about this Bill so freely now, but I am not sure that we have yet got to the stage where we are ready for a great publicity drive on the general question of donating bodies or parts of bodies.[3]

Rycroft had advised the RNIB against accepting responsibility for marketing eye pledges.[4] Too few pledgers was not why the supply of cadaver eyes was failing to meet demand. Few staff at potential collecting sites knew anything about corneal grafting, and the kinfolk of people who had just died were neither asked to agree to enucleation nor advised how a pledge is realized. Another difficulty was inflexible hospital organization, which made it difficult for a surgeon to command at short notice theatre time, staff and a hospital bed for keratoplasty – a procedure that did not count as an emergency and where most potential recipients retain sight in one eye – or find a suitable source for a scheduled operation.

Findings of a nationwide survey of some sixty hospitals conducted by Ministry of Health staff responsible for drawing up the Human Tissue Bill supported Rycroft's scepticism.[5] In 1959, around 550 people had undergone keratoplasty, but 600 names were on waiting lists, which were lengthening. Surgeons operating in general hospitals were mostly relying on the mortuary as collecting site, and often enucleated cadaver eyes 'just in time' for the operation. Ophthalmologists attached to eye hospitals typically relied on external local arrangements established with staff of geriatric hospitals, and institutions for physically and mentally frail people.

Some hospitals had accumulated substantial numbers of pledges; for instance, Cardiff General Hospital had 400 and Wolverhampton Eye Infirmary 1,017. Yet few grafted corneas were those of a dead pledger. The survey found little enthusiasm for honouring pledges, particularly those of pledgers who died in their own bed, which was an inconvenient and inappropriate, unhygienic place in which to enucleate cadaver eyes. Rycroft's team was unique in being able to undertake domiciliary enucleation under sterile conditions. In 1960, Tenterden Lions had presented him with a white van fitted out with two refrigerators, special lighting, instruments and other kit.[6] A typical 'melancholy visit' was paid on a Saturday evening to the home of a ninety-year-old man whose pledge had been recorded in the Queen Victoria Hospital's register. His widow had telephoned the hospital shortly after his death. Switchboard operators had been instructed to contact the ophthalmic ward's sister who contacted the on-duty surgeon who drove the van to the dead man's home, enucleated his eyes and placed them in a moist chamber that was held in one of the van's refrigerators during the return journey.[7]

Despite this discouraging evidence, civil servants had decided to encourage eye pledges, and to prevail upon the RNIB to accept responsibility for the marketing campaign on the grounds that a civil society organization known and respected for supporting blind people would have greater public appeal than a faceless government department or the NHS.[8] The decision indicates a shift in the British government's attitude towards civil society organizations from antagonism (voluntarism was antithetical to state-organized collectivism) to one recognizing that in some instances they were a useful means to an end where government determined the end but wanted to be disassociated from the means of achieving it.[9]

For its part, the RNIB's agreement to accept responsibility for the marketing campaign represented a change of heart, perhaps because the civil society organization is dependent on charitable donations and bequests and feared a hint of disapproval of corpse philanthropy might damage both reputation and income. On 27 September 1961, the marketing campaign was launched. Despite his reservations Rycroft accompanied John Colligan, its director general, to be interviewed on *Today*, the BBC's flagship morning radio news programme. The campaign gained momentum. For instance, the redoubtable Mrs Clemetson persuaded editors of local newspapers to highlight pledging on 7 April 1962, World Health Day, an annual event organized by the World Health Organization, which that year had been dedicated to prevention of blindness.

Representational Dilemmas

The marketing campaign's principal weapon was a leaflet.

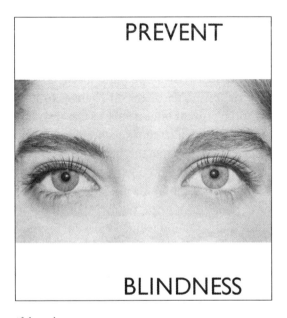

Figure 16: 'Beautiful eyes'

Inside is an invitation to give 'sight to someone who is, at this moment, blind', and, echoing the Eye-Bank for Sight Restoration's original appeal, reassurance that the gift 'will cost you nothing but your understanding and your signature'. However, whereas the Eye-Bank's discourse was sacralizing, the RNIB's was secular: the ethos of the pledge was 'humanitarian'.

The leaflet is without any reference to enucleation and corneal grafting. Its production coincided with a row over *Your Life in their Hands*, a groundbreaking television series, which the BBC had begun broadcasting in February 1958, which showed graphic images of surgery and other 'high-tech' medical interventions. The first series had outraged a vocal section of the medical profession. Medical practice was not a spectacle, one doctor thundered: 'The privacy of the surgical insult to a human body, even a consenting one, should be inviolable, and should never be made the basis of a Roman holiday for the titillation of the public's demand for thrills.' Others insisted that 'operation horrors' were heightening the general public's fear of hospitals, and that some of their more sensitive patients had fainted while watching the programmes.[10]

Many opponents of *Your Life in their Hands* were won over by evidence that the programmes were instilling therapeutic optimism and respect for the medical profession in the general public. Rycroft belonged in this camp; in 1964 a programme devoted to his work had been televised. The following week the RNIB received over 3,000 completed pledge forms. Encouraged by the positive reaction to his 'operation horrors', Rycroft decided to produce a leaflet inviting pledges to the Queen Victoria Hospital eye bank, which was illustrated with two photographs, one before, the other after a clouded cornea had been replaced.

Shortly afterwards, the RNIB entered into discussion with the Ministry of Health's public relations department on changes to its original leaflet's cover. The exhortation 'prevent blindness' was clearly misleading, and was replaced with 'restore sight', which more accurately describes keratoplasty's ambition. But despite agreement that the 'beautiful eyes', as they had become known, had been inappropriate, no one knew what image should replace them. There is no easy solution to the representational dilemma in marketing corpse philanthropy, particularly of the eye. Rycroft was convinced that the public could stomach, even respond positively to, medical imagery, but civil servants judged the photographs on his leaflet far too graphic, and asked him to replace them with a 'before' image of a man with a white stick trapped in the centre of London's traffic, and an 'after' one of the same man seated behind the wheel of a car.[11] Rycroft refused, saying, 'We have living facts to portray and do not need artificial effects.'[12] The

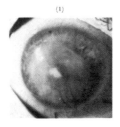 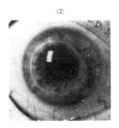

Photograph (1) shows an eye that is blind because the cornea or front "window pane" is opaque. Photograph (2) shows how sight has been restored by a corneal graft operation: a disc of opaque cornea has been replaced by a disc of clear cornea taken from the eye of a person who has just died.

Fresh eyes donated well within 12 hours are urgently needed and all the law requires is that a close relative should give consent. If you are willing to help please complete the form on the next page and appropriate arrangements will immediately be made without involving you in any further trouble or distress.

NATIONAL EYE BANK,
CORNEO-PLASTIC UNIT,
EAST GRINSTEAD,
SUSSEX.

Tel: EAST GRINSTEAD 24111

Figure 17: Rycroft's approach

image finally settled on was a photograph of an English country landscape filled with trees and cows, which was chosen on the grounds that it was the kind of scene most blind people would like, but were unable, to see, although whether or not this was actually the case had not been established.[13] Corneal blindness was represented by the blurred frame and sight restoration by illuminating the central section with sunlight.

Choice of a bucolic scene rehearsed the trope of grafted corneas admitting mostly favourable images onto recipients' retinas. Yet, ironically, when the decision was taken to use it, people dying in rural areas were the least likely to have their pledge honoured.

Technology of Pledging

Three 'donor forms' were printed on the back cover of both original and replacement leaflets that the pledger had to cut out: one was to be given to the pledger's next of kin or someone in the place where they lived; one to their executor – if there was one; one to be sent to the RNIB in an

Figure 18: An image corneal blind people are purported to wish to see

envelope marked 'Corneal Grafting'. On receipt of a completed form, the pledger was issued with a blue pledge card.

The decision to issue the blue pledge card had been taken after other methods of identifying pledgers had been considered and rejected, such as tattoo (effective but outrageous) and brooch (unlikely to be worn). Yet the British public was far slower at acquiring card-carrying habits than its American counterpart, and moreover had a record of resisting attempts to introduce official identity cards. The wartime scheme had been tolerated for reasons of national security, but in failing rapidly to bring it to a close after the war, the Labour government had been accused of 'totalitarianism of the Left'.[14] However, almost a decade had elapsed since a Conservative government had dismantled it, and it was hoped that a pledge card issued by the RNIB would be immune to 'Big Brother' connotations.[15]

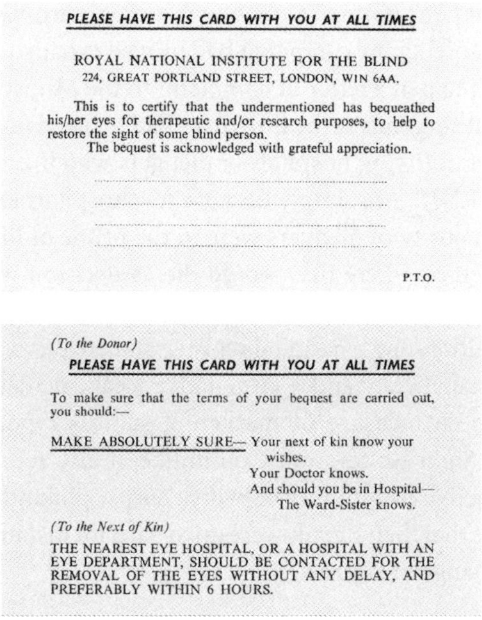

PLEASE HAVE THIS CARD WITH YOU AT ALL TIMES

ROYAL NATIONAL INSTITUTE FOR THE BLIND
224, GREAT PORTLAND STREET, LONDON, WIN 6AA.

This is to certify that the undermentioned has bequeathed his/her eyes for therapeutic and/or research purposes, to help to restore the sight of some blind person.
The bequest is acknowledged with grateful appreciation.

P.T.O.

(To the Donor)
PLEASE HAVE THIS CARD WITH YOU AT ALL TIMES

To make sure that the terms of your bequest are carried out, you should:—

MAKE ABSOLUTELY SURE— Your next of kin know your wishes.
Your Doctor knows.
And should you be in Hospital—
The Ward-Sister knows.

(To the Next of Kin)
THE NEAREST EYE HOSPITAL, OR A HOSPITAL WITH AN EYE DEPARTMENT, SHOULD BE CONTACTED FOR THE REMOVAL OF THE EYES WITHOUT ANY DELAY, AND PREFERABLY WITHIN 6 HOURS.

Figure 19: The Royal National Institute for the Blind's blue pledge card

Pledgers were instructed to keep the blue card about their person at all times, but it was recognized that this was unlikely, and anyway, there was only a slim chance of the flimsy card remaining intact throughout a pledger's life. As a backup measure, the RNIB agreed to inform the hospital closest to a pledger's home that it had received a completed form.

By the end of the campaign's first eighteen months, the RNIB had received more than 25,000 completed forms.[16] H.M. (61) 98, the official circular written in tiny typeface setting out the requirements of the Human Tissue Act 1961, advised hospitals to keep a record of pledges received.[17] But a Mr Smith discovered that hospitals where keratoplasty was seldom performed saw no reason to keep a record. As a gesture of thanks for the excellent care he had received for 'cataract trouble' at the Weymouth and Dorset County Royal Eye Infirmary, he had persuaded the Wyke Regis (Weymouth) Toc H (a fellowship and service club) to organize a local appeal for eye pledges. Leaflets had been sent to the county police, the naval bases at nearby Portland, the UK Atomic Energy Establishment at Winfrith, larger local factories, the Southern National Bus Company, the Southern Railway, all the local opticians, the Established, Free and Catholic clergy, and local branches of the Women's Institute. An 'Eye-bank tableau' staged by the Toc H Women's Group in the town's annual carnival had carried off first prize.[18] Organizing the campaign had proved exhausting. But what had stirred Mr Smith to pen a letter of complaint to the Ministry of Health was his discovery that surgeons at the infirmary had no need of cadaver corneas.

The futility of notifying hospitals of pledgers who lived in their vicinity soon struck the RNIB, not simply because few hospitals kept a record, but also because a majority of pledgers were in the prime of life, and there was no knowing when or where they would die. A decision was taken to stop notifying hospitals, and to devote some of the organization's resources and volunteers to maintaining a national register of pledges.[19] Completed forms were sorted alphabetically and a cumulative total calculated each month. The tally became a measure of marketing success reported at monthly meetings of the Sight Restoration Committee. It also represented the size of an imagined active community of British corpse philanthropists, some of whom made sure that their details were up to date, for instance, by informing the RNIB of a change of address.

Honouring Pledges

Both leaflet and pledge card urged pledgers to tell their family doctor, and 'If you have to go into hospital, tell the Ward Sister.' Admitted to hospital in order to have her gall bladder removed, a doctor's wife refused to mention her pledge lest she conveyed the impression that she feared she would not survive the operation.[20] For their part, many doctors and nurses were reluctant to ask patients if they had pledged their eyes, as the question suggested a poor prognosis. Outpatients of Manchester's hospitals were encouraged to have their pledge recorded in their case notes so that should they die following admission at some later date it might rapidly be honoured.[21] But some people found the practice objectionable, and one patient wrote a letter of complaint to the local newspaper.[22] Greater care was advised to avoid conveying the impression that eye surgeons were casting predatory looks upon seriously ill people.

H.M. (61) 98 additionally advised hospital staff to explain to patients that in some circumstances it might not be possible to honour their pledge. This was partly because demand for cadaver eyes was local and unpredictable, and also the United Kingdom was without an infrastructure that could distribute them nationally. But even in hospitals where keratoplasty was performed, staff on other wards were often unwilling to confront the next of kin of a patient who had just died with the question, 'had she pledged her eyes?' Kinfolk who seized the initiative were sometimes frustrated. In a letter addressed to Anthony Barber, Conservative Minister of Health, a woman complained that no one in the chest hospital where her husband had recently died knew what to do. She was greatly upset, 'not only because some blind person was deprived of having his or her sight restored, but also because my husband was deprived of giving this service which I knew he was anxious to give and I felt that I had let him down.'[23] In September 1964, the Ministry of Health, inundated with complaints along similar lines, asked hospital authorities to send a suitably worded sympathetic letter to disappointed kinfolk.[24]

Where someone died at home the card instructed, 'The nearest eye hospital, or a hospital with an eye department, should be contacted for the removal of the eyes without any delay, and preferably within 6 hours.' But people were often uncertain whom to contact, and found hospital switchboard operators equally ill-informed. The RNIB was inundated with telephone calls from family doctors and kinfolk either seeking information about whom to contact or in the mistaken belief that because its name was printed on the pledge card it was responsible for organizing both hospital and domiciliary enucleation.

Responsibility for domiciliary enucleation typically fell to junior doctors who resented it as an unrewarded extra duty, which had to be carried out almost immediately. Ophthalmologist Patrick Trevor-Roper (1916–2004) proposed offering an 'incentive' of one pound, a sum he believed was sufficient to enlist the support of impecunious junior doctors.[25] Trevor-Roper enjoyed considerable social and professional kudos, and had a thriving private practice. He also operated at Moorfields Hospital, London's leading eye hospital, which led the national field in keratoplasty (the Queen Victoria Hospital came second). But hospital administrators were loath to approve the 'tip' as enucleation already incurred taxi fares or car mileage.

Reducing Waste

'The facts appear to be that there are insufficient corneas available: that you are relying on the Ministry of Health. Heaven help the blind!' a doctor, with justification, warned the RNIB.[26] Education was the Ministry's only policy tool. Out of its offices poured a stream of circulars, typically printed in tiny typeface, containing guidance that local NHS administrators and doctors could observe or ignore.[27] Despite its name suggesting the NHS operated uniformly throughout the British Isles, practice was distinctly local. Ophthalmologists might sail under the same flag, but in many different directions.

As a result, wastage of enucleated cadaver eyes was considerable. Pairs that unexpectedly became available might not be used because at short notice a surgeon was unable to secure an operating theatre or hospital bed, or a patient was unavailable or unprepared for a lengthy stay in hospital. Some surgeons, perhaps unconfident in their skill in excising cornea, insisted on having two cadaver eyes in theatre before starting an operation, one as a backup in case a mistake was made. At the Queen Victoria Hospital only one eye was allocated to each operation; the rule was to excise the cadaver cornea before the recipient's and abandon the operation if anything went wrong.[28] In order to prevent disruption that might disturb his concentration, Rycroft insisted on locking the operating theatre's doors, and communicating with the outside world by passing messages under the door that, hopefully, a passer-by would find.

Cadaver eyes in a moist chamber might be stored for a day or two in a fridge – it was not unusual for staff to find a pair on the same shelf as their lunch.[29] An extended shelf life would allow operations more easily to be scheduled and reduce waste. 'New eyes for old from the deep freeze', was the headline of a newspaper report of experiments conducted in the early 1960s by a team led by Trevor-Roper. One of their research subjects was a woman who had been told eighteen years earlier that she would never see again and who had had her sight restored by a graft of a cadaver cornea that had been frozen for over one month.[30] Rycroft also experimented with freezing; he bought an expensive Matburn Cryobiological Apparatus capable of cooling excised corneas at a rapid and regulated speed, and a Frigidaire deep freezer for storage.[31] Unfortunately, not only was the technique complicated and expensive but grafts of frozen corneas were found to have a high failure rate and the experiment was abandoned.[32]

In H.M. (61) 98 the Ministry of Health had asked regional hospital authorities to consider opening 'eye banks' that would receive cadaver eyes either direct from collecting sites or that were surplus to requirement elsewhere and distribute these to surgeons experiencing a shortage. Appendix II contained an outline plan of one such arrangement. The hospital blood bank was identified as an ideal location, probably because its fridges were dedicated to the storage of human stuff. However, no attempt was made to establish hospitals' response to the request. In 1964, Rycroft renamed his scheme the South East Regional Eye Bank. Each morning its administrator telephoned affiliated collecting sites to ascertain if any cadaver eyes had become available, and then contacted affiliated hospitals to find out where to send them.[33] Enucleated cadaver eyes were secured uppermost in a moist chamber filled with liquid paraffin, which had been packed into a specially made stout copper-lined box half filled with ice. The box lid was held fast by four screws, and picked up either by a Red Cross volunteer car driver or sent by train to the hospital where the eyes were needed, either directly or via East Grinstead.[34] Station staff would telephone the eye bank when the box arrived, and leave it on the platform to await collection. In its first full year of operation the South East Regional Eye Bank handled around two out of every five cadaver eyes enucleated in the country: 200 of the total of 446 were used in the Queen Victoria Hospital's operating theatre; 173 were distributed to ophthalmologists in hospitals as far afield as Southampton and Weymouth; seventy-three were flown overseas to eye hospitals in the Middle and Far East where British ophthalmologists had established a professional relationship.[35]

Rycroft suggested a national network of three or four Regional Eye Banks that would minimize waste and meet growing demand.[36] He was supported by the RNIB because complaints about failure to honour pledges were damaging its reputation, and discrediting its work with the overwhelming majority of blind people whose sight loss was permanent.[37] The NHS might be short of cash but start-up costs were negligible: all that was required was a sink and wash basin, sterile cabinet, microscope and domestic fridge. Indeed, until around 2000, most British eye banks could be mistaken for a kitchen, albeit without a cooker. Equipment was relatively inexpensive: in 1965, it came to around £300.[38] The main expense was the annual running costs of around £2,000, covering the wages of a secretary and several sessions of an ophthalmologist who would evaluate the quality of enucleated eyes.[39]

Ministry of Health staff agreed with Rycroft's proposal and began considering how many and where Regional Eye Banks might best be sited. But speed was not of the essence, and anyway their decisions could only be hortatory. Progress was undermined by senior ophthalmologists who resented Rycroft donning the mantle of leadership and contravening the unspoken rule prohibiting senior medical practitioners from collaborating with central government on plans for changes in medical practice. Before civil servants acted, the Manchester Royal Eye Hospital recast its eye bank as a regional service; Westminster and Moorfields hospitals joined together to form the Westminster/Moorfields Regional Eye Bank, which was promoted as port of call for London-based ophthalmic surgeons.

The Westminster/Moorfields Regional Eye Bank was a collaboration of Trevor-Roper and Thomas Aquinas Casey (1930–1993).[40] London is rich in potential collecting sites. Westminster Hospital (closed in 1994) was one of several large teaching hospitals clustered in the centre. Nonetheless, in its first year of operation, two out of every three eyes had been enucleated by one of the eye bank's team in two London hospices. Moreover, despite its name, Trevor-Roper and Casey, both of whom were keratoplasty enthusiasts, were its chief 'customers'; few cadaver eyes were distributed to other surgeons.

CHAPTER 17

Banking British Cadaver Skin

Skin homografts emerged out of the Second World War with two identities: an actual one as a biological wound dressing, and a potential one as a permanent replacement. In the former skin was handled as if it is homostatic; realizing the latter depended on scientists defeating the doctrinal tyranny. Both identities were confined to burn care, which was suffering its chronic inability in peacetime to capture the support of either policymakers or the general public.

In the early 1960s three new purpose-built regional burns units were opened in Britain. These were 'big-burns units', not 'big burn-units': only the most difficult cases would be admitted. Taxpayers wholly funded two of them. The NHS had been a victim of post-war austerity; its costs rapidly exceeded its architects' forecast, and the little cash the Treasury allowed for capital expenditure was spent on repairing bomb damage and emergency maintenance of hospital buildings. An indication of the depth of government austerity is that capital expenditure pre-war was at least three times higher than 1952–3 levels, if changes in prices are taken into account.[1] In the late 1950s, central government acknowledged that its hospital estate was sorely in need of investment, and the Treasury began releasing funds for an ambitious programme of modernization and hospital building. In 1966, the Leeds Regional Hospital Board opened a new burns unit on the site of the Pinderfields General Hospital, Wakefield. The small town had been chosen because there was little competition from other hospitals for nurses. Staffing is an important consideration: burn care is unpleasant and suffers a high turnover of nursing staff who do most of the work.[2] The South West Metropolitan (Wessex) Regional Hospital Board's Burns Unit was opened in 1967 in one of the single-storey brick wards of the Odstock

Hospital, two miles outside of Salisbury, which the American military had built in anticipation of casualties following the D-Day landings.[3]

In 1963, Queen Elizabeth the Queen Mother laid the foundation stone of the McIndoe Burns Unit, an air-conditioned bungalow on the Queen Victoria Hospital's site, with twelve single rooms that allowed patients to be isolated from one another.[4] Big-money-philanthropists Elaine and Neville Blond had raised £250,000 from family and friends to cover the cost of erecting and furnishing the building. The Ministry of Health provided the site and agreed to cover running costs.

Individuality Markers

Patients admitted to the McIndoe Burns Unit were carefully selected. Its policy was to discourage the admission of small burns and to prevent it from becoming a mortuary for patients admitted with a burn so severe that treatment was futile.[5] It was intended as a research facility, providing research material and subjects for the McIndoe Memorial Research Unit, which had been opened in March 1961 in the hospital's grounds. Initially named in memory of Archibald McIndoe, who had died suddenly the previous year, it was subsequently called the Blond-McIndoe Medical Research Unit to acknowledge the Blonds' generosity. In his speech delivered during the opening ceremony, Conservative Minister for Health Enoch Powell praised big-money philanthropy: 'It will be a poor day for medicine,' he concluded, 'if medical research has to be controlled by a ministry.'[6]

The research unit was the brainchild of plastic surgeon John Watson (1914–2009), who had become McIndoe's protégé shortly after being demobbed from the army in 1947. He envisaged it as a place where laboratory scientists could collaborate with the Burns Unit staff on research into how to counter the antigenicity of skin, which, according to Peter Medawar's doctrinal tyranny, would allow great progress to be made in the field of transplantation. Medawar had called the 1950s the 'golden age of immunology' because it was rich in discoveries in laboratories.[7] His team, which had moved to University College London, was responsible for several in which the skin homograft was the pre-eminent experiment. For instance, they found that skin homografts on mice and rabbits are rejected at different rates: some melt away at a slower pace than others. Medawar postulated that pace of rejection of foreign stuff is determined by how many 'individuality markers' source and recipient share, although he held that even the slightest difference ultimately would result in rejection. In 1953, he announced that his team had discovered how tolerance of foreign stuff might be acquired, a finding coincidentally made by Australian virologist Frank Macfarlane Burnett (1899–1985). Both achievements were recognized in 1960 when the two shared the Nobel Prize for Medicine. Unfortunately, as Medawar was the first to admit, their discovery was inapplicable to medical practice: the investigation involved injecting foreign stuff into embryos in the uterus of a genetically homogeneous-strain of laboratory mice.

In December 1954, Joseph Murray, at Peter Bent Brigham Hospital, Boston, performed the first human kidney homograft that 'worked'. It was a surgical triumph posing no threat to the doctrinal tyranny of skin; it had succeeded because source and recipient were identical twins, and it was common knowledge that body stuff exchanged between identical twins is not rejected. Although few people with end-stage renal diseases have a healthy identical twin, by 1966 Murray had transplanted twenty kidneys where source and recipient were identical twins. Negotiations were not always straightforward: death of one recipient had caused heartache; several successful operations had resulted in a breakdown of relations; sometimes, the healthy twin had refused to donate a kidney.[8]

Attempts to make homovital homografts work included 'denaturing' stuff by subjecting it to extreme heat, cold or chemical treatment, or disciplining recipients' immune systems with drugs or whole body radiation. Investigators had noticed that survivors of atomic bomb explosions over Hiroshima and Nagasaki were more susceptible to illness, an observation which had led them to conclude that radiation somehow weakens the immune system.

Nothing succeeded in extending recipients' tolerance of homovital stuff beyond a few months. However, in 1958, optimism was reinvigorated when French immunologist Jean Dausset identified the first human leucocyte, or white blood cell group, which it was agreed was one of the culprits responsible for rejection of homovital stuff. Dausset went on to propose that white blood cells, like red blood cells, might be classified into various groups.[9] Medawar's individuality markers were now understood as comprising a 'tissue type'. It now looked as if homovital homografts might be tolerated where source and recipient's tissue type matched. 'Compatibility' became an additional criterion of bioavailability of homovital stuff.

The First British Skin Banks

Leonard Colebrook's prototype treatment block was without a skin bank. It was an unthought-of requirement. There was no consensus on how burn wounds should be dressed; each surgeon decided according to their own preferences and resources (scant documented evidence on practice is available). Colebrook's colleague Douglas MacGregor Jackson, who had been appointed to the Birmingham Accident Hospital Burns Unit in 1948, dressed wounds with split-skin mostly the patient's own, cut with a Blair knife, but where insufficient was available, he would instruct a junior member of staff to 'see if you can get some from somewhere'.[10] Sources were mostly donors-on-the-hoof, typically parents.[11]

The Yorkshire Regional Tissue Bank is one of two known to have extracted cadaver skin during the 1960s. It grew out of a collaboration between senior microbiology technician Frank Dexter at the University of Leeds, and Leeds General Infirmary surgeons who wanted to extend the shelf life of arteries for repurposing in grafts. Dexter had experience of freeze-drying bacteria, and also knew how to build freeze-drying apparatus. When too few arteries were found in surgical waste, the team began extracting them from corpses resting in the hospital mortuary. Cadaver bone and skin were added to its inventory of cadaver artefacts, and it was named the Leeds Tissue Bank. It was a modest replica of the Navy Tissue Bank. Unable to afford to copy its elaborate and expensive aseptic extracting ritual in an operating theatre, extraction was undertaken in hospital mortuaries, and the risk of transmitting cadaver infection to recipients minimized by chemical techniques of terminal sterilization.[12] When the Regional Burns Unit was opened, the Leeds Tissue Bank was relocated to a converted linen store on the Pinderfields Hospital campus, and was renamed the Yorkshire Regional Tissue Bank. Its inventory was expanded to include freeze-dried cadaver bone, dura mater and skin artefacts that were distributed to hospitals within the Leeds Regional Hospital Board, which included the Burns Unit.

Watson's plans for the McIndoe Burns Unit included a cadaver skin bank, which would support clinical research into whether or not tissue typing might extend the life of full-thickness skin homografts, and perhaps even transform them into permanent replacements. He had been encouraged by observations that skin homografts survive twice as long on burn victims than on healthy people, which were interpreted as suggesting that terrible injury suppresses the immune system. However, the technology of identifying tissue type was in its infancy. An ambition of the Blond-McIndoe Medical Research Unit was to further its development, and eventually provide a tissue-typing service for British surgeons undertaking experimental grafts of homovital stuff. Medawar agreed to serve as one of its three scientific advisers (the other two were pathologist and immunologist Peter

Gorer (1907–1961) and surgeon David Slome (1906–1995)), and Danish immunologist Morten Simonsen (1921–2002) was appointed to lead the research team. Their involvement attracted financial support from the Medical Research Council.

There are many more tissue types than red blood groups. Hence there was a slim chance of a corpse resting in a local mortuary being compatible with a patient admitted to the McIndoe Burns Unit. Watson hoped that by extracting skin from random corpses, and extending its shelf life by freezing, a wide spectrum of tissue types would be captured, increasing the likelihood that some in the freezer was compatible with a newly admitted patient.

Trainee surgeon Tom Cochrane was charged with the responsibility for creating Britain's first bank of viable cadaver skin. He was awarded the first fellowship in a scheme supported by the charitable foundation of property developer Max Rayne (1918–2003). Technical, logistical and legal obstacles impeded progress. In 1952, Medawar's team had found that soaking skin in glycerol before freezing at a steady controlled pace protects it from damage.[13] But suitable apparatus was unavailable. Cochrane designed a prototype 'biofreezer', which the research unit's chief technician Ron Chambers built out of begged and borrowed parts.[14] It was a 'Heath-Robinson' affair, resembling a giant Thermos flask, about 4 feet (120 cm) high, with the capacity to hold 180 sheets of skin, each measuring 8 x 4 inches (20 x 10 cm).[15] 'The idea is just like a butcher's deep freeze,' Chambers explained to readers of the *East Grinstead Observer*, 'and then we developed the right mixture of liquid nitrogen to freeze the skin.'[16] Watson reached an agreement with Spembly Technical Products Limited to manufacture and market what became known as the STP Biofreeze Unit.[17]

Cadaver skin remains viable up to sixteen hours following death, which left little time to complete the formalities understood to have been stipulated under the Human Tissue Act, for a surgeon to reach the mortuary of local hospitals that had agreed to serve as collecting sites, recover skin from a corpse's back, buttocks and thighs, place it on ice and deliver it to the Queen Victoria Hospital where it was prepared for freezing. Cadaver skin recovered from corpses after the sixteen-hour watershed was freeze-dried, which was technically easier and cheaper than deep-freezing viable skin. The price of an Edwards High Vacuum, off-the-shelf freeze dryer was around £500, around one-tenth of the price of a STP Biofreeze Unit; one freeze-drying cycle cost 20 pence, whereas routine weekly maintenance of the biofreezer came to £2.50, and 50 pence was incurred each time a batch of viable skin was frozen (the latter sum excludes expenditure on tissue typing).[18]

Watson initially planned to stock the bank with the skin of 'coroners' cases'; he was under the impression that the Human Tissue Act had conferred on coroners the authority to agree without knowledge and agreement of next of kin, which was advantageous not only in terms of speed but also because he believed kinfolk would be squeamish and likely to say no.[19] However, not only did this interpretation of the act's notoriously ambiguous wording differ from that of his ophthalmic colleagues, it was also unlike that of staff of the hospitals in London, Brighton and Haywards Heath that had agreed to act as collecting sites. The latter were insisting that agreement of coroner, coroner's pathologist and next of kin was obtained. Watson changed tactics: agreement of next of kin of non-coroner's cases would be sought when they asked for the death certificate, required in law for a corpse's disposal. The task fell to staff in the decedents' affairs office, which was advised to approach only people who had displayed an appropriate attitude towards 'death and flesh'.[20] How this was to be ascertained is unclear. Perhaps the test was their reaction to the skin bank's leaflet, which was illustrated with three photographs of a badly burned child taken before and after skin homografting.[21] The skin bank's dependence on the goodwill of clerical staff was recognized by Watson, who asked the management committees of cooperating hospitals to thank them on his behalf.

The original plan had a surgeon doing the extraction, as the Human Tissue Act required. But a surgeon commanded a sizeable fee, and was available mostly at night when mortuary technicians who were needed to manhandle the corpse were off duty.[22] A decision was taken to train and tip mortuary staff. If challenged, the Medical Research Council's tips for cadaver pituitary glands could be cited.[23] Cadaver skin was suspended in a solution of penicillin in a glass jar until Ron Chambers arrived in the research unit's blue Ford Escort van to collect it.[24] But extracting full-thickness skin is technically challenging, and quite a lot extracted by mortuary staff was rejected.

Researchers found homografts of compatible cadaver skin can survive for up to ninety days before finally melting away.[25] But American surgeon Thomas Starzl (1926–2017), a virtually unknown newcomer to the field of transplantation, had undermined the skin bank's rationale shortly after it opened. In 1963, at a National Research Conference in Washington DC, he announced the recipe of an innovative cocktail of immunosuppressive drugs including a new one called azathioprine, which extended survival of kidney homografts beyond one year. What he called 'the clinical gold rush' immediately took off; hospitals competed to establish kidney transplant centres.[26] The cocktail also delayed rejection of skin homografts, and was more convenient to administer than organizing a bank of frozen viable cadaver skin.[27] During the decade in which the skin bank operated, skin was extracted from around 150 corpses of people whose age at death ranged between twenty-five and ninety-eight.[28]

Changing Landscape

The landscape of burn care was experiencing what has been described as a seismic shift.[29] At its epicentre was the burns unit of a hospital in Maribor, a small town currently in Slovenia, which at the time was behind the Iron Curtain in northern Yugoslavia. Economic privation was responsible for a tragically high incidence of burns and scalds, particularly of children; it also left the burns unit short of cash, particularly of foreign currency with which to buy pharmaceuticals and consumables and pay for the repair or replacement of broken equipment (a barber's razor sharpened on a strap was described as the pearl among its instruments). Each year, around 350 victims were admitted. Isolating patients was impossible. Infection in the overcrowded ward was rampant and evident from the stench. It was exacerbating the extent of injuries, prolonging victims' suffering and frequently killing them.

Zora Janžekovič, a young surgeon placed in charge of the burns unit, realized that the conventional approach to treating deep and extensive burns and scalds, which had been recommended by Blair and Brown between the two world wars, courted infection; it was passive, involving waiting for superficial wounds to heal spontaneously, and only then attempting to close deeper ones with autografts, which frequently failed because of infection. In 1960, she threw convention aside and adopted a proactive treatment regime: shortly after admission to hospital, damaged skin and flesh were excised until the healthy raw surface was exposed, and open wounds covered with split-thickness skin grafts. The procedure is not for the faint-hearted: a great deal of blood is lost. The hospital was without a skin bank: wounds were dressed either with autografts or, where the injury was extensive, homografts, with skin extracted from donors-on-the-hoof. The results were rewarding: deaths from sepsis fell dramatically, wounds healed faster, patients experienced less pain and misery, the average stay in hospital was significantly shortened to around fourteen days, and patients returned to school, work or their usual activity far sooner.[30]

Janžekovič's reputation spread and Maribor became the 'Mecca' of the world of burn care: between 1968 and 1984 she was visited by 237 international surgeons and invited to present her work to numerous professional audiences. In some, but not all, hospitals, early excision and skin grafting became standard treatment; where, when and how depended on the personal preferences of clinicians.

Demand for biological wound dressings took off. In order to maximize the little skin that became available, a new technique called meshing began to be used, in which a commercially manufactured mesher dermatome expands the surface area of skin to more than double its size. Meshed skin is another kind of artefact, irrespective of whether its source is a patient, donor-on-the-hoof or corpse.

Absent Progress

In 1970, an official review of burn care facilities within the NHS found a majority was ill equipped and housed in inappropriate and/or dilapidated buildings.[31] For instance, the Nuffield Burns Unit in Stoke Mandeville Hospital, which Gillies had declared open in 1956, was housed in a converted Nissan hut that had been erected during the Second World War. The last time central government had issued guidance was in 1952. Some changes in organization had been made to comply with the Royal College of Surgeons' insistence that trainee plastic surgeons gain experience of burn care. However, the marriage of burn care and plastic surgery was unhappy: not only do burn victims require treatment by staff of other medical specialties, but few experienced plastic surgeons were prepared to undertake NHS work. Fees from private practice were much more lucrative than the salary offered by the NHS, and, anyway, many hospitals had failed to purchase the expensive instruments required to undertake delicate reconstructive work. Another challenge was accommodating the different patterns of demand within the same facility: drastic burns are unpredictable and seasonal – dedicated beds are often empty during summer months – whereas demand for 'non-urgent' plastic surgery is relentless, and in some hospitals would-be patients were kept waiting for up to ten years.

In 1978, the Department of Health and Social Security (DHSS) published a list of burns units in each of the fourteen Regional Hospital Boards in England and Wales.[32] Three had none, and few of the thirty-nine identified had been purpose built. The list had been drawn up in order to answer a question put by MP Michael Ward (Labour Peterborough).[33] Central government still had no appetite for developing policy on either plastic surgery or burn care. Indeed, civil servants drawing up the department's Sixth Planning Cycle admitted they had no intention of updating the 1952 guidance.[34] Modernization or upgrading depended largely on the hospital in which a burns unit was situated winning a share of the funds the Treasury was releasing for a major programme of hospital building. It had taken local businessmen and dignitaries seven years to raise £400,000 towards the cost of building and equipping a ten-bedded burns unit at St Andrews Hospital, Billericay. When the Duchess of Kent opened it in 1982, the gap in services in the North East Thames Region, which includes a quarter of London, was finally filled.[35] Civilian burn care was a Cinderella service without a charismatic champion.

CHAPTER 18

The Burn-prone Society

Leonard Colebrook's reputation as campaigner for measures to reduce the risk of burns and scalds in children, and pioneer of modern burn care, reached American shores, and in 1950 he was invited by the New York Academy of Medicine to present a map of the American landscape of burns and scalds. As he had found in the United Kingdom, systematic national data were unavailable, and so his map was a sketch drawn from a miscellany of studies. His lecture opened with a timely and telling observation: the number of American civilians dying of a burn or scald each winter, which he put at around 6,000, was equal to the number of American troops killed during the first six months of the Korean War. In other words, the home front was as least as dangerous as the battlefield.[1]

In both the United States and United Kingdom the home environment was more dangerous than that of the workplace. But American homes apparently were more hazardous than British ones because many are built of wood and hence susceptible to conflagration. On the positive side, American children appeared slightly less at risk of a burn or scald than British ones. Yet coincidentally, flammable clothes, one of the chief culprits, was a public issue because an untold number of boys had been badly burned when their brushed-rayon Gene Autry cowboy suits caught fire, turning them into human torches. Between 1945 and 1953, at least 100 families brought lawsuits against manufacturers and distributors of the cowboy suits, and had been awarded around ten million dollars. In 1953, Congress had adopted the Flammable Fabrics Act, prohibiting manufacture and sale of highly flammable clothing. But it had a negligible impact on safety: the mandatory test of flammability excluded only 1 per cent of potentially flammable garments, which were mostly cheap imports, and anyway, enforcement was slack.

Colebrook reckoned that each year at least 70,000 Americans were admitted to hospital following a burn or scald. However, survival rates were increasing thanks to better treatment of burn shock and the discovery of penicillin, which was capable of defeating some infections. The inevitability of skin homografts melting away was generally accepted but their application as biological wound dressing increasingly valued. What had not changed was reliance on donors-on-the-hoof, whose sacrifice was still applauded in morality tales. For instance, the *New York Times* described how ten inmates of Ohio Penitentiary had repaid their debt to society with 640 square inches (more than 4,000 sq. cm) of skin that dressed the wounds of a badly burned nine-year-old-girl;[2] and how fifteen 'skin brothers', motivated by solidarity for a fellow trade unionist, had provided 720 square inches (4,600 sq. cm) of skin for a boiler mechanic caught in a flash fire at Consolidated Edison Company's Waterside plant.[3]

According to James Barrett Brown, these gestures were heroic but unnecessary. On his resumption of civilian medicine, he began championing cadaver skin. As he explained to his colleagues, 'Put in a plain way, a large amount of good, usable skin is going to waste that could be life-saving.'[4] Corpses require neither anaesthesia nor hospitalization for post-operative wound care, sparing donors-on-the-hoof pain, loss of earnings and hospital bills. Moreover, sheets of split-thickness cadaver skin are more effective as a wound dressing than a patchwork of small pieces of skin cut from several bodies. On average, the corpse of an adult can provide 4–6 square feet (3,700–5,600 sq. cm) of skin from flat surfaces – chest, back and thighs.[5]

Brown knew his colleagues regarded cutting corpses as dirty mortuary work that demeaned a professional reputation won by operating on living patients in aseptic operating theatres. The source of skin might be dead but, he insisted, removing split-thickness cadaver skin is technically challenging and demands a surgeon's skill. He published illustrated guides to his methods.[6] Cadaver skin should be clean and its source free of disease that might be detrimental to recipients. Next of kin's agreement must be obtained; their resistance might be overcome by reassurance that neither the corpse's appearance would be harmed nor the funeral arrangements disrupted (although in practice this was not necessarily the case as funeral directors sometimes struggle to contain embalming fluids leaking from areas of the corpse without skin). Brown also suggested showing kinfolk gruesome photographs of a severely burned child, and explaining that death did not preclude people from saving another's life.

Skin separated from the corpse within a few hours following confirmation of death, or up to twelve hours where the corpse is refrigerated, should either be folded and wrapped in damp saline sponges or spread out on a damp roller bandage and wrapped around a test tube. Brown had found cadaver skin suspended in a solution of saline and antibiotics in bottles, and stored in a refrigerator at 4°C, fit for repurposing for up to three weeks (there is no agreement on length of shelf life). Refrigerated cadaver skin unused after its 'best-before' date should be freeze-dried.

Surgeons working in hospitals admitting few burn victims should rely on a 'just-in-time' scheme, where skin is removed as and when required

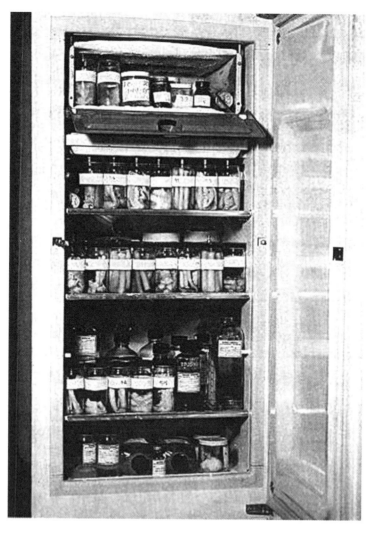

Figure 20: James Barrett Brown's 'skin bank'

from a fresh corpse resting in the hospital mortuary; surgeons working in hospitals admitting significant numbers of severely injured victims should establish a skin bank. 'All that is really needed is a reliable refrigerator that can be kept at 4°C, and the will to make the project succeed.' His refrigerator could hold up to 30,000 square inches (19,000 sq. m) of cadaver skin.

The Burn-prone Society

In the 1960s, Americans were shocked to learn that their society, the wealthiest on the planet, enjoying cheap oil, gasoline and gas, and high consumer spending, was burn-prone. Never mind the threat posed by the communists' arsenal of nuclear warheads: every year it was estimated that around 12,000 Americans were dying as a result of burns and 300,000 were injured, 50,000 of whom required hospitalization.

Realization that fire is a national hazard undermined the entrenched belief that prevention is an individual or local responsibility. Pressure was brought to bear on the federal government to take the lead. Its response was the Fire Research and Safety Act 1968, which proved largely ineffective. In 1971, President Nixon appointed a twenty-member National Commission on Fire Prevention and Control to investigate causes and identify solutions.

The commission's final report, *America Burning* (1973) observed that the public and policymakers had been misled into thinking that a majority of burned or scalded civilians were casualties of fire catastrophes, the official term for a conflagration where twenty-five or more people meet their death that typically occurs in gathering places such as factories, night clubs and cinemas. Fire catastrophes capture news headlines, challenge the resources of first responders, overwhelm medical facilities and staff and often require military assistance. For instance, in 1958, the Navy Tissue Bank provided cadaver skin to dress the wounds of victims of the fire in Our Lady of Angels School, Chicago, in which ninety-five people were killed and ninety-three injured.[7] But they are rare: during the entire twentieth century, 8,530 Americans lost their lives in these circumstances. Moreover, the toll exacted by fire catastrophes has been falling: during the first half of the twentieth century, on average a fire catastrophe claimed 192 lives, whereas in the second half of the century that number had fallen to around forty.[8]

Fire catastrophes typically are attributed to neglect, corruption or wilful flouting of fire precautions in gathering places. The government had addressed these, albeit inadequately, but had neglected Americans' homes, which were filled with dangerous ignitable materials that were either lawfully sold, such as combustible furniture and upholstery, or products and appliances that had become fire hazards through wear or misuse. Economically disadvantaged Americans faced the greatest dangers, forced to live in burn-prone environments, such as crowded run-down apartment houses and tenement buildings where landlords saw no financial advantage in long-term upkeep of their properties. Homeless and physically or mentally frail people are particularly burn-prone. The commissioners also acknowledged the significance of indifference, irrationality, carelessness, desperation, addiction and ignorance.[9]

Modernizing American Burn Care

Congress incorporated many of the commissioners' ninety recommendations in the Federal Fire Prevention and Control Act 1974 that, among other measures, created the National Fire Prevention and Control Administration, a federally funded programme of fire prevention initiatives informing and supplementing local efforts. But some injury is inevitable.

Whereas the Navy's contribution to the United States' struggle against the spread of communism was to create freeze-dried cadaver artefacts capable of withstanding shipment to and use in distant combat zones, Army surgeons had focused on investigating how to improve burn care by establishing a specialist burns centre on American soil, at Brooke Army Hospital, Fort Sam Houston, Texas. Civilian patients including children were occasionally admitted, but during its first three decades a majority of patients had been injured on the battlefields of Korea or Vietnam where rates of scalding, burning and frying were greater than in previous wars as a result of innovations in conventional weaponry that exploited the destructive power of heat. Rapid evacuation by helicopter ambulance to hospital ships *Repose* and *Sanctuary* or military hospitals in South Vietnam and Japan improved chances of treatment of and survival of burn shock.[10] Resources were greatly stretched, for instance, between 1966 and 1978, when the 106th General Hospital in Japan admitted 1,963 burned servicemen, of whom 1,312 were flown by the Air Force to the Brooke Army Hospital. Demand for cadaver skin was so great that the Navy Tissue Bank's supply was exhausted,[11] and Brooke Army Hospital surgeons sometimes resorted to dressing open wounds with the skin of dead dogs and pigs.[12]

The United States Army Institute of Surgical Research disseminated its burns research findings at annual meetings attended by both military and civilian surgeons. The first of a series of annual National Burn Seminars was held in 1959 at Fort Sam Houston. At the eighth seminar, held in Birmingham, Alabama, the evident need for a national burns organization representing the various disciplines involved in burns care was met through the formation of the American Burn Association.[13]

The American Burn Association began organizing civil society and professional organizations in a nationwide campaign to draw the attention of the public and policymakers to the many shortcomings in hospital facilities for treating and rehabilitating victims of burns and scalds. It drew up a national plan organized around hub and spokes. The majority of victims of a burn or scald have superficial wounds that can be treated at home or in a local hospital's emergency department. But ideally, the most severely injured should be admitted to a specialist *burns centre* – a hub – where expertise in treatment and rehabilitation was concentrated, research and training undertaken, and where patients were segregated along the lines first suggested by Colebrook, but with the addition of life support equipment found in intensive care units, which had begun to be opened during the 1960s. At the spokes were *burns units*, with the capacity to treat at least thirty-five less severely injured patients each year, or a general ward where trained staff were part of a *burns programme*, and followed an agreed protocol.

Burn Care is Loss Making

In 1965, Congress adopted Medicare and Medicaid to shelter people who hitherto had been unable to meet medical bills. Medicare is a federal government programme for people aged sixty-five and over; Medicaid is a part federal- and part state-funded scheme for some low-income people. Whereas entitlements under Medicare apply nationally, states can decide who is eligible under Medicaid, and because criteria vary from state to state, the scheme is characterized by a postcode lottery.

Both schemes made more welcome in hospital the typical burn victim, who is low income and/or underinsured or uninsured, but did little to encourage investment in modern facilities. Ironically, the financial viability of a burns centre, with its concentration of medical skills and up-to-date technology, is hostage to any improvement in the survival rates it achieves. Treatment and rehabilitation of a survivor of a drastic burn rank among the most expensive of all types of medical care. Survivors can occupy a hospital bed for months, during which time they will receive intensive nursing care, are wrapped in acres of dressings, consume vast quantities of medication, and on several occasions tie up a surgery team and an operating theatre for a full day.

A nationwide survey of 6,062 hospitals carried out in 1969 found only twelve had a specialized burns centre; twenty hospitals had a burns unit; thirty had a burns programme; 300 treated patients in the intensive care unit; the remaining 5,684 hospitals had no specialized facilities or staff.[14] In *America Burning* the commissioners pointed out that outside dedicated facilities, death rates were significantly higher, survivors' stay in hospital prolonged and their rehabilitation inadequate. They criticized the low priority the National Institutes of Health had been giving to research into burns care and rehabilitation: in the fiscal year 1972, its support amounted to $1.25 million, whereas, for instance, $34 million had been spent on research into kidney disease. This was clearly disproportionate: each year, heat claimed the life of 12,000 Americans, whereas 9,000 died from kidney failure. The National Institutes of Health was instructed to spend up to $5 million on research into burn care and rehabilitation in the fiscal year 1974/5, and invest some of this sum in supporting the creation of twenty-five additional burns centres. But this paltry amount was unlikely to encourage investment in loss-making facilities. In 1976, each day spent in a burns centre cost around $800, and some victims stayed there for up to sixty days. In addition, rehabilitation of one patient could cost more than $50,000.[15] But reimbursement of cost of burn care by third-party payers was seldom adequate, consisting of a hodge-podge of payments for each separate procedure that rarely covered all costs. Burns centres were experiencing unpaid bills at a rate two and half times greater than write-offs for other patient categories.[16]

In October 1972, Congress adopted what is known as the End-Stage Renal Disease Program, through which everyone with chronic renal disease, regardless of age, became eligible for Medicare, which reimburses all costs associated with dialysis and kidney explant/transplantation, including the acquisition of cadaver kidneys, the operation and twelve months' post-transplant care. The programme is extremely expensive, costing Medicare $228.5 million in 1974, $361.1 million in 1975 and increasing every year since.[17]

Why was an exception made for people with end-stage renal disease? Sociologist Richard A. Rettig traced the origins of the unique decision to the early 1960s when the Veterans Administration, the Public Health Service and private providers began opening artificial kidney machine (dialysis) centres throughout the country, a development coinciding with what Starzl had called 'the clinical gold rush'.[18] However, demand outstripped supply: access to dialysis was subject to explicit rationing, and transplantation by happenstance; that is, the admission to hospital of a suitable source whose next of kin had been asked and had agreed to explantation. Vigorous advocacy in the public sphere on the part of representatives of the explant/transplant industries and increasingly vocal patient advocacy groups resulted in late 1965 in the appointment of the Committee on Chronic Kidney Disease (also known as the Gottschalk Committee after its chairman, Carl W. Gottschalk) to advise the Bureau of Budget on how both procedures might be financed so that access might be widened and providers protected from financial loss. Its report, published in 1967, recommended a national treatment programme funded by Medicare. Pressure to extend the programme increased, and included arranging for a man to undergo dialysis, albeit briefly, at a hearing of the House Committee on Ways and Means in November 1971. He was a family man in his prime working years; the message conveyed was, with the government's help more men like him could be rehabilitated and returned to gainful employment. Rettig concludes that, ultimately, politicians reckoned the moral cost of failing to provide care to upstanding American citizens was deemed to be greater than the financial cost of doing so.

Political expediency in the face of vigorous advocacy can account for the exception made by Congress for funding the treatment of kidney disease, but it is also the case that advocacy on behalf of burn victims is encumbered by the over-representation of socially stigmatized disadvantaged people. 'There are people who think what we do is a sin', commented a member of staff of a burns unit. 'Saving burn patients. There are definitely people out there who think that people who get burned deserve what they get and why should we spend all these resources on them.'[19] Moreover, most burn care professionals know that neither their own family nor friends want to hear about their work.

Shaky Financial Context

Just under half of all burns centres in the United States were established in public hospitals that had been founded to serve the poor and indigent, and which unlike private hospitals tended not to 'cherry pick' patients who offer a better and more secure financial return.[20] Philanthropic support was essential. Each year during the 1970s, Shriners raised around $12 million in support of three children's burns centres. The civil society organization 'dedicated to fun and philanthropy' – members sport a red fez – had been established in 1870 as the Ancient Arabic Order of the Nobles of the Mystic Shrine. Its original Grand Endeavour was to provide free hospital care to children, mostly victims of polio but, following the disease's eradication in the 1950s, replaced by burns and scalds, a significant cause of death and disability in children. In 1962 their first children's burns centre was opened in Galveston, Texas.[21] Subsequently, children's burns centres were opened in Cincinnati and Boston.

In December 1976, a twenty-four-bed burns centre was opened on the seventh floor of the New York–Presbyterian Hospital.[22] It had cost $25 million because a helicopter pad was considered essential to allow severely injured patients to be flown in to the centre. It was urgently needed: New York City had more fires and more burns victims than any other city in the United States. It was also over-supplied with hospital beds. Yet none had specialist facilities for treating burn victims of any age. They were scattered around the city's 350 hospitals, and the most severely injured people occasionally flown to a burns centre elsewhere in the country, sometimes as far away as Texas. Fire catastrophes in the 1970s, such as the ones at a Bronx social club and in a chewing-gum factory in Queens, drew calls from community leaders for improvements in local burn care. But it took the huge plane crash at JFK airport in June 1975, in which 109 people died and fifteen survived, for them to be heeded. Two months later, Mayor Abraham Beame (Democrat) led a fund-raising campaign.

'We took a calculated risk in opening the center because of the financial uncertainty of such an enterprise,' explained Dr David D. Thompson, director of the hospital's Medical Center, to readers of the *New York Times*. When it opened the New York Burns Center anticipated an annual loss of $1 million, which the hospital sought to defray by linking it to a plastic surgery facility. 'Under that plan, the deficit for the care of burn patients would be offset by profits from patients having face lifts, nose bobs, bust enlargements and other types of cosmetic surgery.'[23] Put another way, the Burns Center's finances were organized around 'robbing' wealthy self-paying Peters, unhappy with their appearance, to pay for treating burn-prone Pauls as charity cases. Unfortunately, the strategy appears to have failed: in 1984, the Burns Center cost the Medical Center $9 million and had notched up a deficit of between $500,000 and $750,000.[24]

Financing Skin Banks

Funding secured through the End-Stage Renal Disease Program encouraged the creation of a nationwide network of non-profit organ procurement agencies responsible for the explantation of cadaver kidneys, whereas finances of skin banks were shaky, typically dependent on those of the burns centre to which it was attached. Records are incomplete but nonetheless suggest that the number of entities describing themselves as a cadaver skin bank increased from three in 1967, to six in 1972, to fifteen in 1977 and to thirty in 1982.[25] Early ones were similar to Brown's, consisting of a laboratory bench on which strips of cadaver skin were processed for storage in a refrigerator. The first to be equipped with a freezer was opened in 1970 at the Massachusetts General Hospital by surgeon John Burke (1922–2011). Burke is credited with importing Janžekovič's innovation of early excision and wound coverage into the United States; he was also instrumental in the establishment in 1968 of the burns centre at the nearby Shriners Children's Hospital. His team drew on the experience of freezing techniques developed by Cochrane and Watson at the Queen Victoria Hospital, East Grinstead. But whereas the British team's ambition was to extend the survival of homografts by matching tissue type of source and recipient, the Boston team relied on cocktails of immunosuppressant drugs to achieve the same end, and hence avoided the trouble and expense of tissue typing.[26] And whereas Cochrane and Dexter designed and assembled their equipment, their American counterparts adapted items developed for convenience food manufacturing.

In 1973, eight-year-old Sherry White suffered massive burns to her body. Only the sole of her right foot and upper part of her head were undamaged. Charles Baxter (1929–2005), professor of surgery at Southwestern Medical School and head of the burns centre at Parkland Hospital, Dallas, Texas, was responsible for her remarkable recovery.[27] Sherry's extensive wounds were dressed with the frozen skin of six cadavers held in its skin bank, which had been opened in 1972.[28] A team led by microbiologist Ellen L. Heck extracted, processed and froze the skin of corpses, most of which had been identified by Dallas County's medical examiner who contacted next of kin on the skin bank's behalf.

The Southwest Medical Foundation, Crystal Charity (a women's group in North Dallas), the Golden Charity Guild, and District 2X-1 (Texas) Lions (all men) combined forces to raise $240,000 for a new facility at Parkland Hospital, which allowed the skin bank to move its laboratory out of the surgery department into an entire floor above the medical examiner's office.[29] Collecting sites were the Parkland Hospital and its satellite institutions, a funeral director's premises and the medical examiner's morgue. Very occasionally frozen cadaver skin was exported to hospitals outside Texas, but its burns centre was so busy that supply always fell short of demand and each week several requests had to be refused. Recipients were charged $55 per square foot to cover costs – Baxter's were said to be four times lower than cadaver skin from other sources. Subsequently Baxter's team began processing cadaver skin on behalf of smaller banks, and by 1980, around one-half to one-third of all cadaver skin recovered in the United States passed through their hands.

Firefighters were – and remain – enthusiastic fund-raisers. Jack Meara and Joe Hickey, firefighters based in North Bronx, began a fund-raising drive in support of the New York Firefighters Skin Bank. The initial donation of $40,000 was raised at a football game between 'New York's Finest' (policemen) and 'New York's Bravest' (firefighters) – the police won 21:12 – and by fining firefighters a dollar whenever they swore.[30] Opened in 1978, the New York Firefighters Skin Bank was the eighth in the United States of America to boast a freezer.[31] It still occupies a cramped suite of rooms on the twenty-third floor of the NewYork–Presbyterian Hospital and serves the Burns Center on the seventh floor. An annual target of 300 corpses was set. Firefighters donated three vehicles that could accommodate an equipment trolley, which resembles a robust steel hostess trolley. Teams of extractors are available round the clock, consisting of a retrieval technician who is typically a medical student paid a fee per corpse (the experience can stand them in good stead when applying for jobs in surgery), and a technician who prepares cadaver skin for transportation. Bioprospecting territory stretches north to Poughkeepsie, east to the tip of Long Island, and in the five boroughs.[32]

CHAPTER 19

Harvesting the Dead

In 'Harvesting the Dead: The Potential for Recycling Human Bodies', a much-cited article published in 1974, psychiatrist and ethicist Willard Gaylin conjures up the image of a 'bioemporium', a gothic combination of intensive care unit, nursing home, operating theatre, laboratory and warehouse, in which what he called 'neomorts', which are warm to the touch, breathing, pulsating, evacuating and excreting but permanently comatose bodies, are held for repurposing in medicine. Galyin suggests a variety of potential users: medical students learning how to perform embarrassing procedures such as vaginal and rectal examinations or radical operations such as amputation; scientists seeking replacements for laboratory animals, prisoners and 'mentally retarded children' on which to test antidotes to poison or investigate the capacity of novel substances to treat induced diseases; blood banks replacing inconvenient donors-on-the-hoof with 'blood-producing factories'; transplant surgeons requiring a storage facility of organs.[1]

Gaylin was speculating on the implications of official recognition in 1968 of two identities that Americans might assume: the 'brain-dead', sometimes called 'beating-heart', corpse as source of transplantable stuff, and the posthumous pledger. Adoption of the former was clearly involuntary, and had been formulated by an ad hoc committee at Harvard University;[2] the latter was voluntary, and had been set out in the Uniform Anatomical Gift Act 1968, which, as Gaylin put it, gave Americans the right to pledge all 'necessary organs and tissue' en masse simply by filling out and mailing a small card.

Brain Death

A 'brain-dead' person is comatose because their brain stem, the part of the brain that controls vital body functions, such as breathing and heart rate, has been irreparably damaged, typically unexpectedly following a stroke, car crash, suicide attempt or violent physical assault. One study found that without aggressive support of their vital functions, 20 per cent of brain-dead patients experience cardiac death within six hours, and 50 per cent within twenty-four hours.[3]

Machines, called iron lungs, were invented to support people incapable of breathing independently because they had been paralysed by polio. Polio's eradication threatened the machines with redundancy, but an alternative use had been found, which was to support patients in the immediate aftermath of heroic treatments such as cardiac surgery, or who for a variety of different reasons had become comatose. Initially, these patients were scattered around hospital premises, but during the 1960s, hospitals began gathering them together in what became known as intensive care units, which were mostly the responsibility of anaesthetists but where nurses play a crucial role. Demand rapidly outstripped space, equipment and staff. The concept of brain death was originally intended as a management tool providing proof that coma is irreversible and thus further treatment is futile and should be withdrawn.

Shortly after the ad hoc committee announced it had agreed the concept, it was seized on by the kidney transplant community as a solution to the problem of 'freshness' that was hampering its activities. The viability of homovital stuff rapidly deteriorates following cardiac death and cessation of blood circulation. Put another way, a transplant stands the best chance of success where the kidney is explanted from a body in which blood is circulating, is immediately placed on ice and transplanted within hours. In effect, the concept created a 'no-place' in operating theatres in which a 'beating-heart' corpse is disassembled and experiences cardio-pulmonary death.

The concept was, and in some people's minds continues to be, controversial.[4] 'Harvesting the dead' articulates the public's distrust of medicine which had been cultured during the 1960s in a succession of scandals that included the tragedy of thousands of babies with severely malformed limbs delivered of women who had been prescribed the sedative drug thalidomide during pregnancy, and exposés by Henry Beecher in the United States and Maurice Pappworth in Britain of unethical medical experiments.[5] Little wonder then that some people feared the concept provided scientific justification for denying desperately ill patients potentially life-saving treatment, a fear exacerbated and elaborated in 1967 by the first audacious human heart transplant, which encouraged a belief that it was a ruse on the part of surgeons to obtain 'spare parts'.

The Uniform Anatomical Gift Act 1968

Whereas the Anatomy Act 1832 had pretty much resolved matters in Britain as far as Parliament and medical schools were concerned, according to medical historian Michael Sappol, similar legislation had a chequered history in the United States.[6] In 1831, Massachusetts became the first state to legislate on the matter, but popular opposition to the utilitarian ethic of the anatomist and his supporters succeeded in either blocking adoption of similar statutes or, once enacted, in having the legislation repealed. However, the post-Civil War climate favoured medical education, research and dissectors, and by 1913, a majority of states had adopted a law permitting medical schools to appropriate unclaimed bodies for dissection. The exceptions were southern states where demand for corpses for dissection was easily satisfied by prisons housing large numbers of men, disproportionately African Americans sentenced to hard labour under terrible conditions. Elsewhere, following the introduction in the 1930s of taxpayer-funded grants towards the cost of burial and cremation, fewer corpses were unclaimed and, faced with a widening gap between supply and demand, American medical schools began marketing pledging of whole corpses.[7]

Anatomy acts dealt with corpses for dissection; 'donation statutes' were adopted to fill the legal silence on pledging cadaver stuff for repurposing in medicine.[8] Lions staked the field for cadaver eyes. In 1954, the Iowa Lions began buttonholing members of the state legislature, explaining the invaluable service eye pledgers could render people with corneal damage, and in September 1955, it became the first state to adopt a donation statute.[9] In 1960, Governor Nelson Rockefeller, a Lion, signed one that had been sponsored by New York Lions.[10] By 1965, nearly half of all states had adopted a donation statute.

Unfortunately, early donation statutes were at best unhelpful and at worst counterproductive: some states imposed cumbersome formalities such as notarization – the Massachusetts act was particularly onerous in requiring a physician to certify that the pledger was of sound mind and not under the influence of narcotic drugs, and demanding three witnesses to the signature. Others treated the pledge as a testamentary document and insisted it was executed after probate had been granted, by which time the corpse would have been buried or cremated; some states afforded and some denied next of kin the right of objection to enucleation.[11]

In the mid-1960s, an influential group of surgeons, scientists and lawyers began drawing up a model statute that would afford pledges legal support, and at the same time iron out inconsistencies and overturn obstacles in statutes dealing mostly with pledges of cadaver eyes. They were motivated by the reversal in fortune of kidney transplants that hitherto had looked destined to fail because nothing appeared capable of preventing rejection by the recipient's immune system. However, hospitals had begun investing in new kidney homograft units or enlarging existing ones in 'the clinical gold rush' provoked by Starzl's announcement in 1963 of a relatively effective cocktail of immunosuppressive drugs. Early experiments had involved kidneys explanted from donors-on-the-hoof, typically a blood relative of the potential recipient whom surgeons believed was more likely to be immunologically compatible than an unrelated stranger. In 1962, cadavers had become the source of choice when Joseph Murray, at Peter Brent Brigham Hospital, Boston, explanted and transplanted a cadaver kidney that had briefly 'worked'. This was an important milestone in the industry's history because corpses potentially are twice as productive a source as donors-on-the-hoof.[12]

The group's spokesmen were Alfred M. Sadler Jr and Blair L. Sadler, twin brothers and doctor and lawyer, who both worked at the National Institutes of Health, Bethesda. The new legislative field was staked out by a committee convened by the National Conference of Commissioners on Uniform State Laws, a non-profit association of academic lawyers, practitioners and judges, one from each state, who drafted model acts in areas where national uniformity in state legislation is desirable. Pledging of cadaver stuff fell under its purview because the mobility of American people made it essential that a pledge signed under the law of State A is recognized in State B.

The Uniform Anatomical Gift Act 1968 gave adults aged eighteen or over of sound mind the first and deciding voice in posthumous decisions about whether or not medicine might repurpose their corpse. Their intentions had to be recorded by means of a pledge written and signed in the presence of two witnesses. Next of kin were denied the authority to abrogate a pledge (although in practice they are allowed to do so). 'Donees' – physicians, funeral directors, coroners/medical examiners and staff in collecting sites – who acted in good faith were afforded shelter from the threat of prosecution under the quasi-property rule. However, paradoxically, despite denying kinfolk's private interest in a corpse, where no pledge was found the model act provided them with the authority to agree to extraction of cadaver stuff. Kinfolk were ranked in a descending hierarchy of authority: spouse; adult son or daughter; either parent; adult brother or sister; a guardian of the dead person at the time of death. At the bottom of the list was 'any person authorized or under obligation to dispose of the body' who might be executor, or person on whose premises the corpse rests, or a coroner/medical examiner. No one outside of the list could authorize donation, a measure introduced to forestall bodysnatching and illicit sales.[13] Agreement could be sought and obtained by telegraphic or recorded message or by any written document.

It took three years to agree the format of the Uniform Anatomical Gift Act 1968. It was approved by the National Conference of Commissioners on Uniform State Laws on 30 July 1968, and shortly afterwards by the American Bar Association and almost every major medical organization concerned with transplantation. Within three years, a version had been adopted in every state, and in the District of Columbia. Never in the history of the National Conference of Commissioners on Uniform State Laws has a uniform act proved so popular.[14]

Pledge Drives

A model pledge card was attached to the Uniform Anatomical Gift Act; its format was agreed at a meeting held in November 1969 attended by representatives of organizations with a direct interest in cadaver stuff, including the Transplantation Society, National Kidney Foundation, the Navy Tissue Bank, the Eye Bank Association of America and the National Pituitary Agency – although the latter shortly stopped marketing pledges of cadaver pituitary glands.[15] In effect, the wording of the pledge was standardized across the various industry sectors. Appropriate feeling rules were provided: pledgers declared their intention as, 'in the hope that I may help others'. Pledgers were allowed to stipulate the substance of the help by specifying which cadaver fragments could be extracted for the purpose of transplantation, therapy, medical research or education.

The new pledge card slotted smoothly into Lions' pledge drives. In February 1976 – National Eye Bank month – Surfside-Garden City, South Carolina, Lions handed out sheaths of blank pledge forms and eye bank brochures to 3,000 students at local elementary schools.[16] By 1984, Lions in twenty-nine states had distributed two out of every three, a record beaten only by the National Kidney Foundation, which had distributed seven out of ten cards in thirty-five states.[17] However, by this stage the message of corpse philanthropy was organized around organs: the 'gift of sight' was adumbrated under the 'gift of life'.

Organizations with nothing to gain directly or indirectly from corpse philanthropy began marketing it. For instance, in 1970, the United Way of America, a nationwide coalition of service clubs, opened the Transplant Information Center, which was based in New York and charged to lead 'the widely-scattered activities on behalf of tissue and organ donations and for reaching both the public and the concerned professions'.[18] The rallying cry of its public information programme was 'life-saving, health-giving'.

Brain death is a rare diagnosis: one estimate put its frequency at between fifteen and thirty in every 1,000 deaths in the western world.[19] Since 1968, American car manufacturers have been required to fit seat belts, but wearing them did not begin to become compulsory until 1984, and a disproportionate number of people diagnosed as brain dead had suffered a traumatic head injury in a road traffic accident. Moreover, they were typically young. In 1976, the New York State's Department of Motor Vehicles began attaching a pledge form to its permanent driving licence application form.

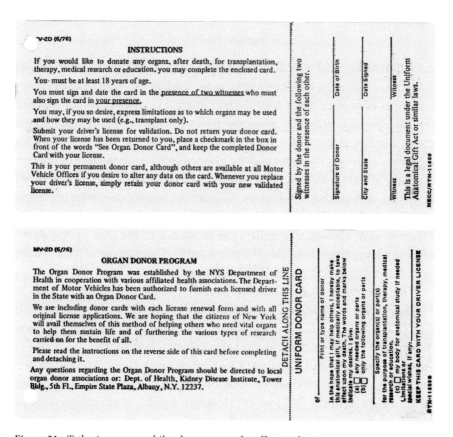

Figure 21: Tethering corpse philanthropy to road traffic accidents

The driving licence is widely used for identification purposes in the United States, and people are much more likely to have it about their person than their pledge card. The measure was soon copied by every state, albeit using a variety of methods of recording pledges. For instance, in some states a sticker was attached to the licence, in others 'ORGAN DONOR' was stamped on the paperwork.[20] Police and highway patrolmen began supporting kidney explantation by notifying their local organ procurement agency of accident victims whose driving licence included a pledge. These efforts were effective: by the 1980s, around half of the cadaver kidneys transplanted had been explanted from a road traffic accident victim.[21]

CHAPTER 20

Horse-trading in the Mortuary

'The sums of money paid for pituitaries although dressed up at times as payment for the trouble taken to collect pituitaries, was an important inducement to collecting amongst people whose wages in 1971 might be £8 a week,' explained Roy Williams, who had worked his way up to become superintendent of a mortuary in Rotherham, Yorkshire.[1] Indeed, in a busy mortuary, a staff member's share of the Medical Research Council's cheque might be worth several weeks' wages. But the tips did not amount to a sum sufficiently large to warrant specification in the council's annual financial report.

Postal workers in sorting offices or collecting and delivering mail were exasperated by the volume of cadaver pituitary glands mortuary staff were sending in parcels that were frequently inappropriately labelled, clumsily packed and leaking acetone – a flammable liquid – and thus failing to comply with Post Office regulations on sending biological and/or hazardous substances through the mail. In 1974, the Cambridge Post Office threatened to destroy them, and the Medical Research Council issued new instructions: parcels had to be clearly marked with 'Fragile with Care' and 'Pathological Specimen', and sent by first-class letter post early in the working week to avoid delay in transit.[2]

The volume rapidly exceeded resources of the overcrowded Protein Hut's laboratory, and a backlog began building up. Anne Stockell Hartree, who had been appointed to oversee manufacture of hGH for the clinical trial, shared the huts' facilities with Department of Biochemistry students and staff who used live animals in experiments.

Hartree's original recipe was a modified version of Maurice Raben's. Cadaver pituitary glands were picked free of bone and gristle, and minced in a Crypto electric meat mincer. The mince was suspended in fresh cold acetone, and the sludge decanted into one-litre, brown, screw-capped glass bottles that were stored in a domestic refrigerator. Periodically a batch of sludge, representing around 2,500 dead people, was taken through the second stage: washing in acetone, draining, spreading out on a large tray and left to dry in the open air. Every so often, the mash would be turned over with a spatula and lumps broken up. It took around twenty-four hours for a grey powdery cake to form, which was replaced in the brown glass bottles, which were returned to the refrigerator. In the third and final stage, hGH was isolated in a solution of the grey powder using precipitation techniques, and the precipitate was freeze-dried. There was an ever-present danger of the cadaver pituitary glands, sludge, mash, powdery cake, precipitate and freeze-dried hGH being confused with or contaminated by glands, sludge, mash, powdery cake, precipitate and freeze-dried hormones of animals, which Hartree and other investigators were preparing for use in research in the laboratory. Freeze-dried hGH was sent to the Wellcome Research Laboratories where it was sterilized and divided into 'stock cubes' sufficient for one injection.[3]

Hartree was not content to serve as overseer of the manufacture of hGH; she harboured ambitions to make her mark as a basic scientist and lead her own research team. Her research mostly involved adapting recipes for the isolation of pituitary hormones of a variety of different species and, like Li and other biochemists, attempting to identify their chemical structure. But her resources were limited, and she had to beg access to or borrow specialist equipment held in other Cambridge laboratories, often in exchange for processing pituitary glands.

In 1970, Hartree threatened to find work in the United States unless her position was made secure, and she was allowed more time and resources for her own research. She was not irreplaceable but if she resigned the Medical Research Council feared supply of hGH for the clinical trial would temporarily be prejudiced. She was appointed to a permanent position on the Medical Research Council's external staff, and in January 1971 her team moved into six small and one large room on the ground floor of 5 Shaftesbury Road, Cambridge, a large detached rambling Victorian house it owned.[4] Permission for substantial alteration to the building was refused because the house was in a quiet residential street. Facilities were cramped and inconvenient and much of Hartree's equipment was outmoded, broken or awaiting installation. The backlog of glands and mince grew.

Hartree had been allowed to keep her space in the Protein Hut for her own research, and several times each day drove one and a half miles backwards and forwards between the two laboratories. For her side of the bargain she agreed to prioritize hGH production, and in order to increase output switched to a modified version of Alfred E. Wilhelmi's recipe, which gave a higher yield of hGH and was gentler than that of Raben, and hence less likely to provoke an inflammatory reaction in recipients, which was frequently reported. But the recipe was held to be less lethal for bacteria and viruses. Wilhelmi had originally been unconcerned about safety because, according to biochemist Albert Parlow who worked in his laboratory, '[his] philosophy was that the material was human protein, and human protein cannot harm human beings'.[5] Nonetheless, in 1973, in light of growing anxieties about infectivity, specifically transmitting serum hepatitis (of which more in Chapter 27), Wilhemi tacked on gel filtration, which is a technologically sophisticated type of sieve used as terminal sterilization. Hartree rejected it as unnecessary; her hGH was safe for clinical use.[6]

In May 1974, an inspection of the Shaftesbury Road premises was undertaken in anticipation of the clinical trial completing in June 1977, after which hGH would cease to be classed as an investigational substance and become a medicine paid for out of NHS budgets. Only licensed medicines qualify, and a licence is granted only where criteria of Good Pharmaceutical Manufacturing Practice are met. Its principles were agreed in 1957 to ensure medicines are safe and consistent, which was not a new idea, but the focus hitherto had been on quality of the finished product, irrespective of manufacturing process. Good Pharmaceutical Manufacturing Practice takes people, premises, plant and equipment, manufacturing processes and paperwork into account, all important considerations where the raw material is biological, and where sources are pooled and one 'bad apple' can contaminate a batch, as was potentially the case with hGH.

During the 1960s, several countries, including the United States, made compliance statutory. The Medicines Act 1968, the British government's response to the thalidomide tragedy, introduced the *Orange Guide*, which sets out current criteria of Good Pharmaceutical Manufacturing Practice that are regularly updated in light of scientific and technical developments.[7] Section 112 of the Medicines Act authorizes a representative of the Medicines Inspectorate to enter and inspect premises where medicines are manufactured. The Shaftesbury Road premises failed to meet criteria. Two practices were singled out for particular criticism: the same equipment was used in processing the pituitary glands of dead humans, sheep, chickens, horses, dogfish and the occasional exotic animal such as baboon; whole, minced and powdered human and animal pituitary glands were held in identical brown glass bottles and kept in an old refrigerator that had an unreliable thermostat. Grey powders of pituitary glands of different species are indistinguishable to the naked eye, and the inspector considered the likelihood of confusion and contamination to be great.[8]

The situation had arisen in part because Hartree had been forced to relinquish her place in the Protein Hut. But it was also the case that the Medical Research Council was reluctant to invest in facilities that would become redundant when the clinical trial was completed. Hartree agreed to use bottles of different shapes and colours to distinguish species and the various stages of the production process. For its part, the Medical Research Council agreed to convert the garage of 5 Shaftesbury Road, which measured 250 square feet (23 sq. m), large enough to house a small British car, into a human hormone laboratory.

The Medical Research Council also decided to take some of the pressure off the Cambridge laboratory by opening a second facility in an annexe of St Bartholomew's (Barts) Hospital, which is adjacent to Smithfield meat market in London. It would focus on isolating hGH in fresh-frozen cadaver pituitary glands. Biochemist Philip Lowry was placed in charge, and the Barts Special Trustees, the charitable arm of Britain's oldest hospital, paid for purpose-built facilities measuring 2,000 square feet (186 sq. m), capable of housing a Rolls-Royce car, a far cry from Hartree's garage with its British Leyland capacity.

The annual target for the fresh-frozen cadaver pituitary glands collection was set at 30,000. Medical criteria of bioavailability were more restrictive than those of the acetone-based one. Pituitary glands of anyone whose death was associated with hepatitis, sepsis, meningitis, encephalitis, multiple sclerosis or who had undergone dialysis or an organ transplant, or who was suspected of illegal drug use, had to be excluded. Moreover, as a further precaution, in January 1974, the Medical Research Council had consulted David Tyrrell, virologist and senior member of its staff, about safety. Tyrrell gave qualified approval of hGH isolated in acetone-preserved cadaver pituitary glands, but was less confident about hGH isolated in fresh-frozen ones.[9] Nonetheless, following reassurance from the United States that no untoward effects had been reported there following injections of hGH isolated in fresh-frozen cadaver pituitary glands, the Medical Research Council decided to tack on gel filtration.[10]

A programme of modernization and rationalization that both central and local government had embarked on around the same time as the collection's inception had reduced the number of hospital and public mortuaries, but had increased throughput in those that remained or had recently been built. In 1972, there were 166 'claimants' of tips. The largest were: Borough Public Mortuary, south London, 1,300 glands; Sheffield Public Mortuary, 1,100;

Radcliffe Infirmary, Oxford, 1,100; Battersea Public Mortuary, south London, 1,100; the London Hospital, Whitechapel, 860; Kingston Hospital, 850; Forensic Medicine, Belfast, 825; Norfolk and Norwich Hospital, 800; Middlesbrough Hospital, 800; City General Hospital, Stoke-on-Trent, 740; North Staffs Hospital, Stoke-on-Trent, 650; Addenbrooke's Hospital, Cambridge, 650; East Ham Public Mortuary, London, 650.[11] Those prepared to cooperate with the frozen gland collection were provided with a freezer in which cadaver pituitary glands, placed in a plastic bag closed with a rubber band, were stored immediately after extirpation. When sixty or so glands had accumulated, the plastic bags were to be collected by Lowry's secretary, who carried them to Barts in a flask of dry ice. Lowry's request for funds to buy a motor van was refused but he was awarded a grant to cover her fares on public transport.

Pay Deal Could Produce Midgets

Mortuary work is sometimes obnoxious and often lonely: staff tend to be marginalized, even stigmatized because of their close contact with death. In the mid 1960s a crisis in recruitment emerged: around half of mortuary staff were approaching retirement age, and potential new staff were being tempted by offers of better-paid employment in the more salubrious surroundings of a funeral home. James Heggie (1907–1990), a pathologist at the North Middlesex Hospital, London, and president of the Guild of Mortuary Administration, proposed elevating mortuary staff's status by requiring them to attend a training programme leading to a diploma in Mortuary Hygiene and Technology awarded by the Royal Institution of Public Health and Hygiene.[12] Successful students would command respect. As Heggie put it: 'The post-mortem technician of the future thus must not be an inferior, ranking lower than a medical laboratory technician. He should be one of their numbers.'[13] The Department of Health and Social Security (DHSS, a giant government department created in 1968 that incorporated the Ministry of Health) agreed, and a new job description was drawn up with a pay scale that placed hospital mortuary staff on a par with laboratory technicians.[14]

The new terms and conditions of employment did not apply to public mortuary staff in local government employ. Moreover, they had almost no immediate impression on hospital mortuary staff's dependence on tips. The new salary scale remained well below the national average wage, and anyone without the recognized qualification, which took at least three years to obtain, languished at its lower end.

In July 1970, the *Daily Express* published an article with the headline 'Doctors' Tangle over "Spare Parts" Cash Offer'. Readers were told how the

Medical Research Council was offering cash 'tokens' for cadaver pituitary glands without ensuring that next of kin agreement to extraction had been secured.[15] Prior to publication, DHSS staff had advised the council to disclaim responsibility for mortuary practices.[16] Its spokesperson is quoted as saying, '[W]e simply organise the collection of the pituitaries and assume that hospital pathologists know their responsibilities and meet the necessary requirements.'

The no-place that the Medical Research Council had established in British mortuaries began attracting other bioprospectors. The most aggressive was Nordisk Diagnostics, a non-profit Danish manufacturer of hGH, which in 1974 began offering £1 for a British cadaver pituitary gland. Not only did this greatly exceed the Medical Research Council's 'tip' and thus threaten the collection, it also led to civil servants being concerned that export of British cadaver stuff could embarrass government were the public to hear about it.[17] However, denying Nordisk entry into British mortuaries risked exposing the Medical Research Council's presence.

Clearly, the cadaver pituitary glands had significant commercial value. The council doubled its tip to 20 pence.[18] It also waved the Union Jack: in a letter circulated to pathologists in January 1976, Frank George Young, in his capacity as chairman of the Steering Committee for Human Pituitary Gland Collection, described the export of cadaver pituitary glands as contrary to the national interest. Unfortunately, the Medical Research Council was unable to increase any further its 'token of appreciation': '[W]e would like to be more generous, but since we are collecting in the region of 100,000 glands a year, you will understand that any increase in this sum would represent a major financial problem to us, and would curtail funds urgently needed for other research.'[19]

Fear of foreign competition proved groundless. Nordisk was sent between 4,000 and 5,000 pituitary glands each year, mostly from Scottish mortuaries.[20] Indeed, in 1976 collection reached its zenith of 74,632, which meant that every eighth British corpse placed in a grave or crematory lacked a pituitary gland, a statistic that excludes cadaver pituitary glands delivered to the Medical Research Council's competitors (the record achieved by the National Pituitary Agency in 1975 was around one in twenty-three).

Skullduggery

The 'human flesh' referred to in the *Sunday People* (overleaf) was the cadaver pituitary gland, and by the time the article was published in August 1976 around 700,000 British corpses had had theirs extirpated.[21] The

Figure 22: Skullduggery exposed

following day *The Times* newspaper reported that Gerard Vaughan, Conservative MP and opposition spokesperson on health, described the collection as 'appalling' and 'a sophisticated kind of body snatching'. He had asked Secretary of State for Health David Ennals, and the British Medical Association to investigate the matter.[22] Ennals subsequently responded by saying that he saw no reason to object to the small fee of 20 pence being paid for a cadaver pituitary gland but that it would be wrong for organs to be sold for gain.[23]

Behind the scenes civil servants were troubled by the suggestion in the *Sunday People*'s article that 'certain parts of the human body were being regarded as mortuary "perks" available to the staff concerned to be disposed of by them to the highest bidder'.[24] But they also anticipated that when the clinical trial ended in June 1977, demand for hGH would escalate to a level that could be satisfied only if the cadaver pituitary gland of almost every British corpse undergoing a post-mortem examination was extirpated.[25] Moreover, home-produced hGH was far cheaper than that sold by foreign manufacturers. KabiVitrum (KABI), a state-owned but for-profit Swedish pharmaceutical company, priced an ampoule of Crescormon® at around £17.50, compared to £1.18 for one ampoule of British hGH. Cost had become even more crucial because the British economy had fallen into the doldrums, and government was implementing a programme of deep public spending cuts. Leave well alone remained the guiding principle.

Matters came to a head again in 1978 when three mortuary technicians at a Stoke-on-Trent hospital were dismissed for 'selling' glands to Nordisk. The scandal coincided with the government signing Council of Europe Resolution (78) 29 that, among other things, forbade the exchange of cadaver stuff for profit. Civil servants decided to consolidate the 'piece rate' incentive into a new, more generous pay scale for hospital mortuary staff.[26] *The Guardian* reported the new terms and conditions, which would come into effect in January 1980, as 'a pay deal that could produce midgets'.[27] But it did not apply to public mortuary staff who were employees of local authorities – for instance, Walthamstow Public Mortuary was administered by Waltham Forest Borough Council's Recreation and Amenities Department. By the end of 1980 the collection had plummeted to 34,380 cadaver pituitary glands.[28] In February 1981, the incentive offered to public mortuary staff was increased to 25 pence per gland,[29] but was almost immediately withdrawn.[30] What a journalist in *Nature* called 'horse-trading' in the mortuary, was drawing to a confused end.[31]

CHAPTER 21

Value for Money in American Mortuaries

This photograph was published in 1966 in *Life* magazine in an article celebrating the National Pituitary Agency's achievement in increasing the height of children who had been failing to grow.[1]

Figure 23: Thirty thousand cadaver pituitary glands

Presented unsentimentally in glass jars, without biographical information about their human origin, the cadaver pituitary glands look like dried beans, more vegetable than human. But once it is realized that the three jars together hold 30,000 cadaver pituitary glands – the population of a small town – the photograph becomes distinctly unsettling. It calls to mind the piles of discarded personal possessions such as shoes and suitcases displayed in museums that convey mass extermination of people as happened in the Holocaust.

In January 1971, members of the House Appropriations Committee were told by Blizzard that recipients of the National Pituitary Agency's hGH had grown a total of 1,600 feet (488 m), three times the height of the Washington Monument.[2] In December 1972, the National Institutes of Health awarded the National Academy of Sciences, Washington DC, $75,200 to consider the case for the agency's continued existence.[3] Its future was being questioned because activities of the National Institutes of Health and its agents are restricted to research. However, hGH demonstrably 'worked' and, strictly speaking, should have been stripped of its status as an investigational medicine, and approval sought for its launch as a prescription drug, which would require approval of the Food and Drug Administration. Its regulatory frame for testing new medicines was a daunting and expensive procedure; it had been created in response to the birth of babies with severely malformed or absent limbs, delivered of women who had taken the drug thalidomide to ameliorate early-pregnancy vomiting, and had been enacted through an amendment of the Federal Food, Drug and Cosmetic Act (known as the Kefauver-Harris Amendment after its sponsors Senator Estes Kefauver and Representative Oren Harris). Moreover, once approved, hGH would become available for general medical use, and paid for out of patients' pockets or by third-party payers, or possibly state schemes providing financial support for children with congenital abnormalities.

A ten-member committee appointed to consider the agency's future undertook a cost/benefit analysis addressing questions such as did it provide value for taxpayer's money, would independent bioprospecting be more efficient than a central collecting, processing and distributing centre, and were commercial preparations of hGH a realistic alternative.

The last was unlikely. Two commercial manufacturers of hGH had recently filed a New Drug Application with the Food and Drug Administration: Calbiochem, a Californian concern and worldwide bioprospector, which withdrew from the market shortly after the committee met;[4] and KABI, which was bioprospecting in mortuaries in Sweden and elsewhere, and using a recipe that Li had given Rolf Luft to manufacture, which was marketed under the brand name Crescormon®.[5] Despite bioprospecting worldwide, KABI was encountering difficulties in obtaining cadaver pituitary glands and was unlikely to fill the gap created by the agency's demise. Moreover, imported Crescormon® was much more expensive than home-produced hGH: one milligram 'stock cube' cost $6 compared to $1.87, increasing the average total cost of treating one child from $689 to $1,750.[6]

The committee found the agency had fulfilled its original remit as a resource for investigators. In 1972, it had distributed research material that the committee valued at around $1,527,600, which represented an excellent return on its grant of $381,276. Moreover, because no charge was levied, savings made by laboratory and clinical investigators could be invested in investigations into other potentially valuable substances secreted by the anterior lobe of the pituitary gland, such as gonadotrophins, or 'fertility drugs': luteinizing hormone (LH) and follicle-stimulating hormone (FSH).

The committee recommended taxpayers continue to support the agency until a mass-produced synthetic growth hormone became available, which looked a long way off since Li's molecule had recently been discredited following the revelation that it was without the full complement of amino acids. And in light of its successful bioprospecting in mortuaries the National Institutes of Health should consider extending its 'product range' to include other cadaver stuff that scientists were repurposing in laboratories, such as ovaries, testes, and thymus and pineal glands.[7]

It might have looked like a bargain in the landscape of American biomedicine, but the National Pituitary Agency's operating costs were considerably higher than those of its British counterpart: in 1972, 92 cents (at 1972 currency exchange rates) covered all the costs associated with extirpating, packaging, transporting and processing one British cadaver pituitary compared to $4.89 for each American one.[8] Some, but clearly not all, of the excess could be explained by the size of tip – around the equivalent of 25 cents in the United Kingdom compared to two dollars in the United States. The committee urged a review of central office overheads but was particularly critical of the agency's lackadaisical approach to processing.

Processing was undertaken under contract in three laboratories: Raben's (where each year hGH was isolated in 20,000 cadaver pituitary glands); that of biochemist Alfred E. Wilhelmi, at Emory University, Atlanta, Georgia (annual share of 20,000–30,000 cadaver pituitary glands); and that of reproductive endocrinologist Brij E. Saxena, at the Cornell Medical School, New York (15,000 cadaver pituitary glands annually). Productivity was low: 192 grams of hGH were extracted out of 62,000 American cadaver pituitary glands, whereas 300 grams were isolated in 56,000 British ones. Moreover, British hGH was more potent than that of the National Pituitary Agency; in other words, more American cadaver pituitary glands were required to initiate the same increase in children's height. Quality of hGH varied from vial to vial. Indeed, impurity was sometimes obvious to the naked eye: instead of being nearly transparent, some 'clinical grade' hGH was the colour of Coca-Cola.[9]

Each processor had been allowed to use their favourite recipe; no attempt had been made to establish which one of the twelve or so in use worldwide was best, or whether it had been modified in light of technical developments.[10] The committee observed that processing in the United States was a sideline of each processor's main activity. For instance, Raben, who had a busy clinical practice, devoted one month each year to fulfil his contractual obligation. Two months ahead of processing, 20,000 cadaver pituitary glands were sent to his laboratory where during her free moments Mrs Vera Grinbergs, his senior technician, examined each one, peeling off excess stuff such as bone fragments. Dr F. Matsuzaki, a Japanese scientist who had been Raben's research fellow, flew over from Tokyo to help Raben and Grinbergs with processing. It took the team of three at least two weeks to grind, dry and mill the glands in preparation for extracting, purifying and recovering hGH and other fractions from the soup.[11]

Motivations and Money

The committee's unfavourable comparison of American and British schemes echoed the conclusion drawn by Richard Titmuss (1907–1973) in *The Gift Relationship: From Human Blood to Social Policy*, which the *New York Times Book Review* had nominated as one of the seven books of special significance published in 1971.[12]

Titmuss attributed the NHS's Blood Transfusion Service's impressive record of keeping pace with growing demand for human blood to the redistributive policies of the British welfare state that encouraged people freely and literally to give of themselves to unknown others. In contrast, American government policy had allowed a mixed economy of blood banks to flourish, which not only failed to keep pace with growing demand but was also responsible for many more American than British people suffering post-transfusion hepatitis. The thesis rang true in the minds of Americans protesting against the government's prosecution of war in Vietnam, and its failure to address abuse of the civil rights of African Americans. However, Titmuss's target was the Institute of Economic Affairs, a British 'think-tank' of neo-liberal economists who had proposed that the NHS's Blood Transfusion Service would be more effective and efficient if blood donors-on-the-hoof were paid.[13]

It was generally accepted that skid row donors-on-the-hoof exchanged blood for cash to buy a heroin fix or bottle of booze. Almost nothing is known about recipients of the agency's tip save a few scraps of information that emerged in the context of scandal.[14] In some mortuaries it was used to supplement an inadequate budget, and was spent on books, equipment or the fees of a professional photographer.[15] But mostly mortuary staff shared it. Blizzard recalled a press story of a private swimming pool built out of the proceeds from 'selling' cadaver pituitary glands.[16] Pituitary pirates still roamed mortuaries: the *Los Angeles Times* covered the prosecution of a thirty-three-year-old autopsy assistant working in the coroner's office at Sylmar, Los Angeles, who had 'sold' more than 1,000 cadaver pituitary glands at $1 apiece to a physician who 'sold' them on to the National Pituitary Agency.[17]

Death of the Hospital Autopsy

In failing to monitor recipes the National Pituitary Agency stood accused of wasting a scarce natural and national resource. The committee recommended concentrating processing on one site. Wilhelmi, who was the most proficient of its three processors, was awarded the contract, and on his retirement at the end of 1976 it was passed to Albert F. Parlow, who relocated to the University of California Los Angeles' Harbor Hospital, Torrance. Parlow used a different recipe that incorporated chromatography and isolated nearly three times as much hGH out of each cadaver pituitary gland as his predecessor.

With hindsight it is evident that Parlow saved the day. In 1975, the National Pituitary Agency's collection reached its zenith of 83,091 cadaver pituitary glands, which meant that one in every twenty-three Americans placed in a grave or crematory was without a pituitary gland. But the following year the collection began falling precipitously: 70,966 in 1976, 58,500 in 1977, and from then on it languished at around 50,000 glands each year.[18]

The records of the National Pituitary Agency are unavailable for research, and it is only possible to speculate why this happened. But evidently fewer autopsies were being performed: in 1972, the corpse of around one in every five Americans who died was opened up on a mortuary table for inspection; by the new millennium the rate had halved to just under one in ten.[19]

The hospital autopsy was largely responsible for the decline. The word 'autopsy' means 'seeing for oneself'. The procedure provided doctors with an opportunity to improve practice by comparing the accuracy of diagnoses made when patients were alive with what their corpses revealed on the mortuary table, to investigate whether crucial signs had been missed or misinterpreted.[20] It also enjoyed pride of place in the medical school curriculum. In the first decades of the twentieth century the hospital autopsy had been adopted as a measure of the quality of hospital care. By the early 1950s, the Joint Commission on Accreditation of Hospitals was demanding a minimum rate of autopsy of one in every five patients who died in hospital, and the American Medical Association was demanding a rate of one in four as the minimum for accreditation of a teaching hospital. Some boasted an autopsy rate of nine out of every ten deaths. Without accreditation, a hospital was unable to offer internships and residencies to junior doctors. Indeed, it was common practice for job candidates to judge potential employers in terms of their autopsy rate, and hospitals strove for a rate as high as possible in order to attract the best ones.[21]

In 1957, the American Medical Association relaxed its rule on hospital autopsies from mandatory to advisory, a decision coinciding with an increasing inability of many hospitals to fill positions in pathology that carried responsibility for the tedious paperwork routine autopsies required.[22] Moreover, pathologists had begun questioning the value of routinely opening up the bodies of elderly patients whose death had been anticipated simply in order to sustain their employer's autopsy rate. At the same time innovations in technology were enhancing the accuracy of diagnoses when patients were alive. The focus of pathology began shifting away from mortuary table onto laboratory bench, a development that created business opportunities in the American health-care system. By the early 1960s, over 20,000 commercial medical laboratories were aggressively competing against one another by lowering prices or offering more tests for the same price.[23] The lethal blow was dealt in 1970, when the Joint Commission on Accreditation of Hospitals eliminated autopsies from its list of requirements for hospital accreditation. Almost immediately hospital autopsy rates fell to under one in six deaths, and the trend continued downwards.[24]

Financial considerations sounded the death knell of the hospital postmortem examination. Their cost had been buried in charges for rooms, but when hospitals' reputations no longer depended on them, their cost began to be charged to the pathology budget, a measure resented by pathologists who objected to paying for a procedure undertaken for the benefit of colleagues whose practice was covered by another budget. Third-party payers began refusing to reimburse the cost of autopsies on the grounds that cadavers are not patients. Medicare and Medicaid, which began paying some patients' bills in 1966, were unwilling to use taxpayers' money to subsidize the training of medical students, many of whom had exceedingly high-income expectations. And because a mortuary has a high fixed cost, the fewer autopsies performed, the more costly each one became. Hospital treasurers were increasingly reluctant to countenance the considerable expense of a non-revenue-producing activity. In 1983, Medicare and Medicaid removed autopsies from the list of reimbursable procedures.[25]

Legislative Consent

Although far fewer hospital autopsies were being performed, the number of medico-legal ones was increasing slowly. As a result, whereas in the United States in 1972 one in five autopsies was ordered by a coroner or medical examiner, by 2007 the proportion had risen to one in two. Commentators are uncertain why this happened. Little was known of the work undertaken by coroners/medical examiners before 2004 when the first national census was undertaken. It found around two out of five of the 2.3 million deaths each year had been brought to their attention, a rate somewhat higher than that in the United Kingdom, and about half (487,000) were deemed to warrant further investigation, and in half of these an autopsy had been ordered.[26] These findings suggest replacement of coroners by medical examiners is a contributory factor: coroners typically are lay people elected to office by their local community and perhaps reluctant to offend constituents who object to the performance of an autopsy, whereas medical examiners are medically qualified people trained in forensic pathology, appointed to their position and therefore without an electorate to consider, and with offices typically better resourced than those of coroners.[27]

As the Eye-Bank for Sight Restoration had found in the early 1950s, coroner/medical examiner offices are potentially rich collecting sites. But bioprospecting had been held back by uncertainty over whether the law allowed coroner/medical examiners whose authority rests on the public interest to agree to extraction and retention of cadaver pituitary glands for repurposing in medicine, and the threat of actions by aggrieved next of kin. Nonetheless the National Pituitary Agency's tip was tempting: in November 1966, a Minnesota newspaper revealed on its front page that staff in the

Hennepin Medical Examiner's Office and Hennepin County Hospital had been extracting cadaver pituitary glands.[28] Local radio and television stations picked up the story. Some next of kin sued for damages. Blizzard was called as a defence witness and his evidence moved the jury to voice its support of the agency. Nonetheless the story found its way into the national press, and Grand Jury hearings were held in several other cities. Adverse publicity was blamed for the slight reduction in the size of the agency's collection over the following two years.[29]

The small print of the Uniform Anatomical Gift Act 1968 afforded coroners/medical examiners some comfort where the corpse was unclaimed. Last of the list of persons on whom legal authority was conferred to agree to extraction of cadaver stuff in circumstances where no pledge is found is 'any other person authorized or under obligation to dispose of the body'. In effect, it allowed people to be treated as if before their physical death they had been socially dead. States began adopting what became known as legislative consent statutes that allowed coroners/medical examiners to authorize extirpation of cadaver pituitary glands (and subsequently cadaver eyes/corneas, of which much more follows in Chapter 22). Whereas what are called presumed consent laws typically treat every corpse as potentially bioavailable, legislative consent applies only to those that fall into possession of a coroner/medical examiner.[30]

Legislative consent statutes were inconsistent. For instance, Virginian law was amended when local physicians persuaded Delegate James Murray of Charlottesville that the needs of the state's dwarves greatly outweighed any 'rights' that might be allowed dead people.[31] In Arkansas the state medical examiner and his assistants were authorized to extirpate the pituitary gland during an autopsy and donate it to the Arkansas Dwarf Association, but with the proviso that neither the dead person nor their next of kin had explicitly objected. In California, the coroner was allowed to hand the cadaver pituitary gland over to 'a public agency for use in manufacturing hormones', which might be either the San Francisco Pituitary Bank or the National Pituitary Agency. In Connecticut, the pituitary gland could be extirpated where the pathologist had reason to believe it would be beneficial to the health of a living person with the proviso that the corpse was not disfigured, next of kin did not object and the deceased person did not belong to a religion that opposed extirpation. Statutes in Ohio and Oklahoma specifically mentioned the National Pituitary Agency. Others permitted any bioprospector to enter into the mortuary. In Nebraska the pathologist was authorized to extirpate the pituitary gland for the purpose of research and treatment of hypopituitary dwarfism, whereas

New York coroners were allowed to extirpate it for isolation of hGH. Medical examiners in Vermont were permitted to extirpate the cadaver pituitary gland for the purpose of manufacturing 'a hormone to be used in the treatment of hypopituitary dwarfs'.[32] Unfortunately, without access to the agency's records, it is impossible to gauge the effectiveness of these statutes.

CHAPTER 22

Financing High-volume Eye Banks

'No one ever called the eye bank when someone died. People were dying every day in local hospitals.' This is Frederick N. Griffith recalling the situation confronting him in 1967 on his appointment as executive director of the Medical Eye Bank of Maryland, which had been established in 1962 by and for ophthalmologists operating in the Baltimore area, and enjoyed support in cash and kind of local Lions.[1] But in its first full year it had secured only thirty cadaver eyes, and in subsequent years its inventory had barely grown.

From his experience as a hospital administrator, Griffith knew that a division of labour is enforced when a patient dies on a hospital ward: senior professional and administrative staff are almost never involved; nurses, nursing aides, ward orderlies and porters prepare and handle dead bodies; junior doctors confirm death and complete formalities, including asking next of kin to agree to a hospital post-mortem examination.[2] It was easier to 'get the post' where the doctor was known and trusted by kinfolk, but this was increasingly unlikely in large impersonal hospitals.[3] Junior medical staff passed around tips on how to be persuasive.[4] Strategy without stratagem, art without artifice was advised; when objections were raised, 'the duty to humanity' approach might do the trick: this is where people are advised that information derived from a post-mortem examination will advance medical understanding of the condition responsible for the death of their kin.

Custom and practice were to offer junior medical staff incentives to increase the number of 'posts'. In *The House of God* (1978), Samuel Shem's satire of medical education in the early 1970s, the chief of medicine presents the intern who has recorded the most posts with a free trip for two to the

American Medical Association's annual meeting in Atlantic City.[5] Griffith's 'incentive' was two tickets for either a Baltimore Orioles baseball game or a Baltimore Symphony Orchestra concert for the intern who notched up the most agreements to enucleation.[6] Baltimore hospital decedents offices cooperated by adding an enucleation request form in the 'death package' of official forms, together with the death certificate and the release of personal belongings form. The competition proved popular: the Medical Eye Bank's inventory grew from 390 in 1967 to 1,725 in 1974.[7] It became America's first high-volume eye bank.

Griffith joined forces with Russell S. Fisher, (1916–1984) Chief Medical Examiner of the State of Maryland, who had previously provided the National Pituitary Agency with legal advice. In 1975, Maryland became the first state to adopt legislative consent in support of eye banking. The statute explicitly protects the medical examiner, staff and the Medical Eye Bank from action brought by people distressed by extraction of kinfolk's corneas.

Autopsies ordered by Maryland's medical examiners were carried out in a central facility. Fisher referred to it as 'his shop'.[8] It was a busy place; Baltimore usually ranks among the winners in the annual competition for homicide capital of the United States, largely as a result of violence associated with young African American men caught up in gang warfare connected with illegal drugs traffic. The 1970s was a particularly lethal decade, with homicide rates nationally reaching record proportions: in 1970 the national rate per 100,000 people was 7.9; by 1980 it had reached 10.2.[9]

The Medical Eye Bank's inventory grew rapidly under legislative consent, and it held many more corneas of younger people. Whereas the typical source of eyes enucleated in hospitals in and around Baltimore was a fifty-three-year-old patient who had died of disease, the typical source of corneas excised under legislative consent was a twenty-eight-year-old man who had died in traumatic circumstances.[10] Demand for younger corneas was growing. Corneas of older people had been preferred because they 'take' more easily, but the introduction in the early 1970s of the specular microscope had raised questions about the evaluation of suitability of corneas. Ophthalmologists had begun asking how many years of useful service a seventy-year-old cornea could provide a thirty-year-old recipient, and some were insisting on a close match in age of source and recipient.[11] The question had arisen because indications for keratoplasty were changing: keratoconus, a degenerative condition that distorts the curvature of corneas, and which typically develops during adolescence, was creeping towards the top of the list. Fewer Americans were suffering infections that hitherto had been a significant cause of corneal blindness; and following the passage of the Occupational Safety and Health Act of 1970 and the imposition of safety measures, in particular, the wearing of goggles, the incidence of workplace injuries to the eye was reducing.[12]

Legislative consent is organized around *in situ* corneal excision – *in situ* being wherever between death and disposal the corpse rests. It became technically feasible in 1974 following the formulation of M-K medium, a preservative solution that extends the shelf life of excised corneas by up to six days, although under routine conditions four days was recommended (the shelf life of an enucleated cadaver eye in a moist chamber stored in a refrigerator at 4°C is between twenty-four and forty-eight hours).[13] Corneas floating in M-K medium are another type of cadaver artefact, easily transported in a standard eye bank Styrofoam container filled with ice. Trim to fit is all that is required in the operating theatre, relieving surgeons of the task of excising the cornea off a whole cadaver eye, a procedure that takes time and is susceptible to error, which is why two cadaver eyes routinely were provided and the second one often wasted.

M-K medium was named after its creators, Bernard McCarey and Herbert E. Kaufman, chief of ophthalmology at the University of Florida and a Gainesville Florida Lion. It was widely available. McCarey published the recipe – supplies were also freely available from the Warner Lambert Research Institute – together with instructions on how to excise corneas *in situ*.[14] Nothing new in the way of surgical instruments was required. The Maryland statute also allowed trained technicians to replace licensed medical practitioners. This concession, combined with morticians' statutes, accelerated the pace at which extraction of cadaver eyes/corneas was de-medicalized, becoming a task for trained lay people.

Revenue of Non-profits

Demand in and around Baltimore was frequently satisfied. Griffith, sometimes assisted by the Eyeball Network of radio hams, began placing free of charge cadaver eyes/corneas surplus to local demand with surgeons operating elsewhere. The offers were eagerly accepted; as Griffith recalled, surgeons took to cadaver corneas 'like ducks to water'. But by covering all costs of extraction, and supply chain operations, the Medical Eye Bank faced financial ruin.

Griffith decided to follow the example of the Rochester Eye Bank, which in the late 1960s had begun billing recipients of cadaver eyes it had enucleated. The introduction of Medicare and Medicaid was changing the business model of American health-care providers, making charges more explicit and more likely to be met. With the help of an accountant, Griffith set $100 as a 'service fee'. But some members of the Eye Bank Association of America objected on the grounds that the exemption of eye banks from taxation is a privilege awarded to non-profits for voluntary service to the community, and if additional revenue was needed it should be found through fund-raising. Jack McTigue, the association's president, countered that federal government had accepted the view that billing for services undermined neither ethos nor tax status of non-profits. Under the End-Stage Renal Disease Program, which Congress had adopted in October 1972 (and of which more is said in Chapter 25), Medicare was reimbursing every cent of 'kidney acquisition costs', which covered explanting surgeons' fees and salaries of other medical and technical staff, cost of preservation materials and equipment, transportation and packaging, administration and overheads.[15]

Opponents of charging laboured under the misconception that the non-profit sector is almost entirely financed by private gifts. This is not the case; it received, and continues to receive, substantial contributions from government. Moreover, a sizeable and growing proportion of its revenue was being derived from charges, dues and investment income. Another misconception is to attribute the preferential treatment of non-profits under American tax law to the nature of services the sector provides, which are supposedly materially different from those provided by the private sector in that they are focused primarily on either promoting and supporting public goods or helping the poor and needy. In fact, eligibility under Section 501(c)(3) of the Internal Revenue Code is neither source of revenue nor purpose served but an assurance that no part of net earnings is distributed to private shareholders or individuals, and that the organization will not engage in political activity.

Opponents further argued that billing effectively monetizes transactions in which cadaver eyes/corneas are transferred from source to eye bank, and eye bank to recipient, and that exchange for valuable consideration was explicitly prohibited under the association's rules. However, a lawyer found nothing in its constitution and bylaws prohibiting cost reimbursement. Without consensus, he advised that decisions are taken locally. The following year, the American Academy of Ophthalmology recognized the association as its standard-setting arm for eye banks; it also observed that charging fees was not prohibited under its regulations, which is not surprising, as fees undoubtedly comprised the main source of income of a majority of its members.

What is it about the funds raised by Lions selling light bulbs or fried chicken to their neighbours that might make them appear a preferable, even more ethical, source of revenue for eye banks (and other non-profits that extract, process and distribute cadaver stuff) than billing recipients? The answer lies in the meaning of money.

Sociologist Viviana Zelizer rejects the commonplace that 'a dollar is a dollar is a dollar', that is, that money is an absolutely fungible, qualitatively neutral, infinitely divisible, entirely homogeneous medium of market exchange that stamps everything it touches with the same quantitative logic, which invariably destroys warm personal ties. This, she argues, is a crude theoretical understanding of its significance, which she attributes to influential German sociologist George Simmel (1858–1918) who in *Philosophy of Money* (1907) insisted that monetary relations are the antithesis of ties of kinship and friendship, and for this reason decried gifts of money on the grounds that these are inevitably tainted by its presence. According to Zelizer, 'money is never just money'; it is invested with many different meanings through a process she calls 'earmarking'.[16] Everyone earmarks money, which is why the difference between an 'honest dollar' earned through hard work and 'dirty money' derived from dubious or illegal practices requires little explanation, and why a $100 cheque given to a nephew as a wedding gift is experienced differently to a $100 cheque that pays a parking fine – the former communicates a warm personal relationship, which is acknowledged in the form of a 'thank you', whereas the latter is an inconvenient reminder of the existence of an anonymous bureaucracy that has the power to institute criminal proceedings against non-payers, and which acknowledges payment by issuing a pro-forma receipt.

Money then is neither culturally neutral nor socially anonymous. Lions earmarked money raised in support of eye banking as confirmation of their identity as Knights of the Blind – it is the booty of battles voluntarily fought on behalf of corneal blind people. Irrespective of whether 'fees', 'handling charges', 'cost recovery' or 'service charges' is written on the invoice, billing disentangled eye banks from dependence on philanthropic support, changed their ethos and entangled them in the politics of the American health-care system where reimbursement schedules are subject to heated negotiations. Indeed, in recognition of the growing importance of government affairs on its members' operations and financial viability, in 1984, the Eye Bank Association of America moved its headquarters from a Texas eye bank to a separate office in Washington DC.

Legislative Consent

In order to manage its growing inventory, the Medical Eye Bank recruited additional staff and instituted a division of labour: a medical director trained and supported technicians who focused on extraction and processing, and other staff undertook administration, 'retailing' and supply chain operations. Other eye banks began seeking its advice on how to become a high-volume entity and charge for services. Its business model offered a solution to the challenges confronting the finances of non-profits in general, including eye banks. Government was cutting back support for health and welfare programmes, and looking to civil society organizations to pick up the slack. The non-profit sector experienced rapid growth during the 1980s, when each year an average of 29,000 new entities left the tax roll.[17] Yet its traditional sources of finance were diminishing. As political scientist Robert Putnam famously put it, American citizens were increasingly electing to bowl alone.[18] Eye banking was a victim of individualism: between 1978 and 1993, Lions Clubs lost one in eight of their North American members.[19] Putnam attributed this development to work and family commitments eating into leisure time, a trend associated with increasing equality between the sexes. If that is the case, it applied mainly to younger Lions: some older men resigned when their association admitted Lionesses, as Lions Clubs finally did in 1987. It might also be the case that the political climate encouraged individualism. In order to raise additional revenue, non-profit entities were increasingly behaving entrepreneurially, competitively, commercially and corporately, adopting activities that were consistent with their original mission, but which generated revenue. For instance, non-profit hospitals opened health clubs, non-profit museums opened shops, and non-profit universities engaged in research alliances with private firms.[20]

When it became apparent that merely telling eye banks what to do was insufficient to change practice, the Medical Eye Bank began seconding its own staff for prolonged periods – six months at least.[21] Griffith testified on their behalf before state legislatures considering whether to adopt legislative consent. He was persuasive: 'I made opposing legislative consent seem like being against mother love.'[22] A version of the Maryland statute was adopted by twenty-four states.[23] However, as usual in legislation relating to death and corpses, there was inconsistency in what was allowed. Some statutes permitted excision without notifying next of kin; others required a 'good faith' effort to contact them, sometimes specifying how much time had to be spent in the attempt. Some allowed the authority to be delegated to a Justice of the Peace responsible for establishing cause of death in certain counties.

By 1980 a majority of the larger eye banks were billing recipients, and just under half were operating under legislative consent.[24] The Medical Eye Bank created a national network of eye banks that had adopted its business model and shared cadaver eyes/corneas surplus to local requirement. In June 1978, the first branch was opened in Orlando, Florida, shortly followed by eye banks in Delaware, West Virginia and California. Members of the network were charged for advice and support, creating a revenue stream consistent with the Medical Eye Bank's mission of promoting sight restoration, but unencumbered by the politics of both philanthropy and reimbursement schedules, and providing the wherewithal to invest in technical innovations and operating efficiencies. A call centre was opened in Baltimore to receive orders from surgeons throughout and occasionally outside the United States in advance of scheduled surgery, and to arrange for these to be met by an eye bank in the network. In relieving local staff of 'retailing' and allowing it to concentrate on extracting, the network's productivity increased. In 1982, it was responsible for the extraction of just under one in eight of the 38,337 cadaver eyes/corneas enucleated or excised by members of the Eye Bank Association of America. The following year its inventory had grown by nearly 30 per cent, to 6,349 cadaver eyes/corneas.[25]

Eye-banking Diplomacy

Corneal blindness is a silent, growing and largely preventable global epidemic. A much-quoted statistic, attributed to the World Health Organization Task Force, is that keratoplasty could restore the sight of around eleven million people, a majority of whom live in a developing country, where impoverished people are unable to afford antibiotics to treat eye infections, and work in dangerous industries without eye protection. Ophthalmologists working outside the United States began seeking the Medical Eye Bank's advice on how to set up an eye bank, which created an international division, adapted its procedures and changed its name to Medical Eye Bank International. Griffith became its roving ambassador, 'teaching how to fish' in places where corpses rest between death and disposal. In 1989, two Egyptian eye banks joined its network; both were in Cairo where around 10,000 corneal blind Egyptians live.[26] That same year, the International Federation of Eye Banks was established. Griffith sat on its board, but other members were recruited from outside the United States in order to avoid the federation being seen as a tool of American diplomacy. Its ethos was unlike that of the International Association of Lions Clubs, which in 1984 had officially adopted the Lions Eye Bank Program, to help local clubs sponsor eye banks.

The collapse of communist governments in 1989–1991, and the shift from socialist to market-orientated delivery of health care, opened up new opportunities for American eye banks. Indeed, Griffith found introducing legislative consent into Eastern Bloc countries 'a piece of cake', because hitherto private interest in the corpse had been legally unrecognized. The Eye Bank of Prague was the first of the international networks to open; it was a beneficiary of a grant of $100,000 awarded by Katherine Graham (1917–2001), publisher of the *Washington Post*, towards the cost of training its technicians. The first recipient of a Czech cadaver cornea was an eight-year-old girl. Lions exploited eye banking to persuade governments in countries that had hitherto been hostile to civil society organizations, especially those associated with overseas organizations, to allow local chapters to be opened. For instance, the Chinese Communist Party's trust was gained through Sight-First China, a five-year programme of blindness prevention work jointly funded by American Lions and the Chinese government. Impressed by their record, Premier Zhu Rongji permitted Lions Clubs to begin recruiting members. In May 2002, both Shenzhen Lions and Guandong Lions Clubs were officially chartered at a ceremony in the Great Hall of the People in Beijing.[27] Clearly, eye banking is a powerful asset in diplomacy.

CHAPTER 23

Regulation is Necessary, but How?

New entities sprang up to meet the growing demand for cadaver stuff. Emulating the Navy Tissue Bank, which was funded by America's war chest, was out of the question: it involved investment in expensive equipment such as freezer, refrigerator, lyophilizer, surgical instruments and bone mill, and revenue to cover rent, staff wages, transport to and from collecting sites and consumables: gloves, gowns and such like.[1] In 1970, orthopaedic surgeon Theodore Malinin, one-time Navy Tissue Bank collaborator, opened one of a few schemes disassembling corpses, which was called the University of Miami Tissue Bank, and which supplied local hospitals and research laboratories with a variety of cadaver artefacts.[2] A majority handled only one or two cadaver fragments and were either established within a hospital by surgeons to replenish their stash or as an independent non-profit or commercial operation. 'Nobody stays around forever in the Navy,' a one-time Tissue Bank technician told me; alumni of the Navy Tissue Bank's 'bone bank school' easily found employment in the rapidly expanding industry or opened their own entity.[3] Funeral directors with training in and facilities for embalming dead bodies were well placed to diversify into extraction, despite some states prohibiting – but seldom policing – the use of an embalming room for other purposes. The entry bar was low: nothing prohibited 'mom-and-pop' or 'garage' entities collecting cadaver stuff in mortuaries where the culture of tipping was flourishing thanks to the National Pituitary Agency.

Ravenis v. Detroit General Hospital (1975) was a wake-up call. Two people who had lost the sight in one eye to infection contracted following keratoplasty were the plaintiffs.[4] The operating surgeon and Harper Hospital where the operations had been performed had followed generally agreed standard aseptic procedures and were absolved of medical malpractice. Detroit General Hospital, the collecting site, was found negligent. The source of the corneas was a sixty-year-old man, a chronic alcoholic. The first-year ophthalmology resident instructed to enucleate his eyes testified before the court that the hospital had failed to formulate criteria of medical bioavailability, that some were communicated informally by word of mouth, and inexplicably the source's case notes holding results of recent blood tests had been missing.

Ravenis also challenged the claim that corpses are disassembled in order to benefit recipients, and threatened the reputation of the burgeoning industry. Oversight was clearly necessary, but what type, and by whom? The Food and Drug Administration looked like an obvious candidate: it was responsible for the safety of food, medicines, biological and medical devices, but it was unclear whether it was authorized to regulate cadaver stuff/artefacts, as nothing in law had them in mind. The exception was hGH, which fell into the medicines category. KABI's Crescormen® had been approved, whereas the National Pituitary Agency was avoiding oversight by insisting its hGH was a research substance and restricting its distribution to investigators whose research proposal had been approved by their peers.

Cadaver artefacts that perform structural work within recipients' bodies might be defined as medical devices, a new regulatory category created shortly after *Ravenis* was decided. Congress had adopted the Medical Device Amendment in 1976 in response to a succession of scandals about significant injuries and deaths attributed to unsafe design.[5] Manufacturers eager to participate in the rapidly expanding medical market had launched a plethora of new devices for diagnosis of disease, and assistance in supporting, repairing or replacing worn out or diseased body parts – computed tomography (CT) scanners, kidney dialysis equipment, heart valves, cardiac pacemakers, intrauterine contraceptive devices, artificial hip joints, contact lenses and so on. Evidence of the harm that faulty design can cause had begun emerging in the late 1960s. In 1970, Theodore Cooper, director of the National Institutes of Health Heart and Lung Institute, estimated that over the past decade implanted medical devices had injured 10,000 people, of whom 731 had died. The only sanction available to the Food and Drug Administration was confiscation after injury or death. Under the 1976 amendment it became gatekeeper of the market: manufacturers had to submit proof of safety and efficacy for its approval before placing a new device on the market. However, the number and variety of medical devices falling under its purview were enormous, with each type demanding a specific method of appraisal. Congress lightened the workload by devising a regulatory scheme organized around three categories of potential harm. Class I held the least risky devices – bandages, tongue depressors, thermometers, and such like – which require negligible oversight. Class II is by far the largest, and includes devices for which standards of safety and effectiveness have been agreed – powered wheelchairs and infusion pumps. Manufacturers of Class III devices, the riskiest category, which includes implanted devices, must submit satisfactory evidence of safety and efficacy. Even here Congress allowed some leeway: manufacturers capable of demonstrating that a new Class III device was 'substantially equivalent' to one that had been approved can escape the burden of pre-market approval.

Food and Drug Administration approval does not immunize manufacturers of faulty medical devices from product liability law, where the financial penalty can be exorbitant, combining redress for injury and punishment. The punitive element is meant to encourage manufacturers to take care; it originated in the 1950s when consumer advocates began challenging the doctrine of *caveat emptor* (let the buyer beware) on which manufacturers had relied. Notable victories include a Consumer Bill of Rights that President Kennedy had proclaimed in 1962 in a message delivered to Congress. In 1964, the American Law Institute issued Section 402A of the Restatement (Second) of Tort, which encouraged courts to recognize consumer vulnerability in the face of corporate power. Despite or perhaps because of their success, consumer militancy continued to grow, encouraged in part by lawyer Ralph Nader who shot to prominence following the publication of *Unsafe At Any Speed* (1965), in which he revealed how the car industry had resisted making safety a priority, and claimed that its negligence was responsible for an annual slaughter of thousands of Americans: he estimated that 51,000 Americans had died from faulty car design in 1965. In 1972, Congress passed the Consumer Products Safety Act, which established the United States Consumer Products Safety Commission.

The burden of proof is relatively light in product liability. Plaintiffs need only show that a defect existed in the product that caused their injury; they need not demonstrate that the manufacturer's conduct fell below a standard of care, that is, what other reasonably prudent manufacturers were doing. Cases brought under product liability law were stalking the American blood industry. Perhaps the most significant is *Perlmutter v. Beth David Hospital* (1954). The plaintiff, Mrs Perlmutter had contracted hepatitis following a blood transfusion, for which she had paid $60, and sought damages from the hospital on the grounds that the blood had not been of merchantable quality. The Court of Appeal rejected the case: blood transfusion occasionally features in medical care, which is a service. However, the decision was controversial; it had been decided by a narrow 4–5 majority, and was ridiculed by legal commentators who argued that the reasoning that blood transfusion is incidental to a service would absolve a restaurant that served food that poisoned diners. Victims of post-transfusion hepatitis continued to seek redress under product liability. In 1955, California became the first state to immunize blood banks from product liability claims by adopting a statute that defined blood banking as a service industry. However, it took more than a decade for other states to adopt so-called 'Blood Shield' statutes, and not all of them did so.

In 1962, the Federal Trade Commission, the government agency that oversees anti-trust law, decided that human blood provided for transfusion is an article of commerce. It was responding to the complaint of two commercial blood banks in Kansas City that local physicians and hospitals were boycotting their blood, and accepting only that provided by their own non-profit scheme. One of the banks had been opened in 1955 in a slum area, and accepted blood from 'skid row' donors. A married couple without medical training owned it, and its medical director was a seventy-eight-year old physician who had no experience of blood banking. The second bank had been opened three years later, and operated along similar lines. Nonetheless, the agency concluded that the boycott constituted an illegal restraint of trade and threatened to impose civil penalties of up to $5,000 each day that it continued.[6]

The commission was fulfilling its statutory responsibility to ensure markets can operate freely, irrespective of the nature of the stuff being traded. Its decision had not been taken lightly: records, transcripts and exhibits presented ran to over 20,000 pages, and cost respondents and taxpayers something approaching $500,000. However, the commercial blood bankers' victory was pyrrhic: blood defined as an article of commerce must carry an implied warranty of fitness, and patients who contract post-transfusion hepatitis could sue under product liability – in effect, the commission exploited the threat of punitive damages to discipline, or even drive out of business, skid row blood banks. Its decision influenced *Carter v. InterFaith Hospital of Queens* (1969), where the court agreed that blood provided by a commercial blood bank is a saleable commodity.

The commission's decision effectively sacrificed patients' health on the free market's altar. It also stymied medical professionals who on the one hand claimed to act in the best interest of patients, and at the same time passionately defended their freedom to trade in the American health-care industry. The solution the American Medical Association proposed was to bundle the cost of blood within charges for medical services. Recipients of 'bad blood' could continue to seek redress on the grounds of medical negligence, where damages awarded are without a punitive element and hence are considerably lower than those awarded where product liability succeeds. However, critics mocked this strategy: the suggestion that patients pay only for medical services and never for body/cadaver stuff is no more persuasive than the contention that restaurants provide 'dining services', not food.[7]

Self-regulation

In cases brought under medical negligence courts decide whether or not the defendant has acted as a reasonable and prudent person according to standards existing at the time of performance. *Ravenis* had drawn attention to the absence of agreed standards not only of criteria of medical bioavailability, but also during handling. Indeed, precautions taken during extraction sometimes consisted of little more than tucking tie into shirt and wearing gloves.[8] There was also a real risk of recipients losing their eyes because of contaminated storage media.

The Eye Bank Association of America began developing medical standards. *Ravenis* also served as a wake-up call to other industry sectors. The Navy Tissue Bank might have scaled down its operation at the end of the Vietnam War but it saw itself as responsible for the reputation of private-sector entities it had encouraged. In August 1976, past and current staff convened a meeting where it was decided to create the American Association of Tissue Banks, a scientific, non-profit, peer-led industry association, with a mission to ensure that body and cadaver stuff was of a uniform high quality.[9] Its first president was Kenneth W. Sell (1931–1996), a former director of the Navy Tissue Bank; James E. Ostrander, of the Naval Medical Research Institute, was secretary. Navy Tissue Bank staff were responsible for administration until 1983 when the association appointed its first full-time employee.

The Eye Bank Association of America's medical standards were officially adopted in 1980, and in 1982 the American Association of Tissue Banks published its own. Inspection and accreditation of facilities were subsequently initiated. Failure to pass muster might result in disciplinary action, even expulsion. Membership might reassure recipients that safety was a key consideration, but it was voluntary; extractors who refused to join were free to determine their own practices.

CHAPTER 24

The Blind Eye Act

'We often find we have to go scrounging round hospitals and mortuaries,' Leonard Lurie, a London-based ophthalmologist, confessed to readers of the *Sunday People* in January 1966. He went on to reveal that on one occasion he was preparing to operate on a patient when he found the hospital mortuary was empty. On the verge of sending his patient home he learned that an elderly man had just died in one of the hospital's wards. None of his kinfolk had been present at the deathbed. Time was marching on, and Lurie was unwilling to wait until they had been contacted. As he put it, 'I dashed off, and removed the eyes and replaced them with a pair of blue artificial ones – and nobody was any the wiser.'[1]

Lurie had no fear of prosecution. Failure to comply with the Human Tissue Act was not a criminal offence. Moreover, there was no consensus on what its ambiguous wording required. For instance, Rycroft complained that in his collecting sites enquiries about enucleation were made at the same time as agreement to a hospital post-mortem examination was sought, which typically was several hours following confirmation of death, often too late for cadaver corneas to be usable.[2] By way of contrast, in Southend hospitals next of kin were contacted while a patient's life was drawing to a close and, if they agreed to enucleation, shortly after the source's death had been confirmed ward staff summoned the on-duty ophthalmologist to enucleate the eyes.[3] But at least both these interpretations recognized the existence of the Human Tissue Act. Some ophthalmologists were under the impression that the Corneal Grafting Act still applied; and others believed that Ministry of Health staff, not Parliament, had fettered their freedom, and hoped the minister might be persuaded to relax the requirement to make reasonable enquiries.[4]

'One must face the fact, gruesome as it is, that just as when you want bread you go to a baker, if you want fresh eyes you go to places where people die,' explained Douglas Gibbs, administrator of the South East Regional Eye Bank.[5] A 'just-in-time' inventory method works only in a general hospital where death is a frequent visitor. But it rarely visited the Queen Victoria Hospital, East Grinstead, where Gibbs was based, and he was struggling to gain the cooperation of staff of nearby places that it frequented. Management committees of four cottage hospitals had refused to cooperate; one member of the Regional Association of Funeral Directors had agreed but backed out when 'people were beginning to talk', and he became fearful that gossip might harm his business; a medical director in the six local mental asylums enucleated eyes of dead patients, but when he fell ill his meagre contribution dried up; the matron of a hospice, staff of four old people's homes and a coroner failed to keep their promise to notify him of suitable corpses; eight large general hospitals together yielded few cadaver eyes.

Hillingdon Hospital, west London, became the South East Regional Eye Bank's most productive collecting site. Thomas Casey was head of its Eye Department, and following Rycroft's death in 1967, also that of the Queen Victoria Hospital. His caseload was heavy: other ophthalmologists referred 'difficult' patients to him because he was reputed to be a skilled surgeon.[6] However, hospital staff were unimpressed, and refused to cooperate with his team, which relied on hospital porters and mortuary staff to notify it when someone died. Their reward was the occasional packet of cigarettes and two bottles of whisky each year.

Lurie used scrounging to mean taking something by stealth; it can also mean exploiting other people's generosity, which is how Casey's team persuaded next of kin of someone who had just died in Hillingdon Hospital to agree to enucleation. The task fell to junior members who were advised to rehearse a fiction of a nineteen-year-old patient in urgent need of a corneal graft. Should next of kin refuse, a more senior doctor would be deployed.[7] The story must have been persuasive. In 1971, 320 eyes were enucleated on the hospital premises, that is, around one-third of all those grafted that year in the United Kingdom.[8] (Statistics collected by civil servants put the increase in keratoplasty from 834 in 1967 to 1,151 in 1970, but this was acknowledged to be an underestimate as an unknown number of operations were unrecorded.)[9] Forty-eight cadaver eyes had been grafted in one of Hillingdon Hospital's operating theatres. Porters at the hospital's main entrance organized distribution of the remainder. Ninety-eight had been dispatched to Heathrow Airport for export by obliging airlines, mostly to hospitals in Jerusalem and Kuwait. Porters sent the remainder either direct to the Queen Victoria Hospital by train from Victoria Station or by taxi to Casey's London home.[10] Eighty of these stayed in the Queen Victoria Hospital; the South East Regional Eye Bank redistributed the remainder to sixteen other British hospitals.

Routine Salvaging

A whistleblower had reported these tactics to MP Shirley Williams. Subsequently, the local MP Laurie Pavitt asked about it in Parliament, and had been assured that the government deplored pressurizing next of kin.[11] When Casey was called to account by civil servants he provided assurances that in future his team's enquiries would be tactful and truthful.[12]

Casey's reputation escaped scrutiny in the court of public opinion, where the first British heart explant/transplant had met a chorus of disapproval. The operations were performed in May 1968. The source was a building worker who had been almost decapitated in an accident on a building site; on admission to hospital he was diagnosed brain dead but his heart was still beating. Although he was unlikely to survive for more than a few days, surgeons in the operating theatre were responsible for his cardiac death. Was this murder, a critic asked: 'If the heart is still beating at the time it is removed from the body of its "donor", has that person in fact died? Does a surgeon who transplants it lay himself open to a charge of homicide?'[13]

Worldwide in 1969, forty-eight hearts were explanted/transplanted; in 1970, seventeen; in 1971, nine. No recipient survived, and a moratorium was announced. Explant/transplantation of cadaver kidneys continued, but numbers were limited by intensive care unit staff's refusal to cooperate. They saw their responsibilities as lying with the brain-dead patient and their distraught kinfolk, whereas those of explant/transplant teams lay with people waiting for a replacement of failing kidneys. There was also conflict over how to care for brain-injured patients – for instance, the routine practice of withholding fluids in order to reduce brain swelling is undesirable from the point of view of protecting organs that might be suitable for repurposing. There was also a clash over the choreography of cardiac death: intensive care staff typically gather kinfolk round the bedside of the comatose patient, and in their presence switch off life-support machinery; surgeons take brain-dead patients into an operating theatre where, out of kinfolk's sight, a no-place is created and following explantation cardiac death ensues. In desperation some explant/transplant surgeons took to touring intensive care units of local hospitals on the off chance of finding someone who was potentially a source of cadaver kidneys. Michael Bewick, who operated at Guy's Hospital, London, acknowledged his visits were unpopular, and that he and his colleagues were known as vultures.[14]

In the late 1960s a vocal lobby had begun pressing for either an amendment to the Human Tissue Act 1961 or its replacement by another statute that, they believed, would stack the cards in favour of explant/transplant surgeons and their patients by allowing 'routine salvaging' of cadaver kidneys unless the potential source during their life had recorded a specific objection, a policy called presumed consent, or opt-out, or contracting-out.[15] Their attempts to place presumed consent on the statute book failed; they were without the support of the government, which was being repeatedly bruised by opposition to the Abortion Act 1967 and was keen to avoid fanning the flames of moral outrage and exacerbating public disquiet provoked by human heart explant/transplants. Instead, the government deployed the classic delaying tactic of convening a group of experts to consider the issue, which became known as the MacLennan Advisory Group after its chairman Sir Hector MacLennan, then president of the Royal Society of Medicine. In its final report published in 1969, the group declared the Human Tissue Act obsolete, but failed to agree on a replacement.[16]

The Broderick Committee, which had been appointed by the Home Secretary in 1965 to review the role of coroners, provided government with another excuse for delay. Brain death typically is associated with circumstances that must be referred to a coroner, but it was unclear whether their lawful possession began at brain or cardiac death. A majority of coroners supported organ transplantation, but if lawful possession began at brain death they were unhappy about the prospect of finding themselves in a situation where agreement to explantation of kidneys might be taken as a signal to switch off life-support machinery. The committee's report, which was published in 1971, failed to resolve their dilemma.

In 1973, considerable disquiet was caused by a report that, despite absence of evidence of a pledge card and without parental agreement, the kidneys of seventeen-year-old Nigel Ford had been explanted forty minutes after his admission to hospital following a collision of his scooter with a van in which he suffered a traumatic brain injury.[17] His parents were on holiday in Spain, and a neighbour informed the policemen who had visited their house that they would object to their son's kidneys explantation. Ford's father accused the explanters of behaving like bodysnatchers. His supporters considered that doctors had unlawfully mutilated Ford's corpse, and wondered whether it was unwise to enter a hospital in case they suffered a similar fate.

What Can I Do About it Now?

In June 1975, the DHSS issued HSC(IS)156, a circular intending to clarify confusion about the wording of the Human Tissue Act, and at the same time increase the number of cadaver kidney explant/transplants. Hospital staff were urged to be on the alert for potential sources of cadaver kidneys for transplantation, and cadaver eyes for corneal grafting, and either approach next of kin themselves or notify a member of an explant/transplant team who would make reasonable enquiries.[18] Section 8 of the circular emphasized that written authority for the removal of organs and tissue '*must*' (emphasis added) be obtained from the person lawfully in possession of the body and recorded in the patient's notes by the person receiving it. Yet Section 17, acknowledging the convenience of removing some cadaver stuff during post-mortem examinations, specifically and exclusively refers to pituitary glands, and suggests that the forms given to next of kin to sign '*might* routinely include' a brief general reference to the removal of 'tissue'. One year later, a civil servant observed that, 'might routinely include' had given a 'take it or leave it impression'. Subsequently, a nationwide survey of post-mortem consent forms found none offered next of kin the opportunity to agree or disagree to extraction of cadaver stuff.[19]

In the *Sunday People*'s exposé of August 1976, a widow is quoted as saying, 'I did not give permission for these parts to be taken out of his body ... I don't like the idea of this sort of thing going on, but what can I do about it now?'[20] The Human Tissue Act provided next of kin with no means of redress. The exposé's revelations about mortuary practices failed to make an impression in the public sphere but behind the scenes caused consternation, to the detriment of the pituitary gland collection. Trade union representatives of mortuary staff, worried that their members were being incentivized to break the law, advised against cooperation.[21] Coroners, who by this stage were responsible for ordering around nine out of ten post-mortem examinations, were increasingly withholding permission. The coroner in Blackpool had begun insisting that next of kin's agreement to extraction was obtained, but policemen on whom that responsibility fell refused to raise an 'emotionally charged' subject.[22] Bernard Knight, (1931–) a leading forensic pathologist, warned against averring the collection was lawful; it operated under what he called 'the Blind Eye Act'.[23]

In an effort to quell anxieties about the law, in August 1977 the DHSS issued circular HSC(77)28 which contained the following appeal: 'It would be tragic if insufficient pituitary glands became available, since it is at present impossible to synthesize human growth hormone or to obtain it from animal sources, and the absence of treatment would leave those children with a deficiency of the hormone as dwarfs.'[24] The circular goes on to suggest hospital post-mortem paperwork include a brief general reference to withholding of cadaver stuff but advises specific agreement is unnecessary.

Unfortunately, HSC(77)28 only exacerbated the confusion: did the statute grant next of kin the right to agree or refuse extraction and retention of cadaver stuff for repurposing in research or therapy during a hospital post-mortem examination but, and despite Knight's warning to the contrary, not one that had been ordered by a coroner? In August 1978, the unnamed author of an editorial published in the *British Medical Journal* under the heading 'Postmortem tissue problems' observed that 'custom and tacit approval rather than the Human Tissue Act' were governing recovery of homostatic cadaver stuff. The recommendation in HSC(77)28 that hospital post-mortem forms include a brief blanket request to allow recovery of cadaver stuff might spare next of kin the 'gruesome details', but strictly speaking did not conform with contemporary expectations of informed consent. It goes on to warn coroners that they are not above the law: 'This point seems to have eluded some administrative bodies: several directives have appeared stating that the Home Office and even the police have looked into the difficulties and that coroners can take such tissues with an easy mind.'[25]

In July 1980, new instructions were circulated, which among other things required every cadaver pituitary gland to be associated with an autopsy number. A pathologist had to sign a declaration that medical bioavailability criteria had been observed, and the paperwork had to be completed in triplicate.[26] But few pathologists were prepared to sign the paperwork in case they were subsequently found to have been in breach of the Human Tissue Act. Fewer coroners were allowing cadaver pituitary glands to be extirpated from corpses in their possession. The situation reached crisis point, and in November 1980 the Home Office – the government department responsible for coroners – convened a meeting where it was concluded that new legislation was the only solution.[27]

The pay deal with mortuary staff was making matters worse. The DHSS made desperate efforts to restore supplies, sending out hundreds of letters and telephoning pathologists and coroners. A radically new approach was adopted: a 'pituitary collection officer' was appointed to bombard staff of busy mortuaries with phone calls, letters and personal visits.[28] Thanks to these efforts the size of the collection began recovering, reaching 38,204 cadaver pituitary glands in 1981. However, it fell the following year to 32,891,[29] despite the DHSS's circular DA(82)9 issued in May that drew attention to the 800 children who were in danger of being 'condemned to stunted growth', and urging hospitals to designate staff of suitable seniority and sufficient in number to provide round-the-clock cover to make the enquiries required under the Human Tissue Act.[30]

A shortfall of home-produced hGH was averted by Hartree's stockpile of around 25 kg of pituitary gland powder, representing the pituitary glands of around a quarter of a million dead people.[31] But when that was used up, the NHS faced the prospect of having to buy expensive imports. In 1983 no improvement in the size of the collection was seen. In February 1984, following adverse press publicity about a mortuary technician accused of pituitary gland 'theft', the chairman of the Advisory Committee for the National Pituitary Collection wrote to pathologists asking them to reassure their 'conscientious and honourable mortuary employees', and to extend to them his thanks for the worthwhile job they were doing.[32] In 1984, which proved to be the collection's last full year, 42,838 cadaver pituitary glands were extirpated.[33]

CHAPTER 25

Creating American Hybrid Extractors of Cadaver Stuff

When Congress agreed to shelter the End-Stage Renal Disease Program under Medicare in 1972, cadaver kidneys joined cadaver pituitary glands as national assets. In addition to funding the National Pituitary Agency to undertake all the tasks associated with extirpation and processing of cadaver pituitary glands, taxpayers now supported organ procurement agencies, non-profit entities responsible for organizing explantation and delivery of cadaver kidneys in a viable condition to transplant teams. Choice of name was curious: not only does procurement mean 'acquiring by care and effort' but also 'obtaining for the gratification of lust'.

Organ procurement agencies were a hybrid of non-profit entities solely dependent on government for financing. Guaranteed funding encouraged a tenfold increase in their number, from eight in 1972 to eighty-two in 1982.[1] The majority were hospital organ procurement agencies, run by a transplant team for their patients; each one was effectively a federally funded stash, albeit holding an inventory with a very brief shelf life. The minority were independent organ procurement agencies, situated outside a hospital, and established by a transplant team and/or civil society organization such as a chapter of the National Kidney Foundation to supply several transplant teams within a specific locality.

Regulation was light touch. There were no rules on the territory that an agency could claim. The only limitation was the willingness of hospitals to cooperate as collecting sites, and to permit a no-place to be established in an operating theatre. The territory of a neighbouring organ procurement agency was sometimes invaded or contested. Hence, whereas the National Pituitary Agency had been established to stall competition in mortuaries, the End-Stage Renal Disease Program encouraged it in intensive care units where a majority of potential 'beating-heart' sources of cadaver kidneys are found.

Expenditure by Medicare on kidney transplants increased much more rapidly than Congress had anticipated, from $31.2 million in 1974 to $258.7 in 1983 when 5,616 kidneys were explant/transplanted, which puts the average cost of each at $460,000.[2] Nonetheless, threats to funding were defeated by lobbying by well-organized patient groups and physicians.[3] The programme was protected in 1983 when Medicare changed its accounting procedures, and abandoned cost reimbursement, where the cost of everything that might work was reimbursed, in favour of a fixed tariff for eligible procedures. The measure was intended to contain expenditure on health care, which was spiralling out of control, and by the late 1970s was consuming 11 per cent of gross national product. A fixed tariff, health policy analysts claimed, would discipline profligate health-care providers by forcing them to manage resources more carefully, and make economies in order to ensure the bottom line of their accounts is written in black ink.

The battle over reimbursement schedules that characterizes the American health-care system intensified. The tariff is organized around Diagnostic-Related Groups (DRGs), and is based on a notional average cost of procedures, and makes few concessions to patient characteristics (age, other health problems and so on), severity of condition or speed of recovery. Burn care finances were badly hit: around two out of five patients admitted to a burns centre were eligible under Medicare. In 1984, the National Coalition of Burns Center Hospitals was joined to protest that the tariff made insufficient allowance for the medical, social and financial complexity of burn care, particularly of elderly people who typically took far longer to recover from their injuries than the tariff recognized. Anticipating a nationwide cumulative loss of more than seven million dollars in 1985, coalition members predicted that the tariff would discourage vigorous resuscitation of severely burned people, and some burns centres would be forced to close, a development that would result in more patients being treated – and no doubt dying – in non-specialist facilities.[4]

Denying Sources a Reward in the Currency of Money

In September 1983, Harvey Barry Jacobs announced that the International Kidney Exchange Ltd. had opened for business. It was a recruitment agency, offering potential sources of kidneys for transplantation two types of monetary incentive. One was aimed at enticing healthy Americans to part with a kidney. Should few be tempted, he proposed bringing citizens of low-income nations to the United States. The other incentive was described as a kind of life insurance because the intended recipients of the cash were kinfolk of someone who had pledged, and on death had had both kidneys explanted. Jacobs had begun contacting hospital administrators, and claimed to have received a few expressions of interest.

Jacobs intended charging recipients a finder's fee of between $2,000 and $5,000.[5] He was not alone in believing that the currency of regard was an insufficient incentive to potential sources, but whereas a majority favouring a financial incentive argued the case in public and policymaking circles, Jacobs viewed it as a business opportunity. Lists of people waiting for a kidney transplant had begun growing by leaps and bounds following the discovery in the late 1970s of cyclosporine, a highly effective anti-rejection drug that improved the chances of an explanted kidney surviving in a recipient's body from 20–30 per cent to 60–70 per cent. Moreover, the likelihood of becoming a recipient depended on where in the United States you lived. Cadaver kidneys might have been declared a national asset, but their distribution was uneven, partly because the productivity of organ procurement agencies varied. Hospital organ procurement agencies were less successful than independent ones at gaining the cooperation of intensive care unit staff. Political scientist Jeffrey Prottas, who has undertaken extensive research in this field, attributed this to hospital organ procurement agencies relying on the goodwill of professional colleagues whereas independent organ procurement agencies viewed gaining support of intensive care staff as a marketing challenge, and employed aggressive techniques such as 'personal selling' (periodic unscheduled visits to critical care units to talk to staff), organized professional education or in-service training, and offered subsidized places at conferences.[6] Despite repeated warnings of shortage, wastage was high: in 1981, around one in five explanted cadaver kidneys was discarded because no suitable recipient could be found.[7] Transplant surgeons mostly decided allocation with some matching tissue type, others ignoring it, some considering length of waiting time, others not. Patient characteristics also influenced decisions: African Americans had – and continue to have – the highest death rate of would-be cadaver kidney recipients.[8]

Jacobs was exposed as having had his licence to practice medicine in Virginia revoked in 1979 following a conviction for fraud for which he had served ten months in a federal correctional facility.[9] The American Society of Transplant Surgeons, the Association of Independent Organ Procurement Agencies, the American Society of Transplant Physicians and the National Kidney Foundation condemned his scheme. But it was legal.

In November 1983, Representative Albert Gore Jr (Democrat) introduced a bill that would prohibit financial reward of sources of body/cadaver stuff, and which provided federal funding for the development of an infrastructure intended to minimize competition and support and coordinate cadaver kidney explant/transplant activities across the nation, and thereby reduce wastage and promote equity.

In February 1984, several witnesses to the House of Representatives Subcommittee on Health of the Committee on Ways and Means hearing on Gore's bill complained of its failure to recognize the detriment that federal funding of organ procurement agencies was having on entities extracting stuff from non-beating heart corpses. Baxter, in his capacity as past president of the American Burn Association, joined other witnesses, including representatives of the American Medical Association, and third-party payers Blue Cross/Blue Shield, and declared the financial burden of the End-Stage Renal Disease Program on Medicare, which then stood at $1.5 billion, excessive: the favoured status of organ procurement agencies should be withdrawn, and they should operate as private-sector businesses like other entities extracting cadaver stuff.[10]

Richard L. Hurwitz, president of the Virginia Organ Procurement Agency and medical director of the Virginia Tissue Bank, protested at the varying esteem in which the repurposing of the various fragments of the corpse was held: 'That a surgeon can turn to the freezer for banked human skin to treat a burn victim or reach up to a shelf to select a bone graft for a patient undergoing a spinal operation is no less a dramatic event (if less newsworthy) than the well publicized routine of organ transplantation.'

Another complaint was that the bill did nothing to address the shortages of cadaver stuff extracted from non-beating heart corpses experienced by other clinical disciplines, such as ophthalmology, burn care and orthopaedics. Robert E. Stevenson, in his capacity as chairman of the American Association of Tissue Banks Standards Committee, informed the hearing that each year skin of 5,000 cadavers was needed to treat the 10,000 people who suffered massive burns or scalds, but as things stood skin banks were incapable of meeting this level of demand.[11] Perhaps realizing that the typical civilian burn victim is poor lobby fodder, Stevenson went on to appeal to the nation's contract with its armed forces by pointing out that there had been insufficient cadaver skin to dress the wounds of the badly burned Marines who had survived the suicide bombing in Lebanon the previous year, an atrocity which, at the time of writing, remains the deadliest single overseas attack on American servicemen since the Second

World War. Duplication of tasks in collecting sites and along the supply chain was another source of inefficiency and waste.

Stevenson urged federal government support for the development of regional, 'multi-tissue banks', run by trained professional staff. He was one of several witnesses complaining that without cooperation a great deal of cadaver stuff that could have been extracted and repurposed was wasted. Despite their unique access to intensive care units, organ procurement agencies rarely sought the agreement of the next of kin of a brain-dead patient to the extraction of anything other than cadaver organs, and, where agreement was volunteered, failed to contact an appropriate entity. For their part, the various entities extracting cadaver stuff out of non-beating-heart corpses jealously guarded their collecting sites, and sometimes competed to be the first to arrive at the deathbed, funeral parlour, medical examiner's office or hospital mortuary. As a result, many corpses were being placed in a grave or crematory with just one or two fragments – pituitary gland, and/or eye, skin, bone or kidney – missing.

Gore had proposed drawing all the entities concerned with cadaver stuff under the same infrastructure. But the Eye Bank Association of America objected strongly to the measure.[12] The proposal failed to appreciate the fundamental differences between eye banks and other entities extracting homostatic cadaver stuff in terms of: law (by 1984 nearly half of the 23,000 corneal transplants performed in the USA utilized corneas excised under legislative consent);[13] work and workforce (enucleation/excision mostly undertaken by trained technicians and morticians/funeral directors); business model (the contribution in cash and kind of civic society organizations).

The formidable lobbying power of Lions was mobilized to write letters of protest to their member of Congress, pressing them to reject the bill. 'We were blindsided,' recollected Jerold Mande, a former aide to Gore who drafted much of the bill, 'put on the defensive.'[14] Gore was forced to compromise. As he explained, 'I have tried to make it clear that the changes we are discussing are changes to improve the vital organ system and not to try to fix the tissue system since the latter ain't broke.' Nonetheless, he went on to clarify, the 'ain't broke' applied only to eye banks; some of the other 'tissue retrieval people', as he called them, did need help.[15]

Evolving hybrids

President Reagan signed the National Organ Transplant Act (NOTA) into law in October 1984. Acquiring, receiving or otherwise transferring any human organ for valuable consideration became a criminal offence punishable by a fine of up to $50,000, imprisonment of up to five years or both. But the statute permits reimbursement of costs associated with extraction, transportation, processing, preservation, quality control, storage and travel expenses.

'Human organ' was defined broadly as kidney, liver, heart, lung, pancreas, bone marrow, cornea, eye, bone, skin and any other human organ specified by the Secretary of Health and Human Services. Blood and blood products were excluded from the definition, despite arguments in favour of their inclusion being rehearsed during the congressional hearings. The reasoning was that sources suffer no harm because blood is replenished.[16] But so is bone marrow. The decision compromised the principle informing the act that buying and selling body/corpse stuff is unconscionable, offends the public interest and thus must be outlawed.

The act provided the start-up costs of organ procurement organizations, non-profit entities replacing organ procurement agencies. Change of name is significant: entities called agency are part of the machinery of federal government, whereas 'organization' situates entities within the private sector. The authority that transplant teams had exercised over organ procurement agencies was weakened. Organ procurement organizations are free-standing, separately incorporated non-profit entities with laypeople in a majority on their board of directors. Moreover, Congress took decisions about allocation out of the hands of explant/transplanters by providing for an independent non-profit entity that would keep a computerized register of people in need of a kidney transplant, and match sources and recipients.

Entities organized around non-renal cadaver stuff could be set up without official approval. But the existence of organ procurement organisations is decided by the Department of Health and Human Services, which at various intervals holds a competition in which the prizes awarded are franchises conferring entitlement to every cadaver kidney explanted within a specified area, called a Donation Service Area (DSA), which is large enough to provide at least fifty cadaver kidneys each year. Competitors have to demonstrate that they have secured the cooperation of a substantial majority of potential collecting sites within a given territory. The franchise is a monopoly: competition within a DSA is prohibited, and Medicare will not reimburse the costs of kidney procurement incurred by any other individual or business, and any hospital found transplanting such a kidney faces the threat of total exclusion from Medicare and Medicaid. However, franchises are tied to performance targets, and failure to meet them results in closure or a forced merger with a more efficient organ procurement organization. This was an important measure – several organ procurement agencies had never explanted a cadaver kidney.

The number of organ procurement organizations grew from eighty-two in 1982 to 100 in 1986. Each was larger than its predecessor in every measure, such as collecting sites and notifications of brain-dead people.[17] Over the same period, many more kidneys were explanted – from 20.3 per million people to 33.5 and wastage fell almost by half. Prottas attributed their greater productivity and efficiency to the appointment of transplant coordinators who are available round the clock, seven days a week. Transplant coordinator is a new profession dedicated to procuring kidneys; Prottas calls them the 'sales representatives', marketing kidney explantation to hospital staff and the next of kin of brain-dead patients in the intensive care unit, which he calls the 'point of sale'.[18]

Another condition of the award of a franchise is membership of the United Network for Organ Sharing, a non-profit awarded the contract to operate the Organ Procurement and Transplantation Network, which had been mandated in the National Organ Transplant Act. The United Network for Organ Sharing had begun life in 1977 as the South-Eastern Organ Procurement Foundation, when transplant surgeons agreed to cooperate in a computerized database that would place explanted cadaver kidneys that were unusable locally.

The act also required the establishment of the Task Force on Organ Transplantation, which was responsible for developing policy that would reduce wastage and increase output. Its report published in 1986 recommended, among other things, that whoever was awarded the contract for the Organ Procurement and Transplantation Network should be invested with the authority to oversee activities of organ procurement organizations by setting standards for acquisition, preservation and tissue typing, and also criteria of allocation. When Congress adopted its recommendations in the Budget Reconciliation Act 1986, the United Network for Organ Sharing became a non-governmental or quasi-governmental body regulating a nationwide network of federally funded non-profit public businesses.[19] The powers vested in it have been described as 'awesome', susceptible to an anti-trust review brought by pro-competition enthusiasts.[20] But, according to Prottas, in its early phase, authority was exercised lackadaisically: under the Republican administrations of Reagan and Bush, rigorous regulation was considered distasteful. Not one organ procurement organization was penalized.[21]

CHAPTER 26

Sharing Pledges and Cadaver Stuff

In the summer of 1971, Mrs Elizabeth Ward placed an advertisement in the personal column of *The Times* which read, 'B.A. *B.T. 15. 12. Group A – A donated cadaver kidney of this tissue type will release an eighteen year old from the wretchedness of his machine.'[1] The eighteen-year-old was her son Timbo whose kidneys had failed and who was surviving on thrice weekly home dialysis. The following day the story was featured on the front pages of the national press. Ninety-eight people replied to the advertisement, but Mrs Ward accepted none of their offers. Her aim had been to publicize the shortage of cadaver kidneys for transplantation, and it had been achieved.

Mrs Ward was a formidable campaigner. In *Timbo: A Struggle for Survival* (1996) she laid out her strategy of combining personal appeals to influential individuals with public pronouncements as spokesperson on behalf of the British Kidney Patients Association, a civil society organization she founded in 1975. Among her correspondence with Keith Joseph (1918–1994), Conservative Secretary of State for Social Services, who led the huge DHSS between 1970 and 1974, is a letter in which she reminds him that his son had been Timbo's classmate at Harrow, an elite school. Wouldn't he move heaven and earth if his son were facing the same plight as Timbo?

Mrs Ward had been favourably impressed by the pledge card attached to the Uniform Anatomical Gift Act 1968, and had sent one to Joseph. Previous attempts at encouraging British people to pledge their cadaver kidneys had been unofficial and had failed. The Automobile Association led the field by offering its five million members a card which they could carry on their person just in case they were killed in a road traffic accident. Worded as a testamentary document, the pledge card was withdrawn when the Medical Defence Union, which provides insurance cover for doctors, warned that it did not constitute valid authorization.[2]

The RNIB's blue-card scheme had demonstrated that persuading people to carry bits of paper or card on their person is an ineffective remedy for the shortage of cadaver eyes. Civil servants had no reason to believe that pledges would be more effective in relation to cadaver kidneys, but, as a sceptic rhetorically asked: 'Will people who have the cards transfer them from suit to suit, have them prominently placed in their pockets, and have their pockets gone through quickly and diligently at the moment of death?'[3] Some of his colleagues were persuaded that marketing pledges might encourage 'a change of attitude on the part of doctors and public alike so that there is less difficulty, reluctance or odium attached to obtaining consent to the removal of kidneys whether or not there is a bequest on the part of the deceased person'.[4] In other words, marketing pledging might make it easier for an enquiry to be made that next of kin might find acceptable.

The government caved in to assuage the lobby pressing for 'routine salvaging' of cadaver kidneys, and in 1972 arranged the production of five million red and blue leaflets, devoid of illustration, emblazoned with 'Your kidneys could help someone to live after your death', a slogan animating organs with agency. The leaflet incorporated a pledge card, which would-be pledgers were instructed to complete, cut out and keep on their person at all times. The wording of the pledge was 'I request', not 'I bequeath', which few people would recognize as a warning that a pledge is not a testamentary document but an offer that might not be accepted. It was also decided that kidney pledgers' identity would be a mobile one, in order to spare the government the inconvenience and expense of securing it to a register.

Disappointed by what she considered a lacklustre marketing campaign, Mrs Ward set about organizing her own, which included persuading petrol stations to display the leaflets on their forecourts, supermarkets at checkouts and retail pharmacies on counters. But the Minister of Defence, William Rodgers denied her request to ensure all members of the armed forces signed and carried a pledge. His reason: the rule would send an uncomfortable message to mothers of soldiers serving in Northern Ireland. What about the mothers of young men on dialysis? she countered.[5]

'Did you know that your eyes could also be used after death to restore somebody else's sight?' was written on the reverse side of the leaflet. Anyone wanting to learn more about eye donation was advised to contact the RNIB, which responded by sending a copy of its *Restore Sight* leaflet with the hazy bucolic scene on the front cover. Mrs Ward might have been dissatisfied with the marketing campaign but it caused consternation within the organization because its blue-card scheme now looked amateurish and its register of pledges redundant.[6] But the charity was loath to divert additional resources away from its work with people who were irremediably or partially blind. Ophthalmologist Patrick Trevor-Roper's advice was, concentrate on encouraging the public to formulate their answer to an enquiry about enucleating the eyes of kinfolk who had just died.

Opportunities for pledging cadaver kidneys greatly increased in 1977 when the Driver Vehicle Licensing Centre began offering applicants for a provisional driving licence a pledge card that was the same size as the driving licence. The typical applicant was one of the 1.25 million young people who each year celebrated their seventeenth birthday, and whose risk of head injury following a road traffic accident was much greater than that of more experienced mature drivers. MP Michael Roberts had borrowed the idea from New York State's Department of Motor Vehicles. It was marketed under the slogan 'Licence to save life', an obvious play on Ian Fleming's secret agent James Bond's licence to kill.

In February 1978, Clementine Churchill, Winston's wife, died, and realization of her pledge was widely publicized. The RNIB was inundated with enquiries, which rekindled its belief in pledging. Following a lengthy discussion, a decision was taken to revamp the blue card and close the register, which by this stage held 85,000 names. The identity of cadaver eye pledgers, like that of cadaver kidney pledgers, was to be a mobile one.[7] The new pledge card was modelled on the cadaver kidney one, and emblazoned with 'I am an eye donor'. But the DHSS turned down the request for financial support because it was planning to replace its kidney pledge card with a 'multi-organ' one that would include cadaver eyes.

'Multi-organs'

In 1980, the DHSS issued six million 'multi-organ' pledge cards that invited people to pledge either *any* part or specify which ones could be extracted out of their corpse from a list of kidneys, eyes, heart, liver and pancreas (the moratorium on heart transplants had been lifted). Unfortunately, the original multi-organ card was flimsy and quickly fell apart, and the following year was replaced with a plasticized one, emblazoned with the slogan, 'I would like to help someone to live after my death.'[8]

Extensive publicity was necessary to counter the detrimental impact on public opinion of 'Transplants: Are the Donors Really Dead?', a BBC

Panorama documentary televised in October 1980, in which it had been alleged that surgeons were removing organs from people who had received a diagnosis of brain death but might have recovered if their body had not been disassembled. The alleged mistakes involved people whose coma had been induced by drugs or alcohol. The programme had been based on an American survey, and its claims were hotly denied in Britain. Nonetheless, in the three months following transmission, the monthly average of eighty kidney explants/transplants fell to thirty-seven, rising only to sixty-five nine months later. However, the figures rapidly improved, and by the end of the year 864 kidneys had been explanted/transplanted, sixty-two more than in the previous year.[9]

The plasticized multi-organ pledge card was promoted on posters featuring actress Susan Hampshire, journalist Anna Ford, athlete Sebastian Coe and footballer Kevin Keegan, and 'radio fillers' lasting forty-five seconds which ended with exhortations such as, 'Carry the card. It does you credit', a deliberate play on the similar size of pledge and credit card intended to encourage pledgers to keep theirs in wallet or purse.[10] For its part, the RNIB produced a customized leaflet dispenser to hold its share of three million leaflets, which it distributed to retail opticians and pharmacies, hospital eye departments and blind welfare outlets – although why these were considered suitable is unclear.[11] It also persuaded the DVLC to add, 'Your eyes could restore someone's sight' to its application form.[12]

Michael Bewick, who claimed to be explant/transplanting more cadaver kidneys than any other surgeon in the United Kingdom, protested that the marketing slogan 'Carry the card' was pointless because on admission to accident and emergency department or intensive care unit, brain-injured people were stripped naked and their clothes and personal effects given to next of kin without being checked for a pledge card. What was needed was either an opt-out scheme or a readily accessible centralized computer register of pledges.[13]

The multi-organ pledge card was criticized for communicating a troubling image of the corpse as a convenient container of spare parts. However, civil servants had believed that risk of revulsion was outweighed by pragmatism: other cadaver organs could easily be added when their transplantation became technically feasible (as happened with cadaver lungs in 1985). Being confronted with a choice forced people to articulate their personal investment in various corpse fragments. For instance, an elderly woman explained hers as follows: 'Yes, I would go along with kidneys, corneas, and livers but my heart belongs to my husband and my husband is dead.' Another long-standing blue-card holder declared: 'I have always, since I was twenty, wanted to donate my eyes (corneas?) but not any other part of my body in the event of my having, say, a road accident.... Reason! In the event of error one could survive without eyes, not without other organs.'[14]

Shares and Partial Shares

The multi-organ pledge card provided a rallying point around which civil society organizations with an interest in the marketing of corpse philanthropy could combine forces. But the RNIB, in a conscious confusion of metaphors, felt its contribution looked like 'small fry' against that of the 'big guns' whose campaigning style was far more aggressive.[15] Some of the 'big guns' were small organizations punching above their weight thanks to the single-minded dedication of one or two people – a good example is the British Kidney Patient Association set up and run by the redoubtable Mrs Ward. Others were championed by a television personality, for instance, the Ben Hardwick Memorial Fund, named after a young boy who died following two unsuccessful liver transplants, which enjoyed the support of Esther Rantzen, presenter of *That's Life*, a popular weekly television programme.

The pledge card's shopping list conveys the impression of organized activity in intensive care unit and mortuary in much the same way as a menu suggests a team prepares food in the restaurant kitchen in response to orders waiting staff take from customers. But in practice it was a marketing fiction: only solid organ explant/transplant teams were organized and cooperating with one another. Extractors of other cadaver stuff guarded their independence in the mortuary.

Towards the end of the 1960s, explant/transplant surgeons worldwide had begun recognizing the wisdom of relinquishing their claim to cadaver kidneys that they had explanted but which were incompatible with the tissue type of any of their patients. Their incentive was the possibility of a compatible one that had been explanted elsewhere becoming available in the future. In 1967, the first scheme redistributing cadaver kidneys to where they were of most use was established at the University of Los Angeles Medical School by transplant surgeon Paul Terasaki, following his development of a reliable method of tissue typing. Two years later, it was awarded a contract by the federally funded Public Health Service transplantation programme to support explant/transplant teams in the western states.

In 1968, Donald Longmore, a surgeon who had participated in the first controversial British human heart explant/transplant at the National Heart Hospital, London, proposed setting up a British scheme for redistributing cadaver kidneys to where they were of most use, which would be supported by Air Call, a radio car service, and part of the Telephone Answering Service Limited. But it was a commercial venture, and denied admission to the NHS. Subsequently immunologist Hilliard Festenstein (1930–1989), at the London Hospital's Transplantation Immunology Laboratory, instituted a pilot scheme. It was without financial support; laboratory staff undertook all the tasks voluntarily. It was a success: participants grew rapidly from eight explant/transplant teams in London hospitals to twenty-one throughout the United Kingdom and Eire.[16] In February 1972, the DHSS agreed to finance what was called the National Organ Matching Service, and merge it into the National Tissue Typing Reference Laboratory at Southmead Hospital, Bristol, to form UK Transplant Service. The entity would store the tissue type of potential kidney recipients on a computer that would be searched for a suitable match whenever cadaver kidneys became available.

Ophthalmologists were without an incentive to share cadaver eyes. A majority agreed with Benjamin Rycroft that the cornea is privileged, and is never challenged by a recipient's immune system. Graft failure was attributed to poor surgical technique, or the poor condition of the grafted cornea, or the recipient's underlying disease. These were often to blame, but research carried out during the 1970s by a team of investigators at the Blond-McIndoe Research Unit, which included Casey and Gibbs, suggested that immunological incompatibility was responsible for up to one in every three corneal graft failures.[17]

In a matter of weeks following publication of the research findings, DHSS staff convened a working group to consider the shape of a national scheme which, in addition to arranging the distribution of tissue-typed cadaver eyes, might also induce British ophthalmologists to reduce wastage by sharing any cadaver eyes surplus to their immediate requirements, such as the second one of a pair. However, the logistical and technical obstacles appeared insurmountable and, anyway, surgeons at Moorfields Hospital refused to cooperate. They were responsible for the majority of the 1,400 keratoplasty procedures performed each year, and were experiencing few difficulties in meeting demand. Moreover, they argued, easier availability of cadaver corneas might encourage ophthalmologists to try their hand at a procedure that demands a high degree of skill and dexterity. It was reckoned that keratoplasty was being performed in around thirty-five British hospitals, and that each year some surgeons operated on fewer than twenty-five occasions. Policy should aim at channelling patients to 'major grafters' like themselves.[18]

Left to their own devices, British ophthalmologists might have continued to work independently. But the Iris Fund for the Prevention of Blindness picked up the gauntlet that thwarted civil servants had dropped. The civil society organization, now part of Fight for Sight, had begun life in 1965 when the League of Friends of the Royal Eye Hospital, in south London, agreed to sponsor experiments that might involve vivisection, research which the RNIB, with its close links to Guide Dogs for the Blind, could never support. In 1976, the Royal Eye Hospital was closed, and St Thomas's Hospital, a teaching hospital situated on the south bank of the River Thames, immediately opposite the Houses of Parliament, became the fund's main beneficiary. Its trustees were big-money philanthropists who attended the annual Iris Ball at venues such as the Lord Mayor of London's Mansion House, which was sometimes patronized by royalty.

In 1982, Susanna Burr was appointed the Iris Fund's director. She canvassed seventy or so ophthalmic surgeons on whether they would welcome a scheme providing tissue-matched corneas.[19] Just over half responded. A majority was supportive, with the proviso that their local arrangements were not undermined.

Burr knew her way around the corridors of Whitehall. She was on first-name terms with Margaret Thatcher, having worked for her in Conservative Central Office when she led the government's opposition. Burr's proposal was framed to appeal to DHSS staff, who were busy implementing the state-shrinking policies of the Conservative government. A four-year trial would make no demands on the public purse; £80,000 would be found through fund-raising appeals to schoolchildren (written by the entertainer Rolf Harris), churchgoers and readers of the *Daily Mirror* (it was their Christmas Appeal), but with the largest donation given by the Save & Prosper Educational Trust (Save & Prosper was an investment trust). Civil servants, who were recent converts to the utilitarian logic of health economics, were persuaded that in reducing the number of graft failures by around 120 each year, taxpayers would save £100,000 in re-grafting costs alone and, furthermore, in restoring sight, the scheme would allow more people to enter gainful employment and reduce the number claiming welfare benefits.[20]

In September 1983, in a ceremony in the Guard Room at Lambeth Palace conducted by the Archbishop of Canterbury, Dr Runcie, and in the presence of Simon Weston, a Falklands War hero who had sustained severe facial burns, the Corneal Transplant Service was inaugurated.

The Corneal Transplant Service was piggybacked onto UK Transplant Service, sharing its premises, computer and taking advantage of the favourable terms it had negotiated with commercial courier services such as

Securicor and St John's Ambulance Air Wing. It was hoped that the connection would encourage transplant coordinators to add cadaver eyes to the list of stuff they asked the next of kin of brain-dead patients to agree to explantation. Not only would these have had their tissue type identified, they would also be younger than those of the typical source.[21] But co-operation was rare; a majority of transplant coordinators feared that mentioning enucleation might provoke such a powerful visceral reaction that agreement to organ explantation would be rescinded.

In 1986, an organ culture facility, funded jointly by UK Transplant Service, the Iris Fund and the Bristol University Department of Ophthalmology, was opened in the theatre suite of the Bristol Eye Hospital, where ophthalmologist David Easty, the Iris Fund's medical adviser, worked.[22] Donald J. Doughman at the University of Minnesota Medical School had developed the shelf-life extending technique in the 1970s. Whereas a traditional eye bank might easily be mistaken for a kitchen, albeit one furnished with a laminar flow hood, there is nothing faintly domestic about an organ culture facility, and anyone wearing everyday clothing is prohibited entry. Organ culture maintains cell viability, which is necessary for tissue typing, and can extend the shelf life of cadaver corneas by up to thirty days. Suffused corneas swell, and must be immersed in M-K medium or another short-term preservative solution for at least twenty-four hours before grafting. Moreover, there is a risk of the medium becoming contaminated, and it must be tested before corneas are dispatched. Orders must therefore be placed at least seventy-two hours in advance of surgery, which is why organ culture never took off in the United States, where a majority of ophthalmologists were tissue typing sceptics, and the business model of eye banking is networks of high-volume extracting entities capable of completing an order for a fresh cadaver cornea shortly after it is placed.

Burr knew progress would be hampered by the Human Tissue Act's insistence that only a licensed medical practitioner can extract cadaver stuff. Enucleation typically was delegated to junior medical staff, who were often unwilling and slap-dash, resulting in a high rate of rejection. She persuaded MP John Hannam (Conservative), secretary of the All-party Disablement Group, to sponsor an amendment to the Human Tissue Act 1961, which would allow trained technicians to enucleate cadaver eyes. During the debate MP Jeremy Hanley (Conservative), recipient of a corneal graft, described how on one occasion he had regained consciousness in Moorfields Hospital to discover that the operation had not taken place because at the last moment a coroner had decided to withhold the cornea he was due to receive.[23] The Private Member's Bill enjoyed government support, and met with no opposition. In 1986 the Corneal Tissue Act was entered onto the statute book.

In its first few years, 428 surgeons registered with the Corneal Transplant Service. But a majority were inexperienced, undertaking fewer than nine grafts each year. 'High volume' surgeons operating at Moorfields Hospital, London, and the Queen Victoria Hospital, East Grinstead, continued organizing their own supplies, occasionally sending those surplus to immediate requirement to the Corneal Transplant Service. Nonetheless, the Corneal Transplant Service grew rapidly: in its first year of operation, it distributed forty-one cadaver corneas; three years later, 326. In 1987, the DHSS decided to accept responsibility for it. In 1988, a second organ culture facility, called the David Lucas Eye Bank, was opened at the Manchester Royal Eye Hospital.[24]

The Corneal Transplant Service's inventory method was similar to that of Cook County Hospital's blood bank: receipt of a cornea incurred a debt that had to be repaid within a year.[25] Repayment was intended to encourage surgeons to be less wasteful of the second unused cornea of an enucleated pair, and to help remedy shortages elsewhere. Additionally, a non-negotiable service charge for packaging and transport was levied, which initially was set at £80 for delivery of the first cornea, and a further £28 for each subsequent one delivered in the same package.[26]

Follow-up studies on both sides of the Atlantic found tissue typing is mostly unnecessary.[27] By the new millennium, ophthalmologists were routinely prescribing immunosuppressive eye drops to ward off any threat of rejection. Where tissue typing had 'worked' was as an unintended ruse to encourage British ophthalmic surgeons to stop nipping into their hospital mortuary.

CHAPTER 27

Iatrogenesis: Disregarding Risk in Plain Sight

Iatrogenesis is Greek for doctor-inflicted injuries, which can range from benign detriments such as operation scar to death. Post-transfusion hepatitis is another. Titmuss, in *The Gift Relationship* (1970), exploited its apparent greater prevalence in the United States than the United Kingdom to support his claim that blood extracted from the forearm of donors-on-the-hoof motivated by altruism is safer than that of those incentivized by cash.

Titmuss formulated his thesis when the pathogens responsible for yellow jaundice had not been identified, and asking potential sources to recall their medical history the only available test. Its effectiveness depended on recall and honesty, which might be less reliable where cash is exchanged for blood, which is why Titmuss's thesis gained currency. However, at the time, no one questioned whether the thesis could be applied to the mortuary where cadaver stuff was frequently being extracted in exchange for cash.

In 1976, American paediatrician Carleton Gajdusek (1923–2008) and American physician and geneticist Baruch Blumberg (1925–2011) shared the Nobel Prize for Medicine for their contributions to the genealogies of two families of infectious disease. Blumberg's achievement was to find a significant clue to the identity of one of the culprits responsible for yellow jaundice. Gajdusek's family was called transmissible spongiform encephalopathies, but, borrowing the famous aphorism of secretary of defence Donald Rumsfeld, the infectious agent was a known unknown. Gajdusek believed it was a slow virus because its typical incubation period is relatively longer than that of so-called conventional viruses.

The family of transmissible spongiform encephalopathies had not been recognized in 1957 when Gajdusek began investigating kuru, one of its members.[1] The disease was epidemic and apparently endemic among the

Fore people living in the remote eastern highlands of Papua New Guinea. Victims experience strange tremors, shaking, shivering, muscle weakness and a general lack of control of their limbs. Between 1957 and 1975, more than 2,500 Fore people, mostly women and children, died. During field trips Gajdusek documented the disease's progression, and wherever possible extracted cadaver brains to take to his laboratory on the National Institutes of Health's campus in Bethesda. His initial hypothesis was that kuru is a genetic disease because it appeared to run in families. But in 1959, American veterinary pathologist William Hadlow visited the Wellcome Museum in London, where he saw a display about kuru, and noticed a remarkable similarity between the spongy texture of the brain tissue of kuru victims and that of sheep that had succumbed to scrapie: both were riddled with tiny holes and looked 'spongy'.[2]

Sheep have economic importance, which is why scrapie had been intensely investigated for around two centuries. It was known to be transmissible. In 1966, Gajdusek and colleagues speculated that kuru might be too, when two chimpanzees began displaying its symptoms a year after a soup of kuru-infected brains had been injected into their brains.[3] Clearly, this was not how the Fore people contracted kuru. But scientists had long suspected that sheep contract scrapie by eating food contaminated with the unknown pathogen. The speculation that the Fore people, who customarily cannibalized the corpse of their deceased next of kin, contracted kuru through this route was confirmed when the disease eventually disappeared some years after they had been persuaded to give up the funerary ritual.

Other family members were identified. The kinship of Creutzfeldt-Jakob disease was suggested by the observation of Igor Klatzo (1916–2007), neuropathologist and colleague of Gajdusek, that its victims' brains are spongy. The disease is named after German neurologists Hans Gerhard Creutzfeldt (1885–1964) and Alfons Jakob (1884–1931), who in the early 1920s coincidentally and separately described six patients with peculiar and previously unrecorded neurological illnesses, and whose brains at post-mortem examination looked spongy. It typically occurs in people aged around mid-fifties and older. Victims usually die within a year of the onset of symptoms, which include loss of language and memory, unsteady gait and excessive sleepiness, symptoms that might be confused with other types of dementia. Its cause was unknown, but in 1968, Gajdusek's team published findings of an experiment that confirmed that Creutzfeldt-Jakob disease could be transmissible.[4]

Blumberg's Nobel award-worthy discovery was made in 1964 while investigating why certain ethnic and national groups are susceptible to particular diseases. He had travelled around the world collecting thousands of blood samples from many different populations, and in his laboratory mixed them together and watched what happened. To his surprise a sample of blood extracted from an American man with haemophilia, who had undergone repeated blood transfusions, reacted against something in the blood of an Australian aborigine, which he named 'Australia antigen' in recognition of its origin.

Blumberg's team subsequently confirmed that the reaction is indicative of serum hepatitis, and in 1970, the Australia antigen was utilized in the first blood test of the disease. In 1973, the contagious agent responsible for enteral hepatitis was identified. But optimism that all the pathogen(s) responsible for yellow jaundice were now known was dashed when a World Health Organization Scientific Group published *Viral Hepatitis* (1973).[5] Among other things, its authors, who were eminent international experts, proposed naming the enteral disease Hepatitis A, and the serum one Hepatitis B. But they also warned that not every case of yellow jaundice could be attributed to either Hepatitis A or Hepatitis B; a known unknown was responsible for a disease they named non-A non-B hepatitis (NANB). A vigorous, scientifically challenging, hunt for its culprit began.

Risk

There is no such thing as unambiguously safe or unsafe body/cadaver stuff. It is putrescible, supports a 'bioburden' of pathogens, and can attract others lurking in the muddled and unstable landscape in which it is extracted, processed, distributed and repurposed. Moreover, scientific knowledge is imperfect and sometimes speculative: known unknown pathogens exist, and unknown unknowns sometimes emerge.

Risk of iatrogenesis is not simply a sum of scientific evidence. As sociologist Marion Nestle observed, other factors influence its evaluation. These include whether or not the risk is visible, invisible or hidden in plain sight; agreed or controversial; controllable or uncontrollable. Nestle's focus is on food, which shares many characteristics of body/cadaver stuff: both can be beneficial in terms of saving and enhancing life; both can cause harm ranging from trivial inconvenient temporary symptoms to death; safety of both requires compliance with appropriate medical criteria of bioavailability, careful handling and effective terminal sterilization (cooking).

Risk is always relative; the ambition is to make it acceptable. But who decides what is acceptable where perceptions, opinions and values might differ, and pressures on demand and threats to profit margins can be influential? Nestle's history of the development of food safety policy in the United States is peppered with warnings of danger, which were sometimes ignored, or deliberately suppressed, or the various parties involved disagreed over the scale of threat to public health, or on necessary precautions (if any) to take, or where agreed measures were not implemented for reasons of self-interest, or cost or convenience, or where regulatory oversight was non-existent, or ill informed, or light touch, or fainthearted, or ill-equipped and under-resourced.[6]

Safety is not an automatic byword; medicine has considered it far longer and with greater urgency in relation to blood than to stuff extracted from cadavers. Indeed, it took *Ravenis*, the medical negligence case decided in 1975, to force American eye and tissue banking industry associations to start formulating and imposing safety standards in extraction, processing and distribution of cadaver stuff/artefacts. These were American private-sector initiatives: only members had to respect the guidance, and joining the associations was voluntary. American extractors outside any trade association, and every British one, were independently choosing how to operate.

Risk Disregarded in Plain Sight

'I just sat up in bed one night late in 1976 and thought "Good gracious – if they are collecting human pituitary glands and preparing growth hormone from it, you're likely on occasions to be using infective pituitaries with CJD". This is veterinarian and scrapie expert Alan Dickinson recalling the moment he realized that hGH might transmit slow viruses to recipients. He went on to say, 'It is crazy to think that it did not occur to somebody before.'[7]

The first suggestion that cadaver pituitary stuff could be contaminated by a slow virus came in 1974, when it was reported that a fifty-five-year-old recipient of a cadaver cornea had succumbed to Creutzfeldt-Jakob disease – the optic nerve connects the eye directly to the brain. Subsequently the source of the grafted cornea was found to have been harbouring the disease.[8] In 1974, and again in 1975, Gajdusek and his team warned pathology and mortuary staff handling stuff extracted from the corpses of dementia sufferers, who might in fact have been harbouring slow viruses, to adopt the precautions then believed to inactivate the slow virus. This was effectively the first indication that social death prior to physical death, which hitherto had represented an opportunity to extract cadaver pituitary glands, should sound a warning to exercise caution.

On 5 October 1976 Dickinson phoned the Medical Research Council to find out what precautions it was taking to minimize risk of transmission of a slow virus. His was the first direct warning that hGH might be contaminated with Creutzfeldt-Jakob disease. Whether or not the National Pituitary Agency received a similar warning is unknown: access to its records has been refused. However, the National Institutes of Health was paymaster of both agency and Gajdusek and his team.[9] Paul Brown, a member of Gajdusek's team, admitted that although by 1973 or 1974 there was reason to suspect that hGH had the potential to transmit Creutzfeldt-Jakob disease, the possibility had not occurred to him.[10] Indeed, when the agency's medical criteria of bioavailability were modified in 1975 and again in 1977, it was specifically to exclude the pituitary glands of people known to have had infectious or serum hepatitis, or to have been a drug addict.[11] Nonetheless, in 1983, Raiti, the Agency's director, in a symposium presentation, listed under 'Current Activities', 'Collect all human pituitary glands'.[12]

Direct warnings were ignored. France-Hypophyse ignored its first warning in 1980 from Luc Montagnier, the French scientist who identified HIV, and its second in 1983 from the Inspector General of Social Affairs.[13] But it is also the case that dangers in plain sight are sometimes unrecognized. Australian pathologist Colin Masters, who visited Gajdusek's laboratory in 1978, speculated that no one had sounded the alarm bell because the danger was 'too obvious to mention'.[14] The official Australian inquiry into iatrogenic Creutzfeldt-Jakob disease attributed absence of recognition of risk of slow virus contamination to Australian scientists and pathologists occupying separate social worlds. Australian scientists had been actively engaged in research into kuru, had met Gajdusek in Papua New Guinea, and Gajdusek had made many visits to Australia. Australian pathologists had heard of kuru, but claimed to have been unaware of Gajdusek's warnings about handling brain tissue.

Dickinson's question was posed at an inconvenient moment: around two months after the *Sunday People*'s exposé had been published. Both were considered at the Steering Committee for Human Pituitary Collection's meeting in December 1976. It was decided to ask him whether Hartree and Lowry's recipes were capable of eliminating slow viruses. Three months later he replied: although he was quite sure that neither the Cambridge nor London recipe could inactivate the scrapie agent, hGH extracted out of acetone-preserved cadaver pituitary glands might be safer than that isolated in fresh-frozen ones. He advised excluding the cadaver pituitary gland of anyone who during their life had had dementia, and proposed testing on laboratory animals hGH produced according to both recipes.

The advice coincided with the publication of the map of another transmission infection route taken by slow viruses. It was reported that two people who had had electrodes inserted into their brains for exploratory tests in a Swiss hospital succumbed to Creutzfeldt-Jakob disease.[15] The electrodes had been sterilized using processes known to kill conventional viruses, but when one of them transmitted the disease to a chimpanzee in Gajdusek's laboratory it was realized that slow viruses are invulnerable to known techniques of terminal sterilization. Gajdusek's team subsequently reported that medical staff were increasingly reluctant to care for patients suspected of having Creutzfeldt-Jakob disease, that surgeons and anaesthetists were refusing to take part in biopsies and autopsies on them, and that some people suffering from dementia had been refused admission to hospital and nursing homes. He believed staff caring for affected people were not at risk, but warned against extracting and repurposing stuff from dementia sufferers' cadavers.[16]

Dickinson's advice was considered at the March meeting, but no action was taken for seven months, when the opinion of two expert virologists was sought. For his part, Cedric Mims at Guy's Hospital, London, observed that any measures aimed at minimizing risk would inactivate hGH, and reiterated Dickinson's advice on medical criteria of bioavailability.[17] Peter Wildy, of Cambridge University and also chairman of the Agricultural Research Council's Working Party on Scrapie, was more cautious; he challenged the wisdom of continuing to prescribe hGH. As he put it,

> We are in the uncomfortable position of suspecting the worst, but not knowing how bad the worst is. Any clinician who uses growth hormone must be made aware of the gruesome possibilities and their imponderable probabilities. It is the clinician who must take the ultimate responsibility for his patients, but it is up to the Steering Committee to ensure that he understands the true position.[18]

hGH is Executed

In February 1985, Raymond Hintz, professor of paediatric endocrinology at Stanford University School of Medicine, informed the Food and Drug Administration that a twenty-one-year-old man had died of Creutzfeldt-Jakob disease. The young man had been a recipient of National Pituitary Agency hGH. Hintz had been troubled by his former patient succumbing to a disease that was virtually unknown in people under the age of fifty, and suspected hGH might be the culprit. One case was considered too few on which to draw a firm conclusion. However, shortly afterwards the Food and Drug Administration was informed that three other recipients of hGH had died of Creutzfeldt-Jakob disease – two American men and a British woman. In April 1985, distribution of hGH was halted by the National Hormone and Pituitary Program, the new name of the National Pituitary Agency, and the two pharmaceutical companies selling hGH in the United States: Swiss-based Ares-Serono, which by this stage had captured 80 per cent of the American commercial market, and KABI, which held the rest.[19] Two months later, the DHSS followed suit. Authorities in Belgium, Finland, Greece, the Netherlands and Sweden did too, but not France-Hypophyse, which continued to manufacture hGH until 1988, and is responsible for ninety deaths from iatrogenic Creutzfeldt-Jakob disease, proportionally higher than in any other country since only 1,300 French children were treated.[20]

Fortuitously, an alternative source of human growth hormone was in the pipeline. In December 1977, KabiVitrum, frustrated by its inability to collect more than around 150,000 cadaver pituitary glands worldwide annually, had approached Genentech, the Californian biotech company, which Robert Swanson (1947–1999), a venture capitalist, and molecular biologist Herbert Boyer, had set up in 1976 to test the feasibility of adapting recombinant DNA techniques to the mass production of human growth hormone. What is sometimes colloquially called 'bug to bottle' technology employs *Escherichia coli* that have had the genetic code of a protein hormone, such as human growth hormone, engineered into their DNA. Each time the bacteria reproduce, which they do continuously and rapidly, a daughter cell containing human growth hormone is produced.

Genentech agreed, and the Swedish government provided financial support.[21] In September 1985, recombinant human growth hormone (rhGH) was licensed: KabiVitrum's in Britain, Genentech's in the United States.[22] Almost immediately demand outstripped wildest predictions. Up until its distribution was halted in 1985, around 10,000 American and 1,800 British children had undergone a course of hGH. A decade later, 30,000 American children had been treated with rhGH.[23] Supplies of hGH had been limited and rationed, but had mostly been provided free of charge; rhGH is available in unlimited quantities but is expensive. In 1990, the cost of one year's treatment ranged from $10,000 to $20,000.[24] As Paul Brown observed, hGH had been officially executed; but among the spectators at the graveside, only Genentech was not in mourning.[25]

Post-mortem Investigations

By this stage around 1.3 million American and one million British cadaver pituitary glands had been extirpated. In an article published in September 1985, Gajdusek, Brown and other members of the team criticized lax criteria of medical bioavailability, and speculated on what the future might hold. At worst, iatrogenic Creutzfeldt-Jakob would reach epidemic proportions; at best, deaths reported so far would represent an unusual cluster.[26] The paper closed with the warning, 'We are again dramatically reminded that human tissues are a source of infectious disease, and that any therapeutic transfer of tissue from one person to another carries an unavoidable risk of transferring the infection.'

Fortunately, fewer recipients of hGH than anticipated have died of iatrogenic Creutzfeldt-Jakob disease, but other transmission routes emerged. By 2006, globally, 405 people were known to have died of iatrogenic Creutzfeldt-Jakob disease: 196 people from infected dura mater grafts (of which more in Chapter 30); four from infected surgical instruments; two from infected EEG needles; two from infected corneal grafts; 194 from infected hGH; four from infected pituitary-derived gonadotropin; three from infected blood products. However, the number of deaths appears to be diminishing.[27]

In neither the United States nor the United Kingdom has there been an official public investigation into the identity of the dead people from whose corpses pituitary glands were extirpated, or how collections of unequalled magnitude were achieved. Next of kin of recipients who contracted iatrogenic Creutzfeldt-Jakob disease have sought redress. But every case brought in the United States has failed. Whereas private-sector extractors can be called to face an accusation of medical negligence, the National Pituitary Agency, as an independent contractor of the National Institutes of Health, is shielded by the federal government's sovereign immunity. British plaintiffs fared a little better: Mr Justice Morland awarded damages to the next of kin of recipients who contracted the disease from contaminated hGH produced after 1 July 1977.[28]

CHAPTER 28

Ask, or Don't Ask: Inconsistencies in Collecting Sites

A reliable and growing supply of cadaver corneas excised under legislative consent facilitated the transformation of keratoplasty from an unpredictable emergency procedure into a routine operation scheduled to meet the convenience of either surgeon or potential recipient. Moreover, it encouraged more surgeons to tackle the operation: in 1979, 11,226 Americans became a recipient of a cadaver cornea; ten years later, it was 38,464.[1]

Nonetheless, legislative consent statutes stood accused of promoting intentional ignorance, particularly in the next of kin of young people who had died unexpectedly in an accident or as a result of manslaughter or murder.[2] However, actions brought by people who discovered and objected to clandestine excision of their dead kin's corneas failed. One reason is that cosmetic detriment is negligible. For instance, in 1985, the action brought against Georgia Lions Eye Bank by the mother of a baby who had died of sudden infant death was dismissed on the grounds that mutilation resulting from the procedure was insufficient to cause her either symbolic or emotional damage.[3]

When the case against Georgia Lions Eye Bank was first heard at the State Court, Chatham County, the judges decided in the mother's favour on the grounds that she had been deprived of her quasi-property right in the infant's corpse. But the Supreme Court overturned the decision: it described the quasi-property ruling as a nineteenth-century fiction dreamt up to resolve a family dispute. Moreover, it approved sacrificing next of kin's private interest in a corpse to the public good. In 1978, before Georgia had adopted legislative consent, around twenty-five corneas were grafted; six years later, that number had risen to more than one thousand.

The Supreme Court rehearsed the same utilitarian argument to defeat another challenge to legislative consent. In 1985, a Florida medical examiner authorized the excision of the corneas of two young men who had died in tragic circumstances. In explaining their reason for denying the parents' suit judges referred to the $138 million the State of Florida was spending on providing blind residents with the basic necessities of life. In 1967, only 500 corneas had been grafted whereas in 1985, following the adoption of legislative consent, 3,000 Florida citizens had had their sight restored.[4] In effect, eye banks operating under legislative consent statutes act as an arm of the state, and can legitimately and clandestinely claim title to the corneas of dead people whose corpse falls into the possession of coroner/medical examiner.

Legislating for Asking

Paradoxically, the Supreme Court's support of treating people who are physically dead as if they are socially dead coincided with adoption of statutes recognizing kinfolks' private interest in the corpse. Routine Inquiry statutes required hospital staff to ask the next of kin of someone who had just died, or was about to die, whether they had pledged or had intended to pledge their cadaver stuff. The law in New Jersey went so far as to require hospital staff to ask patients on admission to hospital whether they had signed a pledge, a question that caused some distress as it suggested survival was unlikely. Required Request statutes were more demanding: hospital staff had to provide next of kin of someone who had just died, or was about to die, with the opportunity to agree to or refuse extraction. The various statutes differed in detail, but few required compliance to be monitored.[5]

'It's incredible,' said Mary Jane O'Neill, executive director of the New York Eye-Bank for Sight Restoration, commenting in 1986 on the impact of New York State's adoption of a Required Request statute in 1982. 'The number of donors is growing all the time now.'[6] Routine Inquiry and Required Request statutes, which had been intended as a solution to hospital staff's reluctance to notify organ procurement organizations that a potential source of transplantable kidneys had been admitted, proved to be more productive in relation to other cadaver stuff, particularly eyes/corneas. For instance, in 1986, New York State experienced a 58 per cent increase in eye/cornea extractions, and a 23 per cent increase in cadaver kidney explantations.[7]

In May 1986, the Task Force on Organ Transplantation, which had been charged under the terms of the National Organ Transplantation Act to examine 'barriers to the improved identification of organ donors and their families and organ recipients', published its report. It recommended that Routine Inquiry should become a condition of participation (COP) in Medicare and Medicaid: only hospitals that had circulated an appropriate written protocol to staff would be eligible for reimbursement under both schemes. The Final Condition of Participation Rule was made legally binding through an amendment to the Omnibus Budget Reconciliation Act 1986, and became effective in March 1988.

The original proposal applied to all repurposable cadaver stuff, but the Final COP Rule was restricted to cadaver organs in order to avoid imposing a heavy burden on organ procurement organizations, which because of the greater urgency surrounding explantation of cadaver kidneys, would have to serve as the first point of contact. Nonetheless, in practice, extractors of other cadaver stuff were sometimes notified of a potential source directly by a hospital or increasingly by an organ procurement organization, initially as a favour but subsequently in anticipation of a fee, which in 1994 ranged from $100 to $1,000 for each notification.[8] Routine Inquiry had created a new income stream.

In 1987 the National Uniform Law Commission published the Uniform Anatomical Gift Act 1987, which had been drawn up to iron out inconsistencies created by state modifications of the 1968 model act. The 1987 version endorsed both Routine Inquiry and Required Request, and recognized the constitutionality of legislative consent in recommending its extension to cadaver organs, albeit fettered by the requirement to 'make reasonable effort' to contact next of kin. There were few takers. It was impractical: according to custom and practice, the body of a brain-dead patient became a corpse when placed on the operating table, which was when medical examiners/coroners took legal possession, and which was too late to obtain their authorization for disassembly.[9] Moreover, explant/transplant teams feared adverse publicity would tarnish their reputation; they were reluctant to deprive next of kin of the experience of moral and emotional benefits associated with the voluntary 'gift of life', a claim that begs the question, why were grieving next of kin denied the solace that might be derived from a 'gift of sight'?[10]

Whereas every state adopted a version of the 1968 Uniform Anatomical Gift Act, only twenty-six enacted a version of the one issued in 1987; eleven had adopted legislative consent. As a result, as the map below shows, treatment of American corpses varied from state to state, and within states, according to in whose lawful possession it was held; it provides confirmation of how dead people and corpses resist rationalization and coherence.

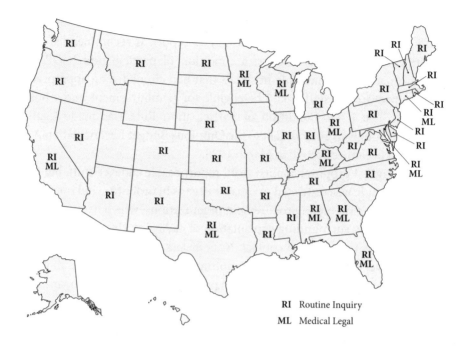

Figure 24: Legislative inconsistencies, 1989

Legislative Consent Discredited

In 1992, national and local entities involved somehow in extraction and repurposing of cadaver stuff, including the United Network for Organ Sharing, joined together to form the Coalition on Donation, a non-profit dedicated to marketing corpse philanthropy and pledging throughout the United States. Its mission is 'to ensure that every individual in the United States understands the need for organ and tissue donation and accepts donation as a fundamental human responsibility'. In 1994, the coalition was selected as one of the campaigns of the Advertising Council, Inc., the leading organization in the United States for public service advertising of social issues, and the fifth largest user of advertising in the country. Jerry and Ketchum Advertising, New York, won the contract for a multi-media campaign. Advertisements appeared on billboards, in newspapers and on radio and television with the strapline, 'Share your life . . . Share your decision.' The campaign garnered over $33 million in pro bono placements.

The campaign's second phase, which began in 1996, featured real-life stories of people who prior to their death had either made known to or had not discussed with next of kin their stance on disassembly. Michael Jordan, basketball superstar and the highest paid advertising icon in the United States, donated his time, voice and image to advertisements encouraging people to have the conversation with their next of kin.[11]

The irony of an African American man renowned for superb athleticism volunteering as the public face of a campaign encouraging corpse philanthropy emerged in November 1997, when a team of *Los Angeles Times* investigative journalists found that in 1996 a majority of the 572 people whose corneas had been excised under legislative consent by staff of the Lions Doheny Eye & Tissue Transplant Bank were young black men, and that seven out of ten were homicide victims.[12] In a critical commentary on the exposé, legal scholar Michele Goodwin drew a connection between black lives lost on American streets to violence or in police custody, and keratoplasty, observing that clandestine, non-consensual excision of cadaver corneas exploits the disproportionate toll drugs wars were exacting on young African American and Latino American men, and was perpetuating the historical distrust these communities have of medicine and other social interventions.[13]

The Lions Doheny Eye & Tissue Transplant Bank is one of the oldest in the United States. It was named after big-money-philanthropist Estelle Doheny, the widow of wealthy oil tycoon Edward Doheny, who had endowed an eye specialist centre serving the Los Angeles area. In 1983, it had joined the Medical Eye Bank's network, which had changed its name to Tissue Banks International to indicate it had diversified into other cadaver stuff, and had expanded overseas.

None of the next of kin had known about legislative consent prior to being contacted by the *Los Angeles Times* journalists. 'I felt I still had to protect my son, even after what he did to himself,' explained Richard Baltierra, whose only son had taken his own life after breaking up with his girlfriend and failing to graduate from college. 'I asked them not to put his body in a body bag. I told them I didn't want him taken to a place and put on a table where they would cut him open.'[14]

Both medical examiners/coroners and eye banks operating under legislative consent are acting in the public interest, but their authority relied on wholly different justifications, which could clash. Yet several Los Angeles County Coroner's Office and Doheny Eye & Tissue Transplant Bank staff members worked in both organizations; some mortuary staff had moonlighted for Doheny; eye bank technicians had their own keys to the coroner's office and had free run of the facility and its investigative files.

Money – or rather its shortage – was at the scandal's heart. It transpired that the Doheny Eye & Tissue Transplant Bank had been paying the Los

Angeles County Coroner's Office between $215 and $335 for each pair of corneas excised. Over the previous five years the Coroner's Office had received more than $1.4 million, two vehicles, and had had its facilities modernized. A member of the medical examiner's staff said the money was 'helpful' but insufficient to plug the deficit created by budget cuts of $3 million over the previous six years, during which policies designed to shrink the size of the state had been pursued. Indeed, in 1991, the National Association of Medical Examiners had realized that budget cuts made offers of payment for assisting in extraction of cadaver stuff more tempting, and had recommended, among other things that any payment be made not directly to the medical examiner/coroner but to an appropriate government agency.[15] In effect, the professional body had recognized that money handed directly to public officials is earmarked as 'dirty money'. Goodwin adds another telling observation: given the predominance of the corpses of young black men taken into a coroner/medical examiner's possession, the exchange of money is reminiscent of chattel slavery.[16]

The *Los Angeles Times* had placed the ethos of legislative consent in the court of public opinion. Almost immediately the Los Angeles County Coroner's Office announced a change of policy: corneas would only be excised with next of kin's explicit agreement. In the month following implementation of the policy, the number of corneas excised on its premises fell by 70 per cent.[17] Less than a year later California State fettered legislative consent with a law requiring its medical examiners/coroners' offices to secure agreement of next of kin before allowing excision of corneas of corpses in their lawful possession.[18]

Next of kin seeking redress in the courts helped turn the legal tide that had supported legislative consent. In *Brotherton v. Cleveland* (1991), Deborah Brotherton had sued when she found from an autopsy report that her husband's corneas had been excised. Stephen Brotherton had died by suicide and his corpse fell into the possession of coroner Dr Frank Cleveland, in Hamilton County, Ohio. Ohio had adopted legislative consent, but it had been fettered by the requirement to establish whether or not the next of kin had recorded an objection. Deborah Brotherton had hers recorded in her husband's medical record, which neither coroner nor eye-banking staff had examined. The case reached the Supreme Court, but its decision was equivocal. *Newman v. Sathyavaglswaran* (2002) concerned the Doheney scandal, and had been brought by the parents of two children who had died in 1997, and whose corneas had been excised under legislative consent when their corpses were in the possession of the Los Angeles medical examiner. The influential United States Court of Appeal, Ninth

Circuit, decided that the state of California had infringed the dignity of the children's corpses when it extracted the corneas without the knowledge or agreement of their parents.[19]

From interviews with eye-bank staff, Goodwin found that in the states where it had been adopted, clandestine non-consensual excision of corneas had fallen out of favour with both lawmakers and physicians by the start of the new millennium.[20] It was relied on by only seven eye banks that together were extracting one in twenty of the American supply of cadaver corneas: one in Florida, two in Maryland, three in Texas and one in Puerto Rico.[21]

Routine Referral

In 1998, a measure known as Routine Referral was adopted as a Condition of Participation in Medicare and Medicaid.[22] It was modelled on Pennsylvania Act 102, which had been signed into law in December 1994 by state governor and organ transplant recipient Robert P. Casey (1932–2000), and which required every hospital in the state to notify in a timely manner the Delaware Valley Transplant Program, the organ procurement organization covering eastern Pennsylvania, of *every* death, actual and anticipated, on its premises. The measure was a radical response to evidence that despite the original Condition of Participation Rule and a Required Request statute, two out of three opportunities for explanting cadaver organs were lost because hospital staff were unaware, ill informed or uncooperative. Under the Pennsylvania Act, it was up to Delaware Valley Transplant Program staff to decide whether or not a corpse was potentially bioavailable and, if so, to approach next of kin. The statute had teeth: failure to report every death would incur a $500 fine paid into the Donor Awareness Trust Fund, which was set up to market pledging. The organ procurement organization experienced a 40 per cent increase in cadaver organ explants in the two years following the statute's adoption compared to an average increase of 3 per cent nationally.

Hospitals are required to cooperate with an organ procurement organization, and with at least one tissue and one eye bank, but because great urgency surrounds cadaver kidney explantation, the organ procurement organization is the first and only entity called – it acts as gatekeeper to the others.

The average number of people who die in an American hospital each year is 400, thus the burden imposed on hospital staff is light. But each year, around 32,000 people die in hospitals within a typical franchise area, so the burden placed on organ procurement organizations is heavy. Moreover, of the 1.2 million Americans who die in hospital each year, 11,000 to 14,000 meet criteria of bioavailability for cadaver organ explantation, whereas at least 100,000 meet those of other cadaver stuff. Nonetheless, the rule provided organ procurement organizations with a new revenue stream fed by charges for referrals.[23] Some expanded in-house capacity; others outsourced the work to a telephone triage agency, a for-profit call centre that might service several organ procurement organizations, and which might also be contracted by wholly different organizations that provide its staff with a computerized scripted protocol. Call centres are often situated where labour is cheap, and can be in a different state or even nation to the organ procurement organization they service, the temporary resting place of the potential source, and that of their next of kin. Fee schedules are arbitrary, and disputes inevitably arise: for instance, the Lions of District 22-C Eye and Tissue Bank and Research Foundation, Inc. was sued by the Washington Regional Transplant Consortium for fees it said it was owed for referrals.[24]

Routine Referral works. The number of dead Americans whose organs were explanted increased from around 4,000 in 1997 to around 8,000 in 2006.[25] Entities belonging to the American Association of Tissue Banks experienced a greater benefit, reporting that sources of cadaver stuff had grown from 6,132 in 1994 to 23,295 in 2003. And almost immediately the rule was enforced, the eye banking industry experienced an increase in referrals of around 10 per cent. Eye banks hitherto without direct access to hospital mortuaries benefited most, whereas some that hitherto had enjoyed fruitful associations with local hospitals reported a detriment.[26] However, the rule greatly increased the authority of organ procurement organizations: they monitor hospital compliance by auditing patient records. Moreover, some exploit their gatekeeping role to bully smaller entities. As an eye bank employee told me in confidence, 'They can squash us like a bug.'

CHAPTER 29

Climbing up the Value Chain

Adoption of the National Organ Transplant Act 1984 was surprising: the Reagan administration favoured small government, private enterprise and competition in deregulated domestic markets, whereas the statute had created a monopolistic network of federally funded, centrally controlled, non-profit, quasi-public entities, called organ procurement organizations. Congress had declared rewarding sources with cash a criminal offence but had permitted entities to seek reimbursement of costs incurred in extraction, transportation, processing, preservation, quality control and storage of body/cadaver stuff; in effect, it had endorsed a monetized understanding of 'services' that is wholly different to either the motivation of Americans who volunteer to serve their community such as members of Lions Clubs, or the redistributive principle that had been infused into the British NHS by its architects.

Demand for cadaver stuff/artefacts was increasing by leaps and bounds. Older people are the largest category of recipient. Medicine has arrived at the view of the ageing body as remediable and malleable, and is continuously adding to its inventory novel procedures, many of which repurpose cadaver stuff/artefacts. For instance, cadaver bone is used in joint replacement surgery, or to stabilize a worn out or crumbling spine, or fix a dental implant; blocked arteries and veins are replaced with those of a cadaver. Medicare reimburses the cost of procedures that restore mobility and independence to Americans aged over sixty-five. But taxpayers are seldom required to bear the cost of facelifts. Nonetheless, more Americans have come to believe that rejuvenation and cosmetic enhancement are acceptable, for others, if not for themselves, and are prepared to pay for them out of their own pocket. As historian of the American cosmetic surgery industry

Elizabeth Haiken observes, cultural acceptance was responsible for 'more surgeons, more dollars, and more surgeries on more parts of the body'.[1] 'Self payers' represent an escape route into an independent (and, hopefully, profitable) practice to physicians and hospitals tied into a financial straitjacket by third-party payers.[2] The so-called aesthetic market is saturated with providers offering procedures at prices within many people's budget.

Save for an exhortation that the various entities should cooperate with one another, federal government never provided the 'tissue retrieval people' the assistance Gore said they needed. Left to their own devices, they adopted a variety of strategies that might meet growing demand.

The first strategy tried was to create up-dated civilian versions of the Navy Tissue Bank, which extract everything from veins to mandible out of a single corpse. These were created in two ways. One was to establish a start-up. In 1980, two surgeons set up the Mile High Transplant Bank, a non-profit in Denver with seed money provided by the Junior League of Denver, a civil society organization of and for women.[3] Their initial intention, and that of the founders of the Eastern Virginia Tissue Bank, a non-profit multi-tissue entity set up in 1982 in Virginia Beach, was to supply colleagues in local hospitals.[4] In 1986 it merged with the Richmond Organ Procurement Organization to create a conglomerate of quasi-public and private non-profit entities with an extensive range of cadaver stuff/artefacts. Its catchment area continually widened, and purchase of a twin-engined 421 Cessna plane allowed its team of extractors to reach corpses resting in seventeen states before putrefaction set in. In 1989, the conglomerate adopted the name LifeNet Transplant Service.

Another strategy was diversification. For instance, seven years after it had opened in 1982, the Skin Bank of the University of Texas Medical Branch at Galveston changed its name to UTMB Tissue Bank to mark the addition to its product range of cadaver bone, arteries and so on. In 1991, the bank moved into new, more spacious quarters in its own building, but still close to its main customers for skin, the University's Truman G. Blacker Burns Unit and the Shriners Children's Hospital Burns Institute, and embarked on a programme of expansion. Its trained technicians also processed cadaver skin extracted elsewhere. Its technical director claimed that, 'physicians and hospitals who use our skin says ours is the best quality they have found'.[5]

'Blood banking has peaked.... We're looking for new avenues to get into.' This is how Dr Todd Kolb, director of the American Red Cross blood programme in Columbia, South Carolina, explained to readers of the *Wall*

Street Journal in 1984 why the American Red Cross, the non-profit entity extracting a significant proportion of America's blood repurposed in transfusion medicine, and blood products for treating people with bleeding disorders, had decided to diversify into cadaver stuff, and had established the American Red Cross Tissue Service, a network of corpse-disassembling entities based in its regional offices.[6] The first entity in St Louis had just broken even, taking in about $200,000 a month. Success had been achieved at the expense of the local Lions eye bank. 'The hospitals said they're doing work only with the Red Cross,' complained Elizabeth Dodd, who ran the local Lions programme.

Perhaps inspired by the American Red Cross's diversification, in 1986, the Community Blood Center, a non-profit entity that had been extracting and providing blood to all the general hospitals in Dayton, Ohio, for two decades, opened the Dayton Regional Tissue Bank. In 1994, its name was changed to Community Tissue Services, to indicate that it had embarked on a programme of acquiring similar entities in other states, the first of which was in Fort Worth, Texas.[7] By 2000, it had a network of entities in Oregon, California and Indianapolis. In retaining their original name, entities held on to a local identity that could be used in marketing, but which obscured their association with an organization that had national ambitions.[8]

Proprietary Processing

Before the early 1990s, surgeons prepared homostatic cadaver stuff for repurposing in the operating theatre. Where preparation was undertaken elsewhere, it was in a laboratory either in a hospital or on its campus, which may have been built or equipped by a civil society organization. However, in much the same way as institutions closed down onsite kitchens in which meals were prepared with fresh ingredients and bought in pre-cooked food, so hospitals closed stashes and in-house cottage-type facilities and began buying in cadaver artefacts from external sources.

As an independent full-service entity, the Eastern Virginia Tissue Bank had been short of cash when it opened for business in 1982, and staff put together its processing facilities. Spare capacity was filled by processing for a fee cadaver stuff extracted by other entities, including one at Ohio State University, Columbus, and another at the University of Iowa.[9] Community Tissue Services began offering a similar service. Both were cottage industries, fashioning cadaver stuff into artefacts, and extending shelf life by refrigeration, freezing or freeze-drying. However, whereas the

Navy Tissue Bank undertook extraction in its operating theatre under aseptic conditions, extraction might be undertaken in non-sterile collecting sites such as hospice, medical examiner/coroner's office, funeral home or hospital mortuary, thus terminal sterilization was a crucial additional stage.

In this business arrangement, processing fees feed the entity's revenue but the extractor retains title to the finished goods, which are returned to them to pack and retail. However, processing can activate a legal property right; lawmakers respect the principle originally formulated during the Enlightenment by English philosopher and physician John Locke that the worker is entitled to ownership of the fruits of their labour, although in these instances the 'worker' is not the technician but their employer. During the 1980s, what might be called second-generation entities dedicated to processing cadaver stuff began to be created. These were participants in the so-called biotechnology revolution that began in the 1970s, which was characterized by start-ups, typically spin-offs from universities financed by venture capital, established to exploit novel patent-protected techniques of preparing human body/cadaver stuff that might open new or widen existing markets. Genentech was one of the first and few to succeed: its achievements include rhGH, which not only released the mortuary's escape hatch, but also offered significant advantages over its predecessor hGH in terms of safety and availability. Genentech belongs to the sector specializing in 'bug to bottle' manufacture of substances with a physiological or metabolic effect, such as insulin. A majority of second-generation processors of body/cadaver stuff were 'engineering' products intended either to promote healing or restore function, or replace or reconstruct malfunctioning or deteriorating body parts.

The business strategy of second-generation processors is to climb up the 'value chain' where a higher price can be justified; it is comparable to that of convenience food manufacturers whose products retail at prices with a healthy margin over and above the cost of ingredients, manufacture, marketing and distribution. Convenience is an extremely attractive property in the modern world; surgeons took to cadaver products like ducks to water.

In 1984, CryoLife, Inc., in Atlanta, Georgia, became the first second-generation processing entity. It was created to exploit patent-protected techniques of terminal sterilization and freezing of cadaver heart valves. It took ten years to become the undisputed market leader, during which time 900 surgeons in 350 American hospitals implanted more than 15,000 of its cadaver heart valves, around 60 per cent of which were used to repair children's congenital heart defects. In 1994, it anticipated sales of 3,400.[10] Relatively high throughput allowed its sales team to offer 'in-stock guarantees', and assure surgeons that operations can safely be scheduled.

CryoLife was also the first for-profit entity; its shares were floated on the New York Stock Exchange. In 1986, New Jersey became home of two others: Osteotech and LifeCell. Osteotech, which focused on processing musculoskeletal cadaver stuff, was launched on the NASDAQ stock exchange in 1991. The following year it took the unprecedented step of marketing under the Grafton® brand a range of highly bioengineered cadaver bone implants that are available in a variety of standardized forms. These 'bio-implants' bear no resemblance to cadaver stuff; they look like nuts and bolts, putty and powder.

In 1986, two investigators at the University of Texas Health Science Center, San Antonio, established LifeCell Corporation in order to take commercial advantage of proprietary techniques of ultra-rapid cooling and ultra-low temperature freeze-drying that remove from cadaver skin the cells responsible for rejection, and deliver it from doctrinal tyranny.[11] Transforming exquisitely homovital stuff into homostatic stuff was a remarkable achievement. The Gulf War (August 1990–February 1991), codenamed Operation Desert Shield, had revitalized the military's interest in freeze-dried biological wound dressings, and support for its research had been found in the United States war chest. In 1992, shares were offered to the public. The following year, AlloDerm® was launched. It is marketed in different widths and thicknesses, sliced and packed like American cheese in sealed foil pouches, with rehydration instructions attached. It can be stored at room temperature and has a shelf life of up to two years.

In 1993, the Mile High Transplant Bank opened a facility that had the annual capacity to process musculoskeletal stuff extracted from 800 corpses. The following year, the Mid American Transplant Services of St Louis, and the Regional Organ Bank of Illinois, Chicago, joined in the creation of AlloSource, a non-profit processing entity. The geographical range of collecting sites was considerably extended when the Colorado Organ Procurement Recovery Services and the University of California at San Diego Regional Tissue Bank subsequently joined.[12]

In 1998, Regeneration Technologies, Inc., in Gainesville, Florida, became the last American for-profit processor to be established in the twentieth century. It was a spin-off of the University of Florida Tissue Bank, processing cadaver bone into precision-shaped cadaver implants for orthopaedic surgery, using BioCleanse™, a computer-controlled, validated, multi-step technique of terminal sterilization. A firm of certified public accountants that had been looking for investments in the North Central Florida Region biotechnology industry provided seed money.[13]

Cottage-industry-type entities were reading the writing on the wall. Some began investing in second-generation technologies, in particular, in 'clean room' processing facilities similar to those used by pharmaceutical and computer industries, which reduce the risk during processing of contamination by airborne microbes, and through contact – technicians are covered head to foot in sterile clothing. Clean rooms are expensive to build and operate: in the first years of the twenty-first century, the annual running cost of a suite with the capacity to process cadaver stuff extracted out of 1,500 corpses was $4 million.[14]

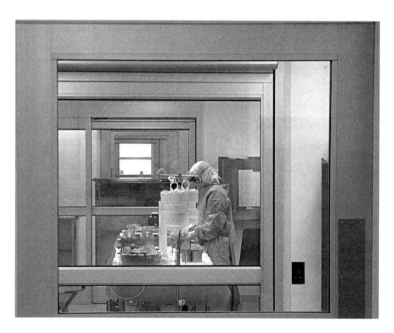

Figure 25: Processing cadaver stuff in a clean room

Competitive Extraction

Death is unpredictable; cadaver stuff is putrescible; employing teams of staff available around the clock every day of the year just in case a bio-available person dies is expensive, which is why entities typically rely on technicians paid either per diem or per cadaver or per cadaver fragment – a fixed rate is attached to each one that recognizes the work and skill required. Technicians might be medical staff moonlighting from their day job, such as surgical residents, nurses, operating or emergency room technicians, morticians or medical examiners, but they must be prepared to drop everything at a moment's notice, and any hour of the day or night or of the year. Christmas is their busiest season; cold weather and suicide boost the death rate.[15]

In 1988, the American Association of Tissue Banks established a certification programme for people without medical training. But attendance was voluntary. Nothing prevented an untrained individual from offering their services, or a 'mom and pop' entity setting up shop. 'My sense was, if certain guys weren't in this business, they'd be in concrete,'[16] observed David Kessler, a former Food and Drug Administration Commissioner. 'If you can find a donor,' Bruce W. Stroever, president of Musculoskeletal Transplant Foundation assured the House of Representative's Committee on Small Business, 'then you can find somebody that will take that donor off your hands.'[17] In 1993, stuff extracted from a single corpse could fetch anything between $3,000 and $5,000.

By the early 1990s, second-generation processing had become the maypole around which extracting mostly danced. Small cottage-type non-profit entities, perhaps organized around a single cadaver fragment, began 'optimizing' their access to collecting sites where corpses of non-beating heart sources rest between death and disposal, such as hospital mortuary, funeral home, coroner/medical examiners' office or hospice. For instance, Orange County Eye and Tissue Bank began shipping cadaver skin to LifeCell, and cadaver heart valves to CryoLife.[18] It was an associate of Tissue Banks International, which had diversified into extracting other cadaver stuff, particularly skin and bone.

By this stage Medicare was reimbursing costs associated with kidney explant/transplants of recipients of all ages, and where the recipient was aged over sixty-five of cadaver heart (from 1987), cadaver liver (from 1991) and cadaver lung (from 1995). Recipients ineligible under either Medicare or Medicaid pay either out of their own pocket or by their third-party payer. Faced with uncertainty over the reimbursement of an expensive procedure, hospitals demand a large upfront down payment before allowing details of a would-be recipient to be entered onto the waiting list.

Organ procurement organizations might be busier, but potential revenue was limited by the number of bioavailable corpses in their franchise area. Moreover, some have more opportunities to explant kidneys than others: for instance, in 1994 and 1995, motor vehicle accidents, the cause of death of one in four potential sources of explanted cadaver kidneys, ranged from 4.4 to about 17.9 per 100,000 population across states and the District of Columbia.[19] And although in theory all costs incurred in relation to cadaver kidney extraction and distribution are reimbursed, in many instances the Standard Acquisition Cost (SAC) set by Medicare falls short because it is based on industry averages, whereas costs vary widely across franchises – for instance, transport costs are relatively high in franchise areas where population is spread thinly. Consolidation of uneconomic franchises and stronger ones was inevitable and initially rapid: the number of organ procurement organizations tumbled from 100 in 1986 to sixty-eight in 1991, and continued falling albeit at a less precipitous rate.

Diversification into extracting other cadaver stuff offered another more predictable revenue stream.[20] Organ procurement organizations began 'optimizing' their privileged access to otherwise healthy but brain-dead sources by presenting next of kin with a list of everything within a corpse that could be repurposed. Unfortunately, as lists lengthened, transplant coordinators began suffering from a new malaise called 'list fatigue', which made it less likely that agreement to the extraction of cadaver stuff towards the list's end would be sought.

A survey conducted in 1993 by the Department of Health and Human Service's Office of Inspector General found that twenty-eight out of the sixty-two organ procurement organizations had diversified into extracting other cadaver stuff, including one that was anticipating gross revenues of about $1 million and an operating surplus of between $50,000 and $100,000 from diversified activities.[21] Another survey found that whereas before 1986 federal government via Medicare was responsible for 90 per cent of the revenue of organ procurement agencies, by the early 1990s its share had fallen to about 65 per cent.[22] Cadaver kidneys were increasingly secondary to organ procurement organizations' earnings from extracting other cadaver stuff. Twelve respondents to the 1993 survey of organ procurement organizations said they were extracting cadaver stuff for processors. One revealed that his charged between $5,000 and $6,000 per corpse, and that the income generated covered a substantial part of operating costs.[23] In 1997, the Regional Organ Bank of Illinois, one of AlloSource's founders, extracted cadaver stuff out of 420 non-beating heart and cadaver organs out of 252 beating-heart sources; two years later, the numbers were 565 and 276.[24] By the turn of the millennium, organ procurement organizations were responsible for around 40 per cent of all the stuff extracted from cadavers.[25]

Organ procurement organizations began acquiring or merging with entities that had access to collecting sites where corpses of non-beating heart sources rest between death and disposal. Take the Midwest Organ Bank, founded in 1973 as an organ procurement agency: it won a franchise, began extracting other cadaver stuff in 1990 and in 1998 acquired the Kansas City Eye Bank. In 1999, the conglomerate changed its name to the Midwest Transplant Network, and adopted the mission statement: 'To provide quality transplantation-related services that will maximize the availability of organs and tissues to the communities we serve.'[26]

In 1987, Osteotech invested $10 million to create the Musculoskeletal Transplant Foundation, an independent non-profit entity that had orthopaedic surgeons sitting on its board. Its staff did not engage directly in extraction; they oversaw entities bioprospecting and extracting in different parts of the United States. Divesting responsibility for activities that had been integral to first-generation entities was advised by gurus of the business strategy of outsourcing, which had become fashionable during the 1980s. In concentrating on 'core competencies', that is, on what an entity does best, it can achieve a definable presence and competitive edge in the market. Activities outside 'core competencies' are best left to those who claim to possess the necessary expertise. Indeed, requiring these entities to compete can reduce operating costs.[27] The downside is that the outsourcer relinquishes oversight of operations, and relies on regulation through contract. However, rumour has it that Osteotech had been motivated not by management theory but by non-profit extractors' refusing to supply a for-profit processor with cadaver stuff.[28]

In 2001, Regeneration Technologies' Chief Financial Officer Richard R. Allen explained the rationale behind its procurement strategy:

> [One] thing that separates [RTI] from . . . other allograft companies is that we work very closely . . . with some recovery agencies around the country. . . . [As a result, we] have more effective control over our supply of the tissue because we are able to help drive and increase the rate of donation in these territories. So we have got probably a very, very solid source as supply goes, and the ability to grow supply in the future.[29]

In 1999, Georgia Tissue Bank, Inc. became a wholly owned subsidiary, and joined a nationwide network of forty non-profit 'affiliates' that extract cadaver stuff under contract for Regeneration Technologies, which sometimes agrees to return some finished goods to them. In 2003, management of extracting entities was delegated to Regeneration Technologies Donor Services, Inc., a non-profit division it controls and supports with an annual operating budget.

The number of American corpses from which organs were explanted increased from 5,099 in 1994 to 5,985 in 2000. Over the same period, the number of corpses out of which cadaver stuff was extracted increased from around 6,000 to around 20,000.[30] Yet despite this threefold growth, competition for cadavers remained intense, offering collecting sites revenue-generating opportunities. For instance, in 2000, the *Chicago Tribune* reported that in San Antonio, Texas, the medical examiner's assistants were paid $50 each time they obtained next of kin's agreement to extraction. It also alleged that county supervisors took bids on the right to extract stuff out of corpses in their possession. South Texas Blood and Tissue Center won by offering to pay $180,000 each year.

CHAPTER 30

Contagious Corpses

In 1991, LifeNet announced that seven recipients of cadaver organs/stuff/ artefacts had been infected with the Human Immunodeficiency Virus (HIV), three of whom had died from iatrogenic Acquired Immune Deficiency Syndrome (AIDS).[1] The source was the corpse of a man who six years previously had been fatally shot in the head during a gas station robbery. Analysis of his blood in two different laboratories had failed to detect that he was HIV positive because tests available in 1985 were first generation, only recently licensed by the Food and Drug Administration, and designed to identify the presence of antibodies to the retrovirus, which can take weeks or even months to appear, which means they are ineffective prior to what is known as seroconversion. Moreover, the tests had been developed for blood donors-on-the-hoof; a Food and Drug Administration's licence did not require evidence of efficacy post-mortem. Yet within minutes of the heart ceasing to beat, blood chemistry begins undergoing changes that may interfere with accuracy and increase the likelihood of false positives and negatives.

Second-generation tests on a stored sample of the source's blood, which work by confirming absence or presence of retroviral DNA, established that he had been HIV positive. Attempts to trace every recipient revealed multiple supply chains of different lengths and complexity, and sloppy paperwork. Every explanted and transplanted cadaver organ is counted out and counted in, and teams at the Medical School of Virginia were known to have transplanted his kidneys, heart and lungs. But the destination of the remaining cadaver stuff was uncertain. If and where his corneas had been grafted has not been reported. According to LifeNet's records, cadaver stuff/artefacts had been shipped to thirty hospitals, but seven of these had no idea if, where and how it had been used. Fifty-eight people were believed to have undergone a procedure incorporating his bone, ligaments, tendons, dura mater or skin, but the exact number was unknown, and not every identified recipient could be found. One surgeon refused to contact his patient, who was unaware that he was a recipient, perhaps because lack of informed consent can constitute an action for battery.

The case revealed how one disassembled contaminated corpse can transmit infection to many patients. The Health Care Financing Administration, the federal agency that administered Medicare, was responsible for the safety of organ recipients, who at the time numbered around 30,000. But official oversight of the safety of around 500,000 recipients of cadaver stuff/artefacts was absent.

Self-regulation

LifeNet was a member of the American Association of Tissue Banks, following its guidelines on safety that include medical criteria of bioavailability, handling practices and requiring staff to attend their training courses. The association had a programme of accreditation, and failure to pass muster could result in disciplinary action or expulsion. However, a private-sector trade association is without the authority to seize suspect cadaver stuff/artefacts or force a substandard facility to close. Moreover, membership of both the American Association of Tissue Banks and the Eye Bank Association of America is voluntary, and was known to fall short of the unknown number of individuals and entities extracting, importing, processing, storing, and/or distributing cadaver stuff in the United States. These varied in size from a surgeon operating a just-in-time inventory – nipping into the mortuary – through 'mom and pop' commercial entities, to a non-profit conglomerate such as LifeNet.

Self-regulation, which industry associations had begun adopting following the *Ravenis* case, had been politically timely. American governments had espoused a preference for voluntary standards, which are cheap to administer and easier to amend than those backed by the force of law.[2] The Food and Drug Administration had also welcomed it because it was hard pressed fulfilling its new responsibilities in relation to the Medical Device Amendment.

The likelihood of official scrutiny retreated even further over the horizon under President Ronald Reagan's programme of small government and sweeping deregulation. Between 1981 and 1987, despite its growing workload, the Food and Drug Administration's workforce was cut by 11 per cent.³ Yet in 1976, when the Medical Device Amendment was adopted, Congress had estimated that 1,000 additional members of staff were needed to implement it, but funding was never sufficient, and by the early 1990s only 600 people had been appointed.⁴

Regulation of the Self

Politicians and policymakers might have greeted self-regulation enthusiastically, but shortly after its adoption there was an unanticipated major threat to the safety of recipients. In the early 1980s, young gay men began dying from a cluster of rare diseases associated with a compromised immune system. Subsequently, heterosexual people, particularly those who shared needles to inject drugs, began suffering a similar fate. It looked as if a hitherto-unknown pathogen was responsible. In August 1982, the disease it caused was named Acquired Immune Deficiency Syndrome (AIDS).

The first report of post-transfusion AIDS was published in December 1982 in the Mortality and Morbidity Weekly Report (MMWR) of the United States Centers for Disease Control and Prevention (CDC), a federal government agency. Its March 1983 report included advice to blood extractors to ask donors-on-the-hoof who identified with one of the groups who appeared susceptible to the syndrome voluntarily to withhold their forearm. This safety measure, known as self-deferral, added a new feeling rule to body philanthropy: consider whether your sexual practices pose a risk to recipients' safety. Civil society organizations protested that the emphasis on homosexual or bisexual men was homophobic, and ineffective because a growing number of heterosexual men and women were contracting the syndrome: the retrovirus, which was identified in 1984, does not discriminate on the grounds of sexuality.⁵

Clearly, dead people are incapable of self-deferral. But there was little public discussion of how they were being or should be treated. Extractors of cadaver stuff were assessing risk by investigating the lifestyle and medical history of potential sources. Information might be sought in medical records, but also by questioning kinfolk, physicians and sometimes police. Absence of a reliable informant became grounds for rejection: social death had become a threat, not an opportunity. But information provided might be unreliable. Indeed, enquiries made by LifeNet about the man whose cadaver stuff went on to infect recipients with HIV had failed to elicit anything that might suggest it would be wise to refuse his corpse. Moreover, information provided is sometimes contradicted by the results of blood tests, placing extractors in the awkward situation of knowing more about a potential source's lifestyle than the latter's immediate circle, and confronting them with the dilemma of whether or not and how to divulge a dead person's secrets that survivors might not welcome.

Regulation by Classification

In 1982, neurologist and biochemist Stanley B. Prusnier proposed that the infective agent of spongiform encephalopathies is not a slow virus, as Gajdusek had suggested, but a rogue protein, which he called a prion. Prusnier's prion research has garnered many awards including the Nobel Prize in 1997, but his hypothesis has provided neither objective test capable of identifying, nor terminal sterilization technique that convincingly eliminates, the prion. Despite the prognostications of Gajdusek and his team, the realization in 1985 that recipients of hGH had contracted and died of iatrogenic Creutzfeldt-Jakob disease provoked few fears about the safety of cadaver stuff, perhaps because distribution of the hormone had been halted, and because both actual and potential number of people affected was dwarfed by that of recipients of blood and blood products who had contracted, and continued to contract, post-transfusion hepatitis and AIDS.

In 1987, another reminder that prions might be lurking in cadaver stuff emerged in reports in the United States and elsewhere that recipients of Lyodura®, a cadaver dura mater product, had succumbed to iatrogenic Creutzfeldt-Jakob disease. Prior to May 1987, Lyodura®'s German manufacturer, B. Braun Melsungen, had been venturing into the mortuary no-place where cadaver pituitary glands were being extirpated, and without next of kin's knowledge and agreement had been extracting cadaver dura mater, the membrane covering the brain and spinal cord. Whether or not mortuary staff were tipped has not been divulged. But what is known is production costs were held down by sterilizing the dura mater in batches: 'They put it in a bucket, a huge bucket,' explained the Centers for Disease Control's Dr Ermias D. Belay.[6] Sterilized dura mater was cut into convenient sizes, freeze dried and marketed as a dressing for wounds created in a variety of medical procedures, from removal of a large brain tumour to dental surgery. B. Braun Melsungen had not sought a licence to distribute Lyodura® in the United States, but it was cheaper than equivalent American products, and American surgeons had been importing it from a Canadian distributor, which shipped it through the mail.

Where recipients of cadaver stuff are harmed, redress for the injury can be sought under medical negligence or product liability, depending in part on whether the defendant is a physician or surgeon or processor. But iatrogenesis caused by an infectious pathogen is also a public health issue. In 1989, the Food and Drug Administration announced that forthwith cadaver dura mater products were medical devices, which meant that anyone extracting, processing or storing it became a 'device manufacturer' and as such was required to register with it, and comply with the medical devices' regulations. Regulation by classification was the administration's only way of prohibiting pooling of cadaver dura mater, which, as happened with hGH, increases the likelihood of 'bad apple contamination'. Extractors and processors of cadaver dura mater products for distribution within the United States were faced with the rigour and expense of registering with the Food and Drug Administration, submitting scientific evidence of safety and efficacy for pre-market approval, having to open the doors of premises to its inspectors, recalling all products that proved unsafe, and the threat of punitive product liability damage awards to infected recipients and their next of kin.

Two years later, the strategy was applied to cadaver heart valves, which, the Food and Drug Administration claimed, warranted the same scrutiny as mechanical heart valves, which belonged in Class III, the riskiest category of medical devices. Six processors including CryoLife formed a consortium to challenge the decision: cadaver heart valves, they claimed, had been implanted in thousands of people without evidence of harm. Moreover, whereas mechanical heart valves can be standardized, every cadaver heart valve differs because no two corpses are alike, and anyway, extractors and processors should not be held wholly responsible for mishap because surgeons often trim the valves before insertion. Put another way, cadaver heart valves are artefacts, not devices, which as the law then stood, were subject to minimal oversight. This argument might hold water but courts dismissed the assertion that the Food and Drug Administration had exceeded its authority. In 1991, the American Association of Tissue Banks established a Governmental Affairs Committee that would negotiate on behalf of its members.

Safeguarding recipients' health on a case-by-case sequestration of cadaver artefacts under the Medical Device Amendment was wholly unsatisfactory: it was arbitrary, excluding a majority of cadaver stuff/artefacts/products. It was simply the wrong approach. As Senator Paul Simon (Democrat) put it, 'You can say a horse is a cow, but it doesn't make a horse a cow, you simply aren't getting the kind of regulation that you need.'[7] But the Food and Drug Administration's staff and resources were stretched, and a budget increase unlikely to be granted by the administration of George

Bush Senior, who believed that mandatory regulation burdens business with unnecessary bureaucracy and stifles innovation. The latter was perceived as a particular threat to the biotechnology industry organized around finding innovative ways of repurposing human body/cadaver stuff, and which at the time was regarded by venture capitalists and stock market investors as one of the fastest growth sectors of the American economy.

Statutory Regulation

In 1991, Governor Mario Cuomo (Democrat) signed into law a statute requiring entities involved in body/cadaver stuff/artefacts/products for repurposing in medicine anywhere within New York State to apply for a licence and submit to rules on criteria of medical bioavailability, and handling and processing practices, agreed by scientific, professional and trade associations. It also required entities where recipients are treated to seek a licence. The measure had been adopted because a significant proportion of New York residents were knowingly or unknowingly HIV positive, and because of the growing number of entities involved in some way in the body/cadaver stuff/artefacts industry. By 1993, New York State Department of Health had licensed fifty-two 'comprehensive tissue services'; 221 'limited tissue procurement services', six eye banks and twenty-four 'gamete banks' (semen and eggs).[8]

In June 1992, Senator Paul Simon (Democrat) introduced the Human Tissue Transplantation Bill, the first attempt to impose mandatory oversight throughout the United States. The bill was organized around the appointment of the National Council on Tissue Transplantation, a non-profit organization that would license entities, collect data on supply and demand, and support the development of private-sector professional standards and inspection. Violation of rules would incur penalties ranging from $10,000 to $50,000 per violation, licence suspension and seizure of stuff considered unsafe.

Neither Simon's bill nor the one he and Representative Ron Wyden (Democrat) submitted in November 1993 was adopted. The American Association of Tissue Banks and the American Red Cross supported the 1992 bill's requirement to screen and test for medical bioavailability, and the tracking system, but argued that mandatory rather than voluntary standards were needed. The 1993 bill, which required federal oversight, was popular, but never received a floor vote and consequently died.[9] Their trajectory was similar to that of thirty bills introduced in the early 1990s in response to increasing frequency of reports of toxic pathogens, including some hitherto unknown unknowns, in food.[10] One of the nastier ones the politicians wanted to eliminate was *Escherichia coli* O157:H7, an unusually virulent member of a family of normal and relatively harmless bacterial inhabitants of the human digestive tract, whose tractable kin had been recruited to manufacture recombinant protein hormones such as insulin and rhGH.

In October 1993, prior to submitting his bill, Wyden presided over a hearing before the House of Representatives Subcommittee on Regulation, Business Opportunities, and Technology that provided an opportunity for enthusiasts of mandatory oversight to set out their stall. Simon maintained that absence of regulation was responsible for a very low entry bar into the industry, which did not deter fly-by-night entities. 'It is like any unregulated field, some people see a chance to make some profit,' he explained; 'they can operate out of their kitchen and put things in the kitchen refrigerator along with the potatoes and everything else.'[11] Americans were not the only ones taking advantage of the low entry bar. The hearing heard how extractors in Central and South America, and former Soviet republics, were marketing directly to American processors and physicians a range of cadaver stuff/artefacts of dubious and unverifiable provenance. Dr Pavel E. Blumberg, President of the Baltic Tissue Bank, was held up as an example; he had telefaxed to potential American 'customers' what Wyden called a 'butcher's list', with prices attached – for instance, 10 square millimetres of cadaver skin cost $7 FOB (meaning, seller pays costs of transportation, etc.) – from St Petersburg airport. Blumberg claimed that the Baltic Tissue Bank, a joint stock company established in cooperation with the Kriolaboratory of St Petersburg Institute of Orthopedy, was one of the largest in Russia, and offered reassurances that stuff was extracted within hours of death from corpses of people aged between sixteen and fifty-five, who had tested negative for HIV, syphilis and hepatitis.[12]

The aim of medical criteria of bioavailability is to reduce the risk of transmitting agents of an infectious disease that a source had contracted when they were alive. But these are not the only potentially harmful pathogens within a corpse. Cardiac death does not eliminate the countless microbes in the intestines, and autolysis, the corpse's intrinsic breakdown that begins almost immediately after the heart ceases beating, provides them with the opportunity to invade normally prohibited parts. Some of these microbes are responsible for what Vladimir Petrovich Filatov had called cadaver infection. Hence, the sooner stuff is extracted after cardiac death the less likely it is to be contaminated. However, witnesses to the hearing testified to delays in extraction, sometimes by a medical examiner/

coroner who failed to complete the paperwork before the twenty-four-hour maximum watershed following cardiac death stipulated by trade associations.

Microbes are everywhere, as Colebrook had established. Thus working conditions are critical factors, but there was nothing to prevent untrained or poorly trained people extracting cadaver stuff in non-aseptic collecting sites. Moreover, the supply chain from collecting site to recipients was lengthening, and the routes taken becoming more convoluted, increasing the likelihood of contamination. Large-scale processors were handling under the same roof large quantities of cadaver stuff that had been extracted from different sources under a variety of conditions. These structural changes are analogous to those in the food-manufacturing industry where for similar reasons reports of outbreaks of foodborne illnesses were mounting. Food producers and manufacturers rely on terminal sterilization, including cooking to kill dangerous pathogens. But vigorous terminal sterilization of some cadaver stuff, such as corneas, for instance, by zapping with gamma rays, destroys its integrity, leading to poor outcomes for recipients.

Interim Regulation

In December 1993, the Food and Drug Administration issued an Interim Rule under the authority granted by Section 361 of the Public Health Service (PHS) Act 1944, which was concerned with preventing the introduction, transmission and spread of communicable diseases by contaminated persons, animals and articles, imported into the United States or transferred between states. It was a stopgap measure, effective immediately. 'Interim' implied that other rules should be anticipated.

The Interim Rule focused on reducing the risk of recipients contracting iatrogenic AIDS, and hepatitis B and C. Stuff should not be extracted from the corpse of anyone falling into the following categories: men who had had sex with another man during the previous five years; anyone who had had sex for cash or drugs within the past five years; anyone who had injected drugs; anyone with haemophilia; anyone who had had sex with one of the above; inmates of correctional systems. It also required records to be kept of source and destinations of cadaver stuff/artefacts/products. Entities had to admit inspectors, who might arrive unexpectedly. Importers had to notify the Food and Drug Administration, and quarantine cadaver stuff/artefacts until release was approved. Violation was punishable by imprisonment of up to a year, and also a fine of up to $100,000 if no one died or up to $250,000 if responsible for death.

The ambition of the Interim Rule was less demanding than the standards set by the two private-sector industry associations. Without a requirement either to register or to apply for a licence, the American universe of entities remained unknown, and an unknown number were escaping the threat of inspection. Moreover, there was no requirement to report when a recipient had been harmed. Yet manufacturers of consumer goods faced a hefty fine for failing to report to the United States Consumer Product Safety Commission when the design of a product on the market proved dangerous.

The Food and Drug Administration was not wholly idle. In 1993, following a tip-off, it ordered the recall and destruction of 5,600 stuff/artefacts/products that had been made out of cadaver stuff imported from Russia. It transpired that the paperwork claiming that their source, 228 Russian corpses, met its medical criteria of bioavailability was false. Subsequently, hepatitis viruses were identified in samples of the shipments. The *Chicago Tribune* exposed details of the case in May 2000.[13] The broker was a Los Angeles dentist, friend of emergency care doctor Dr Valery Khvatov of the Sklifosovsky Institute, Moscow, which had been Filatov's productive collecting site. These 5,600 cadaver 'units' – mostly bones and tendons – had been extracted from the corpses' hips, knees, legs and feet, packed with dry ice in cardboard boxes and shipped from Moscow to New York. Russian law allowed stuff to be extracted without the knowledge or agreement of next of kin, which meant that extractors had not asked questions about sources' social, sexual and medical history. Interviewed by journalists, Khvatov denied any personal gain. 'It's a fairy tale,' he said. 'If I were taking cash, my own side would put me in jail.' His employer though was reimbursed for costs. But revenue accrued along the supply chain: the price charged to American recipients was around six times that FOB.

The bulk of shipments had been bought by AlloTech, a processor in El Paso, Texas. Some had been sold to two San Antonio businesses, Transplant Technologies Inc. and International Tissue Services. The remainder had been distributed to hospitals in thirty states, Puerto Rico and Mexico. Not all of it was traced; around 500 recipients were unidentified.

Media coverage of the case suggested that foreigners are untrustworthy, and that imported cadaver stuff is more likely to be tainted than that extracted from American bodies. But the Interim Rule had made an exception for eye banks operating under legislative consent, where dead people whose corpse fell into the possession of a coroner/medical examiner continued to be treated as if they were socially dead, and evidence of their lifestyle and medical history limited to scrutiny of hospital case notes where these were available, and the poke and sniff test. This now included reading the corpse's surface for signs suggestive of HIV or hepatitis B or C infection, such as tattoo, needle tracks and skin diseases indicative of a compromised immune system. Results from blood tests might suggest cadaver eyes/corneas should have been withheld from repurposing but took several days to emerge, too late for the typical high-volume, short shelf-life business model of American eye banks.

CHAPTER 31

British Prions

In 1985, few British people observed the hasty retreat being beaten from the no-place in mortuaries following the realization that recipients of hGH had succumbed to iatrogenic Creutzfeldt-Jakob disease. However, shortly afterwards, few remained ignorant of the growing numbers of cattle becoming unsteady on their legs, behaving strangely and then dying. Veterinarians had no idea what was responsible for the epidemic until July 1987, when slivers of cattle brain examined under the microscope were found to look spongy, which indicated that the spongiform encephalopathy family had a new member: bovine spongiform encephalopathy (BSE) or, more familiarly, mad cow disease.

The number of infected cattle increased exponentially. Clearly, this was an epidemic. Yet Margaret Thatcher's government dragged its heels until the summer of 1988, when it ordered the compulsory slaughter of herds in which affected beasts were identified. Carcasses were dumped in quarries, drenched with petrol and set on fire – the stench was appalling. The final death toll was more than four million cattle.

Cannibalism was how the Fore people had been infected by prions. Scientists began entertaining the possibility that the cattle had been infected by feed that had been supplemented with prion-contaminated meat-and-bone granules repurposed out of pooled abattoir waste, and rejected carcasses that included scrapie-infected sheep. The protein supplement was manufactured by the rendering industry, which traces its origins to ancient Greece.[1] But industrialization of meatpacking had changed the scale of its operations, and its factories had become 'the goriest expressions of the recycling spirit, hellish places of steam, blood, grease and stink'.[2] The officially approved recipe included a stage in which tallow, which is used mostly as cooking fat, was separated out by hydrocarbon solvents. What was left was minced, sterilized by heat, dried and pressed into pellets. However, during the 1970s, the market for tallow collapsed, the cost of fuel shot up and hydrocarbons were implicated in a disastrous explosion at a British chemicals factory. Margaret Thatcher lent a sympathetic ear to the British rendering industry's pleas for regulations to be relaxed while it modernized. Solvents were excluded, which increased fat content, and savings on the cost of fuel achieved by lowering the temperature of sterilization. Changing the recipe, apparently, and with hindsight, allowed prions to survive.[3] In July 1988, the government banned the feed supplement.

If the prions responsible for scrapie had infected cattle then they had succeeded in crossing the species barrier that hitherto had been regarded as unassailable. The possibility that human meat eaters might be vulnerable to infection from contaminated meat began to be entertained. In November 1989 sale of offal was banned. Home and export markets for British beef and dairy products collapsed. In May 1990, Minister of Agriculture John Gummer (Conservative) famously sought to restore public confidence by encouraging his four-year-old daughter to bite into a beef burger in front of press cameras.

Market Mechanisms

Thatcher's government favoured light-touch regulation, where the invisible hand of the market, and not Parliament or experts, governs commercial practices. It was also committed to shrinking the size of the British state. The NHS was struggling to cope with growing demand, but instead of increasing its share of public spending, which was well below that of equivalent economies, in 1984, the government introduced commercial management practices, as recommended by Roy Griffiths, managing director of Sainsbury's, one of the country's most successful supermarket chains, as a means of effecting efficiency savings. Managers were recruited from industry and the armed forces to identify and implement cost-cutting measures, which provoked considerable resentment among staff, and failed to keep pace with the ever-expanding range of diagnostic and therapeutic options. Nonetheless, by the beginning of 1988, when Prime Minister Margaret Thatcher announced a radical review, underfunding of the NHS seemed to be terminal.[4] Health-care professionals struggled to treat patients in dirty hospitals with inadequate facilities that were an affront to the claims made for the NHS.

So-called 'managed competition', the next cost-containing strategy, was an American import. In the NHS and Community Care Act 1990,

Parliament legislated for a radical restructuring of the NHS that created an internal market. In April 1991, hospitals were freed from their moorings to a regional health authority, which had been responsible for strategic planning of services, transformed into quasi-independent entities, and ordered to compete on price and quality, against one another, and non-profit and commercial entities, to win contracts from so-called commissioners authorized to spend taxpayers' funds on their behalf.[5] Health care remained free at the point of delivery, but public and patients were excluded from negotiations around contracts, which included the type of procedures that would or would not be provided.

Market mechanisms prescribed to treat the NHS's financial woes had recently been rejected as a remedy for the shortage of kidneys for repurposing as transplants. Parliament had adopted the Human Organ Transplants Act 1989 in response to a scandal involving three surgeons who had paid people, mostly Turkish men, to travel to their private London hospital, where one of their kidneys was explanted. The surgeons were found guilty of serious professional misconduct, and the government took swift action to criminalize anyone offering cash to potential kidney donors-on-the-hoof.

The scheme the surgeons had operated was similar to the one proposed by Jacobs, which had resulted in the adoption in the United States of the National Organ Transplant Act 1984. However, whereas the statute reorganized the American infrastructure responsible for explanting and distributing cadaver organs, no change in the UK Transplant Service was considered necessary. Instead, the government created the Unrelated Live Transplant Regulatory Authority (ULTRA), and invested in it the authority to investigate the motivation of living people volunteering a kidney for transplantation in order to prevent exploitation.

The Nascent British Cottage Industry

The Human Organ Transplants Act applied to stuff falling under the statute's definition of 'organ', which was 'any part of a human body consisting of a structured arrangement of tissues, which, if wholly removed, cannot be replicated, by the body'. Nothing replicates in a corpse except bacteria and fungi, but whether or not the statute applied to cadaver stuff was neither considered in Parliament nor tested in court.

The statute allowed reimbursement of expenses incurred by both donors-on-the-hoof and entities extracting, transporting, processing, or extending the shelf life of stuff. This was fortuitous because the contracting process of the internal market required each type of service to be costed, a demand intended to replace profligacy with tight budgetary control.

Since its inauguration in 1983, the Corneal Transplant Service had been seeking reimbursement of costs incurred in extracting, processing, storing and distributing cadaver eyes/corneas, but for the first time British entities involved in other cadaver stuff had to work out how much to charge. The maths was challenging for the Yorkshire Regional Tissue Bank, the UK's only full-service entity distributing a variety of cadaver artefacts. John Kearney, who had replaced Frank Dexter on his retirement in 1985, considered two equations.[6] One involved dividing total annual costs of the bank by number of cadaver artefacts distributed, in which case everything would have the same price irrespective of the amount of work embodied in it. The other was to use as a guideline the price of commercial alternatives, but this was possible only where these were available, and, if that was the case, were likely to be relatively expensive.

In 1993, seventeen years after the American Association of Tissue Banks had been established, the British Association for Tissue Banking was inaugurated. It was Kearney's idea and he was elected first president. The nascent British cottage industry gathered under its wing. Members included established and newly minted entities. The former included the Corneal Transplant Service, and the Heart Valve Bank at the Royal Brompton Hospital, London, which had been opened in 1962 by cardiac surgeon Donald Ross (1922–2014), who in 1968 had led the team that performed the first British heart transplant, and which claimed to be the largest in Europe. Not every established bank joined: Moorfields Hospital's eye bank continued to cherish its independence. However, the NHS blood transfusion service, which was diversifying into extracting and processing body and cadaver stuff, was an enthusiastic member. Safety was the association's watchword: in 1993, it published its first set of guidelines, which were also the first British ones to set out medical criteria of bioavailability and describe how to minimize risk of contamination during extraction, handling and processing.[7]

The Scottish Blood Transfusion Service had been the first to expand its inventory. In the late 1980s, it began creating bone artefacts out of femoral heads. Demand for bone for repurposing was growing, albeit at a pace slower than in the United States. In 1991, it was reckoned that each year in the United Kingdom bone was being repurposed to secure the prosthesis replacing around 5,000 knees and 23,000 femoral heads. Some surgeons were using autografts, and subjecting their patients to two incisions. Others kept a stash of fresh or frozen surgical waste – mostly femoral heads removed during hip replacement surgery. However, in 1993, a British Orthopaedic

Association Working Party found ill-equipped stashes of bone from sources that had neither given consent nor had their medical bioavailability checked.[8] The latter is unsurprising: it is no easy matter to question elderly people about to have their hip replaced about their sex life.[9]

Diversification into surgical waste proved relatively easy for blood service entities: staff knew how to ask awkward questions; femoral heads could be stored in hospital blood departments' refrigerators; its drivers deliver blood and blood products to hospitals almost daily; its laboratories adhere to the principles of Good Manufacturing Practice. In 1993, the East Anglian Blood Centre, Cambridge, followed the Scottish lead. The regional health authority provided £250,000 start-up funding for the East Anglian Tissue Bank, which envisaged becoming self-financing.[10] Its freezer also held cadaver heart valves extracted by pathologists, and other cadaver stuff extracted by its technical staff. Cryobiologist David Pegg knew that the Human Tissue Act 1961 permitted only licensed medical practitioners to cut into corpses, but had come across paperwork incentivizing extirpation of cadaver pituitary glands by mortuary staff, which he intended citing if challenged.[11] Subsequently, the North London Blood Centre, Edgware, began collecting femoral heads, and sending teams of medical students and technicians to hospital mortuaries to disassemble bioavailable corpses.[12]

In an ageing population femoral heads are plentiful: in the year 1993–4, 1,456 were collected in Scotland.[13] But the shortage of cadaver skin was critical. A round table discussion of burn care experts held in Geneva in 1987 had concluded that Zora Janžekovič's method of early excision and grafting was the best treatment of a drastic burn; it was also cost effective: patients suffer fewer infections and their stay in hospital is shortened.[14] However, NHS burns care facilities were mostly neglected, outdated and understaffed. A working group of the British Burn Association and Hospital Infection Society drew up proposals for modernization, but recognized that implementation depended on resources being found, which was unlikely.[15] Ten years later, this pessimistic forecast proved correct; the financial viability of NHS burns units was shaky.[16]

In 1991, the Sheffield Skin Bank was opened in the Northern General Hospital.[17] It was the first new skin bank to be attached to an NHS burns unit for nearly three decades, and had been funded out of existing resources, with laboratory space, equipment and technical expertise provided by the university Department of Medicine. Around half of the cadaver skin extracted was frozen; the remainder was glycerolized, a relatively simple technique involving soaking skin in a solution of sugar alcohol. The Dutch

National Skin Bank, which the Dutch Burns Foundation had opened in 1976, had appropriated the technique from the Dutch flower industry, where it was used to extend the shelf life of cut flowers. Glycerolized cadaver skin packed in small sterile containers stored in a refrigerator at 4°C has a shelf life similar to that of fresh milk. Another advantage claimed for it is that it 'attenuates immunogenicity'; in other words, it slows the pace of rejection, which remains inevitable.

In 1996, cadaver skin banks were attached to two other NHS burns units. In March, Falklands War hero Simon Weston opened the Stephen Kirby Skin Bank at the Queen Mary's Hospital, Roehampton, south London, which housed the South Thames Regional Burns and Plastic Surgery Unit. It had been named after Stephen Kirby, who had died in the hospital in 1994 from complications following drastic burns sustained while rescuing his family trapped in a tent that had caught fire on a camping holiday in France. Cadaver skin was unavailable to dress his wounds. Twelve family members and friends had volunteered theirs, despite being warned that the wounds would be sore and take around ten days to heal.[18]

Kim Kirby, Stephen's widow, her family and friends, supported by the *Eastern Daily Press*, decided to raise funds for a skin bank. The British Diplomatic Spouses Association and Currie Motors donated a 'retrieval car', which allowed extraction to be undertaken in hospitals throughout the southeast corner of England where transplant coordinators were co-operative. Frozen or glycerolized cadaver skin was either used locally or distributed to any surgeon who requested some.[19]

The Frenchay Skin Bank, which was officially opened in April 1996, was an affiliate of the Euro Skin Bank, the name adopted in 1992 by the Dutch National Skin Bank, when its financial footing had become sufficiently secure for it to expand its operations, and offer burns units in other European countries unmeshed glycerolized cadaver skin at the competitive price of $0.55 per square centimetre.[20] Funds had been raised through an appeal launched by the *Western Daily Press* following the death in September 1994 of Susan Guest, six weeks after a horrific barbecue accident. She was seven months pregnant when she had poured methylated spirits on smouldering coals, and was turned into a human fireball.[21]

In 1996, the Sheffield Skin Bank closed. It had been unable to keep pace with demand, largely because it relied on surgeons to extract skin, who were often unavailable when it was notified of a bioavailable corpse.[22] Its facilities were transferred to the local blood centre, and its senior staff joined a small select chorus lobbying for greater official support of the National Blood Service's diversification into cadaver stuff.[23]

Unfortunately support was unlikely to be forthcoming. The National Blood Service was in the throes of upheaval. In 1993, the National Blood Authority had been created to take responsibility for blood collection and distribution from the fifteen regional health authorities in England and north Wales, which were about to disappear, and formulate a national tariff for blood and blood products. Following recommendations of management consultants Bain & Company, the authority began a programme of rationalization and consolidation in which some centres were downgraded and destined for closure, and the remainder reorganized into three zones. Few supported the measures. Surgeons began complaining of shortages of blood for transfusion, and warned that the reorganization threatened the safety of recipients. In April 1998, Secretary of State for Health Frank Dobson (Labour) told MPs that in a review he had ordered into the authority's record, senior staff stood accused of disregarding day-to-day exigencies in order to satisfy the political and contractual demands of the internal market.[24] The following month, Dobson sacked the authority's chairman.

British Prions

In April 1995, court hearings began in the class action for damages brought by the next of kin of recipients of hGH who had died of iatrogenic Creutzfeldt-Jakob disease, and some recipients who were without symptoms but feared the same fate. Prions were wreaking havoc outside the courtroom. In May 1995, nineteen-year-old Stephen Churchill became the first person to die of variant Creutzfeldt-Jakob disease, another new member of the family of spongiform encephalopathies, which is characterized by, among other things, the younger age of victims and a shorter incubation period than that of its older relative. By March 1996, variant Creutzfeldt-Jakob disease had claimed ten British lives. The Secretary of State for Health, Stephen Dorrell (Conservative) admitted the existence of a probable link between variant Creutzfeldt-Jakob disease and bovine spongiform encephalopathy; in other words, people might contract the disease by eating contaminated meat. Sales of beef plummeted. The European Union blocked imports from Britain, not only of beef but also of many beef by-products, including lipstick and cough medicine.

'When mighty Roast Beef was the Englishman's food, it ennobled our brains and enriched our blood.' Henry Fielding's lyrics lost their jingoistic meaning and became tragically ironic when microbiologist Stephen Dealler, who had been investigating mad cow disease in his laboratory, found that an injection of a small amount of blood might be sufficient to communicate prions. He went on to speculate that one pint of blood might contain about one thousand times the necessary levels, and warned, 'This makes it [blood transfusion] the second most infective way of spreading the disease after direct injection into the brain.'[25] The National Blood Service forbade extraction of blood from the veins of anyone whose kinfolk had had Creutzfeldt-Jakob disease or who had been a recipient of hGH.

The Chief Medical Officer's Committee on Microbiological Safety of Blood and Tissues for Transplantation included the same advice in guidance it published in 1996. It also reiterated the National Blood Service's current medical criteria of bioavailability intended to reduce the risk of contamination by viruses and retroviruses, which advised testing sources, taking a lifestyle and medical history, and following the principles of Good Manufacturing Practice.[26] The Food and Drug Administration's Interim Rule had a similar objective, but whereas failure to comply was a criminal offence, the Chief Medical Officer's guidance was without teeth. The NHS and Community Care Act 1990 had lifted Crown immunity, which is why the guidance included a warning of the remote possibility of negligence falling under product liability legislation. It had been 'loosely based on a draft paper produced by an informal working party', and reflected government inertia in the face of the Council of Europe's 1994 stipulation that member states, which include the United Kingdom, ensure sources of body/cadaver stuff repurposed in medicine are tested for transmissible disease 'in compliance with the law and practice of the country concerned'. Additionally, it had recommended official licensing of entities along the length of the supply chain.[27]

People with bleeding disorders such as haemophilia, who were suffering the consequences of government delay in acting to minimize the risk of blood products being contaminated by viruses and retroviruses, began demanding protection from prions.[28] In March 1998, Tony Blair's government announced that from 1999, the National Blood Service would manufacture blood products out of plasma imported from countries that, apparently, were free of variant Creutzfeldt-Jakob disease. This was an expensive measure: each year its laboratory was processing 600 tons of plasma, then worth £57 million ($91 million). It had been recommended by the Committee of Safety of Medicines, which the previous November had had to recall batches of blood products when three sources succumbed to variant Creutzfeldt-Jakob disease.[29] Four months later, another expensive precautionary measure was announced: from 1999 onwards, blood collected for transfusion would undergo leucodepletion, or leucoreduction, where a majority of white blood cells is removed (white blood cells are where prions responsible for variant Creutzfeldt-Jakob disease are suspected to lurk).

In August 1999, the Food and Drug Administration in the United States banned the extraction of blood from the forearm of anyone who had spent more than six months in the United Kingdom between 1980 and 1996. No victims of variant Creutzfeldt-Jakob disease had been identified in the United States (the first was in December 2003). Its absence was attributed to the availability of safer and cheaper sources of protein, such as soybean and cottonseed, with which to enrich animal feed. But it is also the case that the huge American meatpacking industry rejects thousands of animals and produces vast quantities of abattoir waste, and that much of it is repurposed in animal feed, particularly for beef cattle reared in commercial feedlots.[30] In December 2003, bovine spongiform encephalopathy was diagnosed in a Canadian-born dairy cow, and the US Department of Agriculture introduced a programme of random testing that found only two infected animals out of 759,000 in eighteen months. Subsequently, it was decided to radically reduce the programme, a measure defended by the National Cattlemen's Beef Association and opposed by consumer organizations.[31]

In December 2003, the press announced the death of the first person to contract post-transfusion variant Creutzfeldt-Jakob disease.[32] Both source and recipient were British. When the blood was extracted in 1996 the source had shown no symptoms of the disease, but had died of it three years later. Prions have succeeded in overturning the long-held belief, originating with Titmuss, that British blood is relatively safe because it is freely given by donors-on the-hoof, motivated by the redistributive ethos of the post-war welfare state. Infectious pathogens do not respect sources' motivations.

CHAPTER 32

Compassion and Commerce

'Burn victims lie waiting in hospitals as nurses scour the country for skin to cover their wounds, even though skin is in plentiful supply for plastic surgeons,' claimed a team of investigative journalists in a series of articles the *Orange County Register* published in April 2000.[1] 'The skin they need to save their lives is being used instead for procedures that could wait: supporting bladders, erasing laugh lines and enlarging penises.' Surveys subsequently conducted by the American Burn Association and the American Association of Tissue Banks confirmed that occasionally shortages occurred, and that treatment was sometimes delayed, or wounds dressed with sub-optimal amounts.[2] Yet, the journalists had heard that in hospitals where a burns centre or unit was facing a shortage, in the mortuary external entities might be extracting cadaver skin for delivery to a processor.

A majority of first-generation skin banks had lost the battle for survival. The American Burn Association listed forty in its 1986 *Directory of Burn Resources*, which was published to help surgeons experiencing a shortage of cadaver skin to identify potential sources. The edition published in 1999 had twenty-four.[3] *Orange County Register* journalists identified eleven. Extracting cadaver skin is a relatively lengthy procedure, and dispatching a team of extractors, especially to a distant collecting site, for a 'skin only' source is expensive. Moreover, compliance with the Interim Rule had increased costs, which reimbursement tariffs inadequately recognized.[4]

Exceptional circumstances were responsible for skin bank survival. For instance, in 1999, the UTMB, which had been processing around half of the cadaver skin extracted in the United States, closed.[5] The Shriners bought its equipment and established a skin bank at their nearby Galveston Children's Hospital, which it had been serving. A majority of the children admitted to its burns unit are flown in from Mexico, where they have been burned or scalded at home by unsafe domestic heating or cooking appliances, or their parents' hazardous work such as manufacturing fireworks.[6] Children are treated for free – the hospital does not have a billing department. In 2008 the running costs of the skin bank, which was processing and freezing skin extracted from around 500 corpses, were $1 million. First responders continued to support the New York Firefighters Skin Bank, raising funds to pay for new equipment such as a dermatome and a freezer. But its continued existence is largely due to the fact that the New York Presbyterian Hospital views the burns centre to which it is attached, the busiest in the country, as a feather in its cap, and an essential community service in a crowded city prone to fire disasters, a policy no doubt confirmed by the terrorist destruction of the twin towers of the World Trade Center on 11 September 2001.[7]

None of the several branded biosynthetic composites of carcass or cadaver skin and synthetic and wholly synthetic burn wound dressing rivalled the effectiveness of split-skin cadaver skin. Moreover, they were more expensive, and it was difficult to convince regulators and third-party payers to reimburse their cost. Burn care does not admit the collection of systematic data they required. Every burn is different, which makes organizing a randomized controlled trial of comparable groups of patients a challenge. Moreover, sceptics doubt the evidence of American experts, who may have a financial interest in products, and are less likely to report negative findings.[8]

Cadaver Skin Shifts Business Paradigm

> For the amount of skin that I could use to help half a burn victim, I can now help people with bladder incontinence or reconstruct eyelids for 200 people. . . . If I crash my Harley, and I know that I can help 200 people with my skin, that's a pretty good deal in my book.[9]

This is LifeCell's vice president John Harper defending the processor's business strategy in the *Orange County Register*.

LifeCell's paradigm-shifting acellular technology had created far greater and potentially more profitable opportunities than those provided by the burns care market. Its product range includes Repliform®, which is used in surgery to treat urinary incontinence, which around 75,000 Americans – mostly women – undergo each year, and which typically is listed in third-party payers' reimbursement schedules. Liquid Cymetra® is injectable acellular cadaver skin that can strengthen weakened vocal chords; it is also a dermal filler, a 'cosmeceutical' which competes against products such as Botox® in the 'lunch-hour facelift' market where each year several million people pay out of their own pocket to have the contours of a youthful face restored.

From a strictly utilitarian viewpoint, sacrificing one person's life for the sake of the well-being of 200 people might make sense. But repurposing cadaver skin to enlarge penises or in 'cosmetic surgery with a syringe' is a mockery of the feeling rules promulgated in the marketing of corpse philanthropy, under its 'Donate Life' logo. 'We didn't think of these applications,' Harper protested. 'Surgeons came to us and told us how they had discovered different uses. It's not within our power to tell the doctor what to use it for.'[10] Cosmetic enhancement has been described as the last bastion of free enterprise medicine; it is a free-for-all, mostly carried out in private office operating suites, sometimes performed by individuals denied access to privileges to perform the same procedures in a peer-review setting, hospital or ambulatory centre.[11]

The United Network for Organ Sharing won the contract to run the Organ Procurement and Transplantation Network, which had been created under the National Organ Transplant Act to set standards and rules regarding the distribution of cadaver kidneys. But no one has authority to direct how cadaver stuff is repurposed. Moreover, there is no reason to expect LifeCell's business practices to be organized around the public good; it is a company with shares publicly traded on the NASDAQ stock exchange that must comply with corporate law, and is under shareholder pressure to increase its market position and maximize profits and dividends.

The Meaning of Profits

'I thought I was donating to a non-profit. I didn't know I was lining someone's pocket,' complained a woman who two years previously had agreed to the extraction of stuff out of her brother's corpse. *Orange County Register* journalists had shared her belief that non-profit entities, the public face of the cadaver stuff industry, are guided by an ethos wholly unlike that of entities with stock trading on Wall Street, but they had been disabused following interviews with hundreds of people and reviews of thousands of pages of documents: some are subsidiaries of for-profit entities; some have mutually beneficial agreements with one or several for-profit processors; some remunerate chief executives with generous salaries. Indeed, the unique nature of their raw material – corpses – did not immunize entities against the seismic structural changes that the American non-profit sector had begun experiencing in the 1970s, and which by the turn of the millennium had blurred any difference in ethos and business practices between non-profit and for-profit entities operating within the cadaver stuff industry, except for how and where tax returns are filed.

'Why is there a change in price between Boston and Nashville and Des Moines and Los Angeles?' asked Richard Kagan, director of the burns centre at the University of Cincinnati, staff surgeon at the Cincinnati Shriners Hospital for Children, medical director of the Ohio Valley Tissue and Skin Center, Cincinnati, and president of the American Association of Tissue Banks. 'The service charge should reflect the service, not the market.'[12] In 1984, when the National Organ Transplant Act had been adopted, the pressing issue was whether or not sources of body/cadaver stuff, or their next of kin, should be remunerated in cash; the financial status of intermediaries, which at the time were all non-profit entities, was deemed irrelevant save for the obvious necessity of allowing them to cover operating costs in order to stay afloat. As protection against accusations that reimbursement of costs represents valuable consideration, and hence cadaver stuff is being bought and sold, entities had claimed a place within the service industry category, and had billed recipients accordingly. But processors employing professionally trained sales representatives or agents to promote their branded products, which are set out in glossy brochures, now dominate the industry, and some had adopted controversial market-driven practices such as offering lower prices for larger orders and to secure customer loyalty, which create wide discrepancies in charges for equivalent cadaver products.[13]

In September 2000, medical ethicist Tristram Engelhardt, of Baylor College of Medicine, Houston, advised those present at the annual meeting of the American Association of Tissue Banks against shying away from discussing the profit motive; it is a means of spurring medical advancement.

> If you simply say, 'Hey, do you know we're going to make money off of your tissue,' that could be deceptive. You say, 'We're going to use it as all other institutions do – that is, there will be increased wealth from this, which will pay salaries of people involved and in some cases even stockholders.'[14]

'No one objects to the fact that the surgeon who implants a tissue should get a salary, and then similarly, you could take that concept and just work it back through all the other parts of the process,' explained George G. Grob, deputy inspector general for evaluation and inspections in the United States Department of Health and Human Services, in May 2001 in front of a hearing of the Senate's Committee on Governmental Affairs, which was considering whether legislation was required to address the issues set out in the *Orange County Register*, and which five weeks later had been reiterated and elaborated in an exposé published in the *Chicago Tribune*.[15] Grob had led an inquiry commissioned by the department's secretary Donna Shalala to consider whether the reputation of Medicare was being besmirched by the fifty-eight out of fifty-nine organ procurement organizations that had some sort of involvement in other cadaver stuff.[16]

America's anti-trust laws forbid any distinction between non-profit and for-profit entities. Moreover, investigations had found that members of the public either accept that a profit was being made by intermediaries between source and recipient, or would prefer not to be told that this was the case. But excessive profiteering troubled them. However, 'there was no way any of us could get a handle on exactly what is excessive,' explained Grob.[17] 'One could look at the profits a company makes, but big companies earn more profits than smaller ones do, for example, and it is very difficult to second-guess the cost that a company incurs because of all the overhead that goes into the company's operation.' New processes, which are continually introduced, make calculations out of date.

First Person Consent

Informed Consent in Tissue Donation: Expectations and Realities, the report of the inquiry led by Grob, advised periodic public disclosure about entities' finances.[18] But it focused on the 'donation conversation' conducted over the telephone between so-called requesters and kinfolk of someone who had just died, in which, it found, inadequate or inappropriate information about how stuff is extracted and repurposed was provided.[19] The problem was 'requesters' were often poorly trained, which was not surprising: entities typically were relying on 'external requestors' such as a hospital nurse, chaplain or social worker, or someone employed by a telephone triage agency.[20]

The Condition of Participation Rule, which required hospitals to notify an organ procurement organization of actual or imminent deaths in order to be eligible under Medicare, had provided no guidance on what information should be provided or how. The American Association of Tissue Banks, the Association of Organ Procurement Organizations and the Eye Bank

Association of America collaborated on the production of guidelines requiring that requestors provide kinfolk with the following information: the names of whatever is being sought; an explanation of how it would be repurposed; a general description of method of extraction; an explanation of how medical bioavailability is established.[21]

Paradoxically, at the same time that the principle of respecting people's private interest in a kinfolk's corpse was being acknowledged, it was being rejected in another measure that was being adopted in order to increase the supply of cadaver kidneys.

In 1994, Pennsylvania Act 104 introduced 'first-person consent' or 'donor designation', which confirmed the pledge as a legally binding document, a measure that effectively avoids investing what would be a controversial property right in the corpse. It had been advanced in the Uniform Anatomical Gift Acts of 1968 and 1987 but where kinfolk objected, explanters had rarely used it for fear that bad publicity and reputational damage might ensue. However, the Pennsylvania Act legally enforced honouring of pledges. It also included a provision requiring the Pennsylvania Division of Motor Vehicles to set up a register of details of pledges that drivers had recorded on their licence, and allow round-the-clock access to local organ procurement organizations.

Implementing the measure had been difficult. 'The feeling among some of our staff was that we simply couldn't go against a family's wishes, even if the person had indicated the wish to be a donor on his or her driver's license,' recollected Brian Broznick, executive director of CORE, the organ procurement organization serving the western part of Pennsylvania and much of West Virginia.[22] But the measure was declared a success: by the end of 2001, twelve other states had adopted a version.[23]

In an effort to encourage nationwide adoption of first-person consent, and iron out inconsistencies in state laws, the Association of Organ Procurement Organizations had asked the National Conference of Commissioners on Uniform State Laws to revise the 1987 version of the Uniform Anatomical Gift Act.

The Uniform Anatomical Gift Act 2006 endorsed first-person consent. Section 8 states, 'An anatomical gift that is not revoked by the donor before death is irrevocable and does not require the consent or concurrence of any person after the donor's death.' Section 11 provides immunity against civil actions brought by kinfolk and in criminal prosecutions where the extractor had acted in good faith in the face of family objection. Yet, although kinfolk are unable to revoke a pledge, where none is found, the act allows them to decide; their private interest in a corpse is reinstated.

The commissioners withheld their predecessor's support for legislative consent. The Eye Bank Association of America did not object because the industry had accepted the necessity to contact kinfolk in order to review potential sources' lifestyle and medical history.[24]

The model act was controversial. It required critical care staff to sustain the life of dying pledgers while the local organ procurement organization evaluate their bioavailability, effectively giving its staff the authority to decide when a ventilator could be turned off. Critical care staff protested that this was unethical. 'If you promote organ donation too much, people lose sight that it's a dying patient there,' a physician is quoted as saying in the *Washington Post*. 'It's not just a source of organs. It's a person.'[25] Moreover, it conflicted with the Patient Self-Determination Act 1991, which gave living persons the legal right to accept or refuse medical treatment, including procedures that might be undertaken when they are dying. The statute encourages people to set out their wishes in relation to how they die in an advance directive, such as living will and durable power of attorney.

Confronted by heated protest, in 2007 the model act was amended to allow medical staff to follow a dying person's wishes, even if doing so resulted in the loss of potentially transplantable organs. However, despite the revision, the category error remained on statute books. Irrespective of their legal status, pledges are admittance cards to an imagined posthumous existence, whereas advance directives are instructions about end of life care of living people.[26] Put another way, autonomy, the right to decide what happens to one's body, is possessed by competent living people, and cannot be claimed by dead people, who are without agency.

Meanwhile, requestors were being trained to follow a 'donation conversation' script that takes agreement to extraction as the default position. In what is called the 'standard approach', requesters presuppose kinfolk might object, and that their task is to provide information that might make them change their mind. The script of the donation conversation went something along these lines: 'Some families choose the option of donating their loved one's organs. I am here to help you make the decision that is best for you and your family.' In the new 'presumptive approach', requestors start out by presuming that, if given the chance, everyone will agree to save a life. 'I'm here to provide you with the opportunity to donate your loved ones organs' is how they introduce themselves.[27]

The new script had been written with cadaver organs in mind. But it was inevitable that the presumptive approach would be adapted in requests of other cadaver stuff. However, despite the size of the 'gift' growing, and a huge expansion in cadaver artefact/product range and proprietary processing and commercialization, marketing of corpse philanthropy devised in the 1940s by Madison Avenue advertising agency staff for the Eye-Bank for Sight Restoration is little changed. 'True and wonderful stories' continue to rehearse the way that next of kin's gifts have transformed the lives of upstanding citizens. Moreover, there is little evidence of the message being heard. The majority of people know something about how cadaver organs are repurposed, but few know anything about why other cadaver stuff is sought, and requesters frequently meet with surprise.[28] Ignorance greatly decreases the chances of recently bereaved kinfolk responding positively to a telephone call asking them to 'help change at least fifty people's lives'.[29]

CHAPTER 33

A Roadmap for the Future

'I see this industry as operating like a bunch of Wild West gunslingers that are just shooting from the hip, doing it anyway they want to do it, with no laws or regulations they are just making it up as they go.'[1] This is Steve Lykins, the father of twenty-three-year old Brian Lykins, who died on 11 November 2001, four days after undergoing reconstructive knee surgery in a Minnesota hospital, testifying before the hearing on the dangers of tainted tissue held by the Senate Committee on Governmental Affairs in May 2003. Cause of death was *Clostridium sordellii*, a close relative of the bacteria that cause gas gangrene. The Centers for Disease Control and Prevention and the Minnesota State Department of Health traced the infection to a cadaver cartilage processed and sold by CryoLife. Nothing in the source's medical history indicated that he was not medically bioavailable. But he had committed suicide at an unknown time, and his corpse had been unrefrigerated for nineteen hours after discovery. The extractor's identity has not been revealed, but whoever it was had begun extractive work four and half hours later. *Clostridium sordellii* is generally confined to the intestines, but evidently, between death and disassembly, bacteria had had ample time to escape the bowel and invade hitherto inaccessible parts of the corpse. In other words, putrefaction had set in, and his corpse should have failed a poke and sniff test.

In August 2002, thirty days after Lykins's parents had filed an action for damages from CryoLife, the Food and Drug Administration shut down a substantial part of its operation, and ordered the company to recall all cadaver products processed since October 2001, and to withhold from the market and destroy everything in its warehouse. 'We found significant violations from our regulations,' explained Dr Mary Malarkey, director of the agency's Office of Compliance and Biologics Quality. Without a legal requirement to report incidents of iatrogenesis – adverse events – 'shoe-leather epidemiology' had identified twenty-seven recipients who had developed serious infections, thirteen of whom had contracted *Clostridium sordellii* from products sold by CryoLife. Brian Lykins was the only one who died.[2] Shares in CryoLife, then the United States' largest processor of cadaver stuff, supplying about 15 to 20 per cent of the market, fell by 42 per cent, and the New York Stock Exchange suspended trading.[3]

CryoLife specialized in cadaver fragments such as heart valves and cartilage that cannot withstand aggressive terminal sterilization without suffering structural damage. Lykins's family was one of those of several recipients harmed by a Cryolife cadaver artefact to seek damages under product liability laws. But the cases were heard in different states, and decided in different ways. In *Cryolife v. Superior Court of Santa Cruz County* 2003, the defence successfully argued that Cryolife was not a manufacturer but provided a service, and furthermore was protected by California's blood shield law.[4] In states without a blood shield law, the defence argued that the company had not been negligent because it had not contravened Food and Drug Administration's regulations. That was correct. The Interim Rule, which had become the Final Rule in January 1998, focused on medical criteria of bioavailability, but was silent on what constituted safe practices. Moreover, CryoLife was not a member of the American Association of Tissue Banks, and so could not be accused of failing to comply with its standards.

The *Proposed Approach*

Cryolife was not unique: the Food and Drug Administration had evidence of other entities placing tainted cadaver artefacts/products on the market. In February 1997, comments were invited on the *Proposed Approach to the Regulation of Cellular and Tissue-Based Products*, which David A. Kessler, the retiring commissioner of the Food and Drug Administration, described as 'a framework to replace a patchwork of regulations we've developed over the years'.[5] Whereas the Interim Rule had been introduced in 1993 to deal with pressing concerns about medical criteria of bioavailability of cadavers, the *Proposed Approach* was intended as 'a road map for the regulation not only of therapies that exist today, but therapies that will be invented in the future'.

The *Proposed Approach* perforce was politically circumspect; it sought to safeguard the public's health without courting criticism for burdening entities with stringent and inflexible rules, or raising the price of stuff/artefacts/products, or frustrating the ingenuity of biotechnology entrepreneurs who were being lauded for creating a fast-growing sector of the American economy. It was a compromise that acknowledged the ever-present risk of disease transmission, while at the same time conceding that the threat of harm is greater in some instances than others – for example, highly processed freeze-dried cadaver bone powder is much less likely to transmit an infection than fresh-frozen cadaver bone; that considerations of efficacy are sometimes as important as those of safety – for example, where cadaver heart valves are used in the reconstruction of life-threatening abnormalities; and that in the context of procedures that are not medically indicated, and which people pay for themselves, the agency must respect the high value American society places on individual freedom and allowing consumers to take an informed risk.

Whereas the Interim/Final Rule had been restricted to cadaver stuff/artefacts/products, the *Proposed Approach* applied to everything extracted out of both bodies and corpses for repurposing in medical therapies. Everything is 'human cells, tissues, and cellular and tissue-based products' (HCT/Ps), a new capacious category of stuff into which cadaver heart valves and cadaver dura mater could escape from the medical device category, and which has plenty of space for novelties that might be discovered by bioprospectors or innovators of techniques of processing, terminal sterilization and extending the shelf life, or shifting products higher up the value chain. The Food and Drug Administration was also following the lead of New York State, which had included human cells such as semen, oocytes and placental/umbilical cord blood stem cells when it adopted mandatory licensing and oversight following scandals such as one about a sperm bank run by two men who turned out to be the only sources. However, the category excludes cadaver organs, blood and blood products, which fall under different established regulatory regimes.

'I am incredulous that in the last two years, the Food and Drug Administration has made virtually no progress in strengthening its regulatory requirement for the industry,' said Senator Susan Collins (Republican) in May 2003.[6] In 2002, Senators Collins and Richard Durbin (Democrat) had introduced the Tissue Transplant Safety Act in order to spur the agency into action. Collins was speaking as chairwoman of the hearing before the Senate Committee on Governmental Affairs, prior to reintroducing the bill, renamed the Brian Lykins Human Tissue Safety Act, which required the agency to issue a final version of the *Proposed Approach* within ninety days.

The Food and Drug Administration had not been standing idle. In January 2001, it had required entities involved in extraction, processing, storage, labelling, packaging or distributing cadaver stuff, and/or investigating bioavailability to register and provide a list of their products. However, individuals such as a medical examiner and mortician moonlighting as extractor under contract or agreement with, or other type of arrangement for, a registered entity were not required to register. The agency placed responsibility for their conduct in collecting sites on registered entities on the grounds that this was an appropriate way of 'easing the regulatory burden on individuals while ensuring the protection of public health'.[7]

By 2004, a total of 1,302 entities had registered, including 134 eye banks (of which ninety-three had been accredited by the Eye Bank Association of America), and 166 'conventional tissue banks' involved somehow in cadaver stuff such as dura mater, heart valves, skin and bone (of which seventy-five had been accredited by the American Association of Tissue Banks). The largest group by far consisted of 510 entities handling oocytes, semen and embryos, reflecting the considerable growth of the American (in)fertility industry.[8]

The number of registered entities was far greater than had been anticipated, indicating that a large number had been operating under the regulatory radar. The task confronting the Food and Drug Administration was immense. It had repeatedly complained that the Interim Rule was an unfunded mandate, and was having to borrow resources from other programmes such as blood and plasma regulation, in order to fulfil its responsibilities.[9] In 2002, it reckoned it required an additional $4.35 million to implement the *Proposed Approach*, and had requested a 10 per cent increase in its budget.[10] However, politicians repeatedly fettered its authority and denied it resources sufficient to carry out its work.

'Donor Deferral Criteria'

In May 2004, the agency invited comments on *Eligibility Determination for Donors of Human Cells, Tissues, and Cellular and Tissue-Based Products*, its proposed rules on criteria of medical bioavailability and their investigation. Anyone infected with a 'relevant communicable disease' must be 'deferred' (meaning rejected). 'Relevant communicable disease' is an elastic category, frequently discussed with a reference to Rumsfeld's aphorism about known known, known unknowns and unknown unknowns. The category soon welcomed new entrants, such as rabies, West Nile virus, and severe acute respiratory syndrome (SARS), foreign known knowns that without permission had entered American territory.

Pathogens responsible for 'relevant communicable disease' were not the only reason why the potential pool of bioavailable corpses was shrinking. The rule that corpses of anyone who had been incarcerated in jail or mental institution for more than seventy-two hours in the year prior to dying should be excluded had a substantial effect in the United States, which leads the world in producing prisoners (it has less then 5 per cent of the world's population but almost a quarter of the world's prisoners and at any one time has around 2.3 million people behind bars).[11]

The Food and Drug Administration's *Proposed Approach* insisted on a 'medical historian' searching for clues of what industry insiders call 'sleaze' in a 'documented dialog' with someone knowledgeable about a potential source's medical history and lifestyle. The United States is a mobile society, and sometimes no appropriate person who can answer the questions can be found, and the corpse has to be rejected. Where the source is an adult, kinfolk might be unaware of their lifestyle. Therefore, the first question a medical historian might ask is, 'Will you be able to give me detailed information about his or her life or is there someone else that I should also speak to?'[12] Several different people might be interviewed. A reassuring sign is when different sources provide the same answer. Unanswered questions or inconsistent information is taken as a warning not to disassemble a corpse.

The Eye Bank Association of America protested that this requirement would result in the loss of corneas excised under legislative consent, and would create a shortage. But the Food and Drug Administration was not persuaded: no one knew how many corneas were excised under these conditions, and anyway, at the time around 30 per cent of corneas each year were surplus to American requirement and being exported.[13]

Before extraction can proceed, the corpse must be examined for signs suggestive of 'sleaze', such as needle tracks between fingers and toes that indicate possible injectable drug use, and signs of diseases. Blood must be taken for a variety of specified tests, including nucleic acid-amplification tests (NAT), a new generation of objective tests capable of detecting the presence of virus DNA during the initial 'window period' of seronegativity, that is, before antibodies have developed. In 2002, the Food and Drug Administration had approved NAT tests for identifying HIV and hepatitis C (HCV) in corpses.

'Good Tissue Practices'

Scrupulous attention to medical bioavailability is insufficient; between source and recipient the bioburden of hundreds of viruses, bacteria, fungi and prions carried by cadaver stuff can proliferate. In November 2004, the Food and Drug Administration invited comments on 'Current Good Tissue Practice for Human Cell, Tissue and Cellular and Tissue-Based Product Establishments: Inspection and Enforcement', which sets out the measures entities must take in order to minimize the risk of contamination. These are analogous to the Hazard Analysis and Critical Control Point (HACCP) (pronounced 'hassip') system that originated in the late 1950s when the

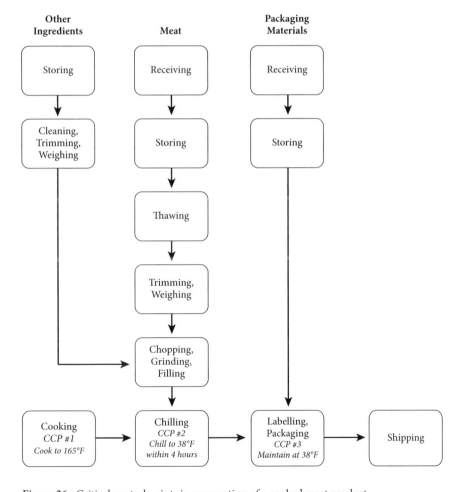

Figure 26: Critical control points in preparation of a cooked meat product

National Aeronautics and Space Administration (NASA) asked the Pillsbury Company to develop a safe food system for astronauts in outer space.[14] Their food had to be completely safe; the consequences of food poisoning in a zero gravity environment are too horrible to contemplate. Pillsbury identified each step – called a critical control point – in which microbial contamination might occur, and put in place scientifically validated methods of preventing contamination.

Good Tissue Practice requires that entities scrutinize and document every stage of their operation, from where and how cadaver stuff is received from an extractor, processed, sterilized, stored, packed and shipped, and institute scientifically validated procedures for minimizing contamination. Facilities must be clean and in a good state of repair. However, the industry relies on outsourcing; each stage might be undertaken by an independent entity with its own staff and premises. Hence, each cadaver fragment must be given a distinct identification code that allows tracking and traceability.

HACCP had demonstrably worked in outer space, but the Food and Drug Administration is without the resources to inspect, sample or analyse more than a fraction of the body/cadaver stuff for which it is responsible. Instead, it relies on the diligence of all entities concerned, which are required to inform it whenever something goes wrong. Failure to comply is a criminal offence, punishable by fines and imprisonment, closure of operations, seizure and destruction of stock, reputational damage, collapse of share price and actions for damages by injured recipients. In anticipation of critics claiming that the burden of complying with and monitoring of compliance of regulations was excessive the agency provided estimates of how much might be saved each year by the adoption of Good Tissue Practice: between $61,000 and $1.4 million was spent on replacing corneal grafts that had failed; $8 million on replacing bone grafts that had to be removed because of infection; $29.6 million on heart valve replacement that transmitted a fungal infection, which was also responsible for the death of 176 recipients.[15]

CHAPTER 34

Repairing the Past

As the new millennium approached, American entities were finding routes up the value chain, and the Food and Drug Administration was considering how risks to recipients' safety might be reduced. On the opposite side of the Atlantic, the modest ambition of British entities, particularly those sheltering under the National Blood Service's logo, was being thwarted by an admixture of prions, chaos created by successive governments' attempts to contain expenditure on the NHS, and shock waves from what became known as the retained organs scandal.

The scandal had no regard either for entities extracting cadaver stuff for repurposing in medical therapies or for the safety of recipients; it was fuelled by a succession of revelations about how, without the knowledge or agreement of kinfolk, cadaver stuff extracted during post-mortem examinations had been withheld for repurposing in medical education or research. It forced the public and policymakers to overcome death denial and investigate what happened behind closed mortuary doors. The trigger was a documentary televised in March 1996, which examined criticisms of the Bristol Royal Infirmary's paediatric cardiac service. Three months later, parents of children whose surgery at the hospital had resulted in either death or disability formed the Bristol Heart Children's Action Group to provide mutual support and call for an independent public inquiry. In June 1998, Secretary of State for Health Frank Dobson (Labour) announced that one would be held.

The Bristol Royal Infirmary Inquiry began hearing evidence in October 1998. In September 1999, Robert H. Anderson, professor of morphology at Great Ormond Street Hospital in London, explained how a dramatic improvement in the survival of babies born with severe heart abnormalities had been achieved through the careful scrutiny of children's hearts extracted during and withheld following post-mortem examination. He went on to single out as particularly valuable the collection at the Royal Liverpool Children's Hospital, known locally as Alder Hey Hospital, which dated from 1948.

Parents were outraged and dismayed to learn that their child's corpse had been without its heart when the mortuary had released it for burial or cremation. Professionals defended the practice with a paternalistic assertion that seeking permission would have exacerbated parents' grief, and a

Figure 27: The tenor of media coverage of the retained organs scandal

utilitarian argument that the ends – better outcomes for children requiring cardiac surgery – justified silence and half-truths. Subsequently, it transpired that retaining cadaver stuff extracted during post-mortem examination for repurposing in medical education and research was the national norm, and that all age groups were affected, and the Bristol Royal Infirmary Inquiry decided to extend its investigation into mortuary ethos and practices.

Press coverage of the Bristol Royal Infirmary Inquiry was muted in comparison to the sensational reporting of the discovery in the filthy basement of an annex of Alder Hey Hospital of fragments of the corpses of around 850 children and foetuses in around 2,000 plastic containers. Dutch pathologist Dick Van Velzen was responsible for this grotesque accumulation. In September 1988, he had taken up a joint appointment as hospital pathologist and professor of paediatric pathology at the University of Liverpool. Despite cutting open numerous corpses, he provided neither crucial final post-mortem reports that his colleagues required to give parents in order to explain why their child had died, and, in some instances, advise them how to prevent affected children being born, nor carried out research for which he had been awarded substantial funds. These amounted to £250,000, awarded by the Foundation for the Study of Infant Deaths for research into the causes of 'cot death', and £110,000 by the Wellcome Trust for research involving foetal and neonatal eyeballs. Instead, he had 'stockpiled' children's physical remains in chaotic, unsuitable and wretched conditions. Van Velzen was dismissed in December 1995, and his licence to practise medicine in the United Kingdom was withdrawn in June 2005.

In November 1999, PITY II (Parents who have Interred Their Young twice) was established, to support one another, demand Alder Hey Hospital return their child's cadaver stuff, and fight for compensation. In December 1999, an independent confidential inquiry into Van Velzen's bizarre misdeeds was appointed. Subsequently, more protest groups were formed, organized either around the retention of children's cadaver stuff at specific hospitals, or cadaver organ retention in general, the largest of which was the National Committee Relating to Organ Retention (NACOR).

This context enabled other scandals to gain purchase in the public sphere. In response to the growing furore, the British government set in motion seven investigations. Three, including the one undertaken by the Bristol Royal Infirmary Inquiry, considered how and why cadaver stuff extracted during post-mortem examinations routinely had been withheld without the knowledge or agreement of next of kin. In order to gauge the extent of the practice, the Chief Medical Officer of England ordered a census of size and composition of cadaver stuff held in 2000 in hospitals. In January 2001, he invited parents, professionals and policymakers to a national summit in London to make representations that would inform official guidance. Under the Scotland Act 1998, health care had been devolved on to the Scottish government, and in September 2000, the Scottish Executive established its own review.

The other investigations examined specific scandals. The Bristol Royal Infirmary Inquiry went into why children's hearts had been withheld, and the Royal Liverpool Children's Inquiry, which was instituted in December 1999, analysed the circumstances that allowed Van Velzen to behave so bizarrely. In February 2000, an investigation was ordered into amputation and retention of the hands of twenty-five out of fifty-one people who in August 1989 had drowned in the River Thames, when a dredger cut through the side of the pleasure boat the *Marchioness* on which they were partying. Their corpses had been taken to Westminster Public Mortuary, and were in the lawful possession of the coroner, whose responsibilities included identification. The fingers had been amputated for fingerprinting; relatives had questioned why this had been necessary, and why other non-invasive methods of identification such as dental records had not been relied on.

In July 2001, the Chief Medical Officer of England asked Her Majesty's Inspector of Anatomy to investigate why, in the face of his widow's expressed objection, in February 1987, the brain of Cyril Isaacs had been retained following a post-mortem examination that a coroner had ordered. In September 2001, the Scottish Executive instructed the committee reviewing the retention of cadaver stuff in Scottish hospitals to investigate the 'bone-snatching' scandal, which had been provoked by the revelation in the *Sunday Herald* that during the 1960s, thighbones extracted from the corpses of more than 2,100 Scottish children had been used in a research project into the danger of radiation following nuclear tests in which the Medical Research Council and the United Kingdom Atomic Energy Authority had collaborated.

Between May 2000 and May 2003, eight weighty reports were published.[1] Not surprisingly, a consensus on the Human Tissue Act 1961 had been reached. It had been tilted towards marginalizing next of kin's private interest in a corpse and supporting that of extractors. Where children were concerned, it had allowed a flagrant disregard for the parental sense of duty of care that is in force whether their child is living or dead. As one aggrieved parent put it,

When a child died that child is still the parent's child – not a specimen, not a cause, not an unfortunate casualty of a failed procedure, but someone's baby, someone's child. In life the parent is responsible for every aspect of a child's well being. In death that responsibility should not be taken away.[2]

Yet many parents who gave evidence to the various inquiries said that had a request for retention been sensitively and honestly made they would have agreed, because knowing stuff extracted out of their child's corpse was put to good use might have eased some of their grief.

More generally, the reports agreed that the statute was out of date. Its defence of utilitarianism was contrary to Article 2 of the Council of Europe Convention on Human Rights and Biomedicine, agreed in Oviedo, Spain in 1997 by, among others, the British government, and which insists that 'the interest and welfare of the human being shall prevail over the sole interest of society or science'. Moreover, absence of evidence of refusal, rather than evidence of agreement, as the statute's test of authority to extract cadaver stuff was out of step with the current expectations of openness and sensitivity in medical encounters, and the principles of the Human Rights Act 1998, which had come into force on 2 October 2000.

Without a legal stick with which to punish failure to make the reasonable enquiries the statute required had sustained medical professionals' sense of entitlement to cadaver stuff. The *Isaacs Report* was the only one to mention carrots: it drew attention to the mortuary culture of tips, of how an interesting brain might fetch £10 and how the various circulars issued by the Department of Health and Social Security extirpation of cadaver pituitary glands had made routine the practice of opening up the skull. But the collection's remarkable size was still unrecognized. According to the Chief Medical Officer's Census, published in January 2001, between 1970 and 1999 one in fifty of the three million or so British corpses that had undergone a post-mortem examination was placed in the grave or crematory incomplete; at the cadaver pituitary gland collection's zenith it was one in eight.[3]

Repairing the Past

The various inquiries agreed that imprecise terminology had helped keep people in the dark about what exactly is extracted out of corpses during a post-mortem examination. Commonly used terms, such as 'organ' and 'tissue', hold no precise meaning for a doctor let alone the layperson. The former sounds meaty, the latter insubstantial, although in practice the opposite might obtain, for instance, 'brain tissue' might mean whole brain. Ambiguity creates anxiety about what makes a corpse incomplete, and whether it matters. Some parents were demanding the return of everything that had been extracted from their child's corpse, including the tiny amount held on glass slides prepared to allow a histopathologist to scrutinize cells under a microscope. Their requests presented hospitals with a daunting and time-consuming task, which frequently proved impossible to complete because cadaver stuff had been scattered around premises, or was without any indication of its source, or had been mislabelled or disposed of as clinical waste. Some parents ceremonially buried or cremated cadaver stuff that had been returned to them, and on a few occasions, where stuff was returned piecemeal, held several funerals. This behaviour was ridiculed behind the scenes, but it registered an objection to children's cadaver stuff being regarded as biotrash, thrown into a yellow waste bag and burnt in an incinerator together with soiled dressings.

The Chief Medical Officer's Census was acknowledged to underestimate the national inventory of cadaver stuff. Nonetheless, English hospitals and universities reported finding on their premises 54,000 cadaver fragments, corpses of stillborn babies and foetuses. Brains accounted for nearly half; hearts a sixth. Of the sources, 17,800 had been adult; 9,800 had been children aged between one and fifteen years.

In April 2001, the Retained Organs Commission was established to oversee the process of returning claimed and decide the fate of unidentifiable and unclaimed cadaver stuff, and restore public trust. Hospitals had their work cut out: they were fielding tens of thousands of enquiries. The commission standardized procedures, provided a Freephone helpline, undertook casework, held public meetings, met families and representatives of the various support groups and held workshops with professionals and hospital managers.

Organ Donor Register

'Anguish of mother after child's death sparks organ donor storm', is the strapline under the *Daily Express* front-page article reporting on Van Velzen's misdeeds. The various scandals had a detrimental impact on people's corpse philanthropy: in 1992, organs had been explanted out of the corpses of 876 people; in 2001, that number had fallen to 777. Moreover, the proportion of older sources grew: in 1992, three out of ten were aged over forty-nine; in 2001, it was four out of ten. Eye banking also suffered. Indeed, in 2001, Frank Larkin, ophthalmic surgeon at Moorfields Hospital, warned that if its inventory continued to shrink, patients who hitherto could be treated on a first come, first served basis, might have to spend time on a waiting list.⁴

Marketing of corpse philanthropy was intensified. In October 1994, the UK Transplant Support Service Authority (UKTSSA), the new name (briefly held) of UK Transplant, celebrated the launch of the Organ Donor Register, a computerized record of pledges, by releasing 78,000 balloons, each one symbolizing either source or recipient of a cadaver organ since 1972. The decision to create the database had been taken following a five-year campaign by the parents and sister of a young man whose pledge had not been realized following his death from a brain tumour. Advertisements were placed in national and local newspapers, and application forms delivered to nineteen million households.⁵

UKTSSA was the principal port of entry into the computer. The Driver Vehicle Licensing Agency (DVLA) became a 'feed organization' and began sending details of pledgers to UKTSSA. In March 1993, it had incorporated the text of the pledge, which hitherto had been on a separate card, into the full driving licence. Yet tethering the pledge to a driving licence was recognized as an ineffective method of narrowing the ever-widening gap between supply and demand of cadaver kidneys: between 1970 and 1990 the number of people suffering a fatal brain injury in a road traffic accident had fallen by 70 per cent, because since 1973, British motorcyclists have had to wear a crash helmet, and since 1983, British car drivers and front seat passengers have had to wear a seat belt, and hence in a collision are unlikely to be thrown through the windscreen.⁶ Other organizations volunteered to feed the register: the Family Health Services Authority, which registers patients with an NHS general practitioner; Boots, a nationwide chain of pharmacies, which added a pledge to the application form of its customer loyalty scheme; the Passport Office and NHS Business Services, which issues the European Health Insurance Card.

Pledgers were invited either to tick the box beside 'any part of my body be used for the treatment of others', or limit extractors by ticking boxes next to kidneys, corneas, heart, lungs, liver or pancreas. In 1998, MP Bob Blizzard (Labour) requested the addition of a box for skin.⁷ He was speaking on behalf of the Stephen Kirby Skin Bank, which had been advising would-be skin pledgers to write SKIN on their pledge card, but the computerized register was ignoring it. The Department of Health refused; research had found that many people believe skin extraction mutilates the corpse, and it was reluctant to add fuel to the retained organs scandal.

CHAPTER 35

Globalizing the Gift

On 25 January 2005, Joseph Pace, a fifty-four-year old unemployed widower living alone in the Kensington neighbourhood of Philadelphia, died in Northeastern Hospital. His step-daughter paid Liberty Crematory $3,160 for a simple cremation. But before carrying out her instructions, its owner Louis Garzone contacted Michael Mastromarino, chief executive officer, and executive director of operations at Biomedical Tissue Services, Ltd., Fort Lee, New Jersey. Mastromarino had been paying Garzone $1,000 each time he allowed him and his team of extractors to disassemble on his premises a corpse destined for disposal in a crematorium.

Pace's corpse was one of 1,077 Biomedical Tissue Services disassembled without either evidence of a pledge or knowledge and agreement of kinfolk, and with an egregious disregard for recipients' safety. In December 2005, revelations of Mastromarino and his team's crimes made front-page news on both sides of the Atlantic when it emerged that one of the corpses was that of Alistair Cooke, a New York-based British-born journalist and broadcaster, famous in the United States as host of Public Broadcasting Service's *Masterpiece Theater*, and in the United Kingdom for *Letter from America*, a weekly radio essay that the BBC had broadcast for fifty-eight years.

A New York police detective, investigating an allegation that the previous owner of the Daniel George Funeral Parlor in Brooklyn had defrauded a client, got wind of the scam when the new owner mentioned a 'bone program'. The interest of the Brooklyn district attorney's office was piqued, and its rackets division began investigating. It transpired that New York Mortuary Service, a cut-price funeral home in Spanish Harlem, which Cooke's daughter Susan Cooke Kittredge had picked out of the Yellow Pages to handle her father's cremation, was one of seven in New York, New Jersey and Pennsylvania that as a sideline to the disassembly racket had been billing two separate parties for their funerals: state welfare departments awarding 'funeral' grants' in cases of need; and kinfolk.

Biomedical Tissue Services masqueraded as a legitimate independent extractor of cadaver stuff. It had been registered with the Food and Drug Administration. But 'it was harder to sell a hot dog on the street than it was to recover transplant tissue' is how a freelance extractor described the undemanding registration process.[1] An online form asked for names of entity and director, and which cadaver fragments would be extracted. Neither criminal record nor qualifications was sought, which is why the Food and Drug Administration was unaware that Mastromarino's licence to practise dentistry in New York had been revoked following prosecution for drug-related offences.

The paperwork accompanying shipments of cadaver stuff was ostensibly valid, but the data fiction. For instance, documents accompanying shipments of Pace's cadaver stuff stated he had died following a heart attack, whereas his official death certificate stated that he had died of cancer, and that HIV and hepatitis C were contributory factors. Another corpse was that of James Herlihy, a former Philadelphia Naval Yard worker who had been incarcerated for six months when he was transferred to a hospital just before he died – incarceration is grounds for rejection. Several sources had had a drug-related cause of death entered on their official death certificate. It was also claimed that 'Dr Hixson' had answered questions about Pace's medical and social history, but he was fictitious. Pace's corpse had remained unrefrigerated for fifty-two hours before disassembly, and should have failed a poke and sniff test. RTI Donor Services was responsible for organizing the screening blood tests, but the blood sent was not Pace's but that of a healthy individual.[2]

Biomedical Tissue Services endangered public health. In January 2006, the Food and Drug Administration ordered Mastromarino to cease operating, and withhold all the cadaver stuff in his entity's stockroom. It also ordered his customers to recall or withhold from distribution every artefact and product that had been made out of cadaver stuff they had bought from him, and to inform and offer to test recipients.

In March 2008, Mastromarino pleaded guilty to bodysnatching, reckless endangerment and enterprise corruption – an admission to being a mobster. He and his ex-wife agreed to forfeit $4.6 million of his ill-gotten gains to be shared by kinfolk of sources. Whether or not – and how much – cash has been distributed is unrecorded. In July 2013, Mastromarino died in prison from bone cancer, having served five years of a sentence of up to fifty-eight years. Several kinfolk of his sources joined together in a claim for damages for emotional injury resulting from loss of sepulchre, a common law cause of action that had begun gaining currency in certain jurisdictions, including New York, as an alternative to the quasi-property rule.[3] The right of sepulchre evokes the ancient Christian belief that proper and timely burial in consecrated ground is essential to resurrection. But its contemporary feeling rule is that proper dignified disposal of a corpse is a source of solace for the bereaved. The action failed. Indeed, in their defence, processors, who are legally responsible for the safety of cadaver artefacts/products they distribute, numbered themselves among Mastromarino's victims.

Journalists used graphic language to convey what disassembly entails. The images conjured up were somehow more shocking where a corpse had a recognizable face. But in being socially alive despite being physically dead, Cooke was unlike the majority of the millions of nameless ghosts haunting these pages. Moreover, Mastromarino's methods were unexceptional. Wherever possible, extractors seek to 'maximize the gift', a slogan sometimes rehearsed in the marketing of corpse philanthropy where it is meant to be interpreted as honouring the dead person's pledge, but which might also be understood as a way of ensuring a financial surplus remains once costs are covered.[4]

Raising the Flag in American Collecting Sites

'He who has the most bodies wins', is how an extractor described the industry's ethos in the new millennium.[5] In 1998, Medicare had stacked the cards in the favour of organ procurement organizations by imposing the Condition of Participation Rule requiring that hospitals notify one of their number whenever a patient died, or was about to die. The rule helped Musculoskeletal Transplant Foundation hoist its flag in hospital mortuaries. 'We are an OPO-oriented tissue bank with 95 per cent of our donors resulting from referrals from OPOs and some 70 per cent of our donors actually recovered by OPO personnel', is how in 1993 Bruce W. Stroever, its president, described the non-profit's business model.[6] In 1995, the Musculoskeletal Transplant Foundation had secured its financial independence from Osteotech, its parent, which had shares trading on a stock exchange, but continued to supply raw material under contract. It also diversified into processing.

In 2004, Musculoskeletal Transplant Foundation paid an undisclosed sum for American Red Cross Tissue Services, which despite an annual turnover of $40 million was losing money.[7] The acquisition included processing facilities that extended its product range to include cadaver skin and heart valves, and collecting sites in small hospitals and funeral homes in states where the Musculoskeletal Transplant Foundation had registered scarcely a presence. More than 2,200 corpses were added to its annual total of around 5,000, giving it a significant share of the extractive side of the industry. In the previous year, around 70,000 bioavailable corpses had been identified in the United States: 6,057 had their organs explanted;[8] around 40,000 had their cadaver eyes/corneas extracted;[9] 23,925 were disassembled.[10] Disassembled corpses might have provided organs and eyes and hence have been counted twice, or even three times. But there was also under-counting: statistics on disassembled corpses provided by the American Association of Tissue Banks fall short of the actual total because only one in four organ procurement organizations was a member.

In announcing the acquisition of American Red Cross Tissue Services the foundation forecast sales of its cadaver artefacts/products in the United States and Europe would reach $300 million the following year. Turnover would have been significantly larger if its 2005 bid to acquire Osteotech had succeeded.[11] However, this setback was only a hiccup in expansion, and it continued to provision the processor. That same year, it affiliated LifePoint, Inc., an organ procurement organization disassembling corpses in sixty-two South Carolina hospitals. The deal is LifePoint sends cadaver stuff to Musculoskeletal Transplant Foundation for processing, and receives a proportion of the finished goods for retailing to its participating hospitals, allowing it to reassure people in its locality that stuff extracted out of their kinfolk's corpses might help neighbours and friends.[12] A similar contract was signed with LifeLine, an organ procurement organization operating in sixty-four hospitals in thirty-eight Ohio counties and in Wood and Hancock Counties in West Virginia,[13] and Iowa Donor Network, an organ procurement organization extracting cadaver stuff in 123 hospitals throughout the state.[14] In one year, Musculoskeletal Transplant Foundation had added 249 hospitals to its 'recovery network'. Although the total number is unknown, it is probably safe to hazard that a significant proportion of the 5,700 hospitals registered with the American Hospital Association have been drawn into its orbit.

Allosource and Community Tissue Services were also consolidating into non-profit high-volume entities, albeit somewhat smaller than Musculoskeletal Transplant Foundation but nonetheless capturing many productive collecting sites and a significant share of the American market for cadaver stuff/artefacts/products. These entities might be non-profit but their ethos is little different from that of commercial entities. However, whereas for-profits distribute some of their financial surplus to bankers, investors and shareholders, non-profits keep all of theirs, rewarding staff and investing in facilities such as up-to-date suites of clean rooms, and supporting research into new cadaver products and processes – patents expire and entities require capital to produce a constant stream of innovative products that will comply with Food and Drug Administration regulations.

Non-profit status does not prohibit 'strategic alliances' with for-profit entities in which they agree to perform a variety of services for one another. It is in the industry's food chain that a difference is found: non-profits are less likely to be eaten; entities that have been floated on a stock exchange are vulnerable to acquisition. In 2008, Kinetic Concepts, Inc. paid shareholders $1.7 billion for LifeCell, which subsequently operated under the Acelity umbrella. That same year, Regeneration Technologies merged with Tutogen, a New Jersey processor, which in 2000 had gained a listing on the American stock exchange, to form RTI Biologics. The merger consolidated a strategic alliance: Tutogen has a German subsidiary, which had been marketing both entities' products in Europe.[15] In 2010, international manufacturer and distributor of medical devices Medtronic Inc. acquired Osteotech by distributing $123 million to its shareholders.

Passporting

Mastromarino's racket was one of many relatively small independent extracting entities bioprospecting in the hinterland of collecting sites in which large entities had raised their flag. Regeneration Technologies, its biggest customer, paid $5,650 for Pace's bones, ligaments and tendons. LifeCell, second in line, paid $2,364 for his skin.[16] Tutogen Medical, Inc., Lost Mountain Tissue Bank, the Blood & Tissue Center of Central Texas, and SpinalGraft Technologies, Philadelphia, bought some small fry. The tariff is organized around Standard Acquisition Costs (SACs) that cover extractors' operating costs such as services of a telephone triage agency, designated requestor, historian and technicians who disassemble corpses, and also the cost of supplies such as sterile drapes, packaging (including storage solutions) and transport to and from the collecting site; they also allow extractors an extra amount that can either be held as a financial cushion or invested in equipment and facilities. SACs vary throughout the United States: in 2004, the SAC of cadaver skin ranged from $250 to £500 per square foot (930 sq. cm). Corpses provide on average 3.5 square feet (0.33 sq. m), that is, each one can provide skin worth from $875 to $1,750, which suggests Biomedical Tissue Services' prices were relatively high.[17]

In *Heads, Shoulders, Knees and Bones* (2011), Philip Guyett's self-published book, he describes how in 2001 he used his training and experience as an autopsy assistant to set up Donor Referral Services, an independent extracting entity disassembling corpses on mostly funeral parlour premises, and which was registered with the Food and Drug Administration. He had no difficulty in finding takers, which included what he called 'a trailer-park tissue bank'. But following the Mastromarino scandal, and armed with finalized rules, Food and Drug Administration's oversight intensified, and in 2006, the entity closed.[18]

Overseas bioprospecting also intensified, with American entities sometimes acquiring a foreign one from which to export to the United States cadaver stuff extracted locally, or to which American finished goods can be exported. Foreign extractors, with the exception of those in the United Kingdom, can register with the Food and Drug Administration, but it is without the resources required for diligence overseas. In 2012, an investigation undertaken by members of the International Consortium of Investigative Journalists revealed how staff at a Ukrainian public mortuary had disassembled corpses and sold various cadaver fragments to RTI Biologic's German subsidiary, which had exported them for processing in the United States.[19] Kinfolk's agreement had not been sought, although it is a requirement of Ukrainian law, which means that the sources' medical bioavailability had not been investigated.

The scandal provoked the Pentagon, a customer of RTI Biologics, to review the provenance of cadaver artefacts/products repurposed in American military hospitals.[20] But traceability is sometimes impossible; information is not necessarily available. Between source and recipient, cadaver stuff/artefacts/products may undertake lengthy and complex journeys, with stopovers in different countries and American states, taking in a variety of entities that can number between twelve and twenty different organizations, and may even reach into the hundreds where stuff extracted out of a single corpse is highly processed and sold in small packs to many customers.

Passports may also be incomprehensible. During the First Gulf War, thirty-four countries sent blood and blood products bearing bar codes that could be read only by the source nation's devices, and labels written in various languages.[21] The International Community for Commonality in Blood Banking Automation (ICCBBA), an international non-governmental organization associated with the World Health Organization, is encouraging governments to adopt ISBT 128, a standardized system of labels and bar codes, that originated with the sponsorship of the United States

Department of Defence, and the American Red Cross. Ironically, ISBT 128 makes international travel of body and cadaver stuff far easier than that of the people from whom it was sourced.

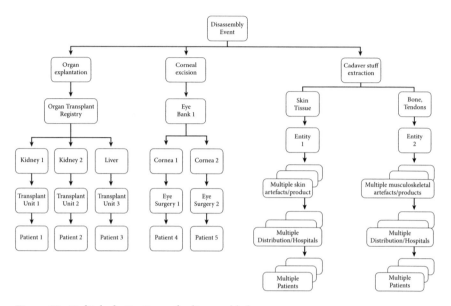

Figure 28: Multiple destinations of a disassembled corpse

Seeing the World through American Eyes

The United States exports cadaver artefacts/products. Indeed, around 800 of the 25,000 recipients of cadaver artefacts/products made out of cadaver stuff extracted by Mastromarino's team were Canadian, Korean, Italian, Greek or British; 2,000 have never been located, and some may live outside the United States. These scant data are rare: records of composition, volume, value and destinations are publicly unavailable. The exception is American cadaver eye/corneas, where exports are counted: in 2001, 13,497 American cadaver eyes/corneas were implanted overseas; by 2014, it was 28,901. Overseas travel is facilitated by commercial storage media, which extend the shelf life by up to fourteen days where the corneas are stored at 4°C.

In 2005, the industry experienced a temporary hiccup when compliance with the Food and Drug Administration's insistence that a potential source's lifestyle and medical history must be investigated forced it to abandon legislative consent. The number of corpses out of which eyes/corneas were extracted fell from 41,131 in the previous year to 38,532. First-person consent came to its rescue: by 2014, eyes/corneas were extracted out of 65,558 corpses, just over half of which were those of pledgers.[22]

Exports are typically cadaver eyes/corneas suitable for surgery but refused by American surgeons for reasons such as mismatch of age of source and recipient, or which become available during a holiday period, or where neither surgeon nor potential recipient is available, or just simply where supply exceeds immediate demand. Keratoplasty is almost always a scheduled procedure typically performed in an ambulatory care centre or ophthalmologist's office. Increasingly, only the damaged portion is replaced, lowering risks of rejection and infection. Surgeons order cadaver corneas that have been pre-cut to a specified size. These are relatively expensive but shorten the procedure by an average of twenty-one minutes, and also reduce wastage: the likelihood of mistakes is lower, and only one is ordered.[23] Their popularity took eye banks by surprise. For instance, the capacity of the National Eye Bank Center, a large processing facility in Memphis, close to the distribution hub of Federal Express, which Tissue Banks International opened in 2005, has been doubled to meet demand.

Exports are handled by the various networks to which eye banks belong, such as Vision Share, a mutually beneficial co-operative of eye banks created in 1998 by Lions Clubs.[24] Tissue Banks International's global reach is extending; it levies a full or nominal fee on the grounds that cadaver corneas provided free of charge might be wasted.[25] Most exports goes to hospitals in the Middle East, followed by Mexico and Central and South America. It is entertaining to speculate on whether and how vision might change where foreigner recipients literally see the world through American eyes.

CHAPTER 36

Consolidation without Cooperation

The various inquiries into the retained organs scandal concluded that the Human Tissue Act 1961 had encouraged rather than constrained the medical profession's belief in its entitlement to cadaver stuff, and that the regulatory framework urgently needed a comprehensive overhaul. Failure to replace its ageing structure was attributed to central government's focus on containing NHS expenditure. Replacements were hastily published, which lacked the force of law.[1] In July 2002, the Department of Health and the Welsh Assembly government published *Human Bodies, Human Choices*, which sets out a variety of measures that might be included in a new law.[2] Whereas the Human Tissue Act 1961 was drafted by Ministry of Health civil servants, with input from neither medical professionals nor the public, and was considered in Parliament on Christmas Eve, all and sundry were invited to comment on *Human Bodies, Human Choices*, and many did.[3] The first draft of the Human Tissue Bill was introduced into Parliament on 3 December 2003. After much argument, lobbying and adjustment in and outside Parliament, the Human Tissue Act 2004 received Royal Assent on 15 November 2004.

The statute covers far greater territory than its predecessor, which dealt primarily and inadequately with extracting cadaver stuff for repurposing in therapies. Firstly, in repealing and replacing the Anatomy Act 1984, it regulates pledging, acquisition and disposal of corpses for repurposing as 'teaching material' for medical students. Secondly, it applies to both body and corpse stuff falling under its definition of 'relevant material', which is 'material, other than gametes, which consists of human cells'. Sperm, eggs and embryos are excluded because they fall under the Human Fertilisation and Embryology Act 1990, and so were hair and nails, otherwise the statute would have to cover hairdressers, barbers, chiropodists and manicurists. Kidneys and corneas fall under the definition, which is why the statute repealed and replaced the Corneal Tissue Act 1986, which allowed extraction of cadaver eyes/corneas by people without a medical licence, and the Human Organ Transplant Act 1989, which prohibited paying kidney donors-on-the-hoof. Thirdly, the statute regulates 'scheduled purposes', which are a collection of disparate activities including therapy, dissection and public display of cadavers and cadaver stuff. Regulation of the latter had been covered since the Anatomy Act had been placed on the statute book in 1832, but it had come to public attention while the bill was being considered when Gunther von Hagens staged in London his controversial *Body Worlds* exhibition, which displayed 'plastinated' anatomized corpses manipulated to appear engaged in an activity, such as playing chess or riding a 'plastinated' anatomized horse carcass. The statute also permits nine named national museums to de-access and repatriate human remains over 100 years old.[4] This measure implemented the agreement drawn up in July 2000 by British and Australian governments that wherever possible British museums would respond positively to requests by Australian indigenous communities for the return of culturally important artefacts incorporating cadaver stuff; it constitutes official recognition of the cultural, spiritual, religious and personal significance people attach to corpses and cadaver stuff, which in some instances should override entitlement claims of education, science and medicine.

'Appropriate consent' is the golden thread running through the statute; it must be sought and obtained from sources of 'relevant material' for 'scheduled purposes', but only where the stuff is that of a cadaver; extraction of body stuff continues to fall under common law, where failure to afford people the right to say 'yes please' or 'no thank you' is susceptible to a charge of battery. Considering the statute's immediate history, the emphasis is on agreeing to a corpse being incomplete on burial or cremation, whereas in the United States, following the Orange County exposé, it fell on commercial exploitation and/or repurposing of cadaver stuff in medically unnecessary applications.

The statute enshrined first-person consent in English law, and the Organ Donor Register became the instrument through which it is provided. People can also nominate a representative to decide on their behalf whether or not to allow cadaver stuff to be extracted. However, where neither pledge nor nominated representative exists, the statute reinstates people's private interest in a corpse: persons standing in a 'qualifying relationship' to the dead person can agree to or refuse extraction.[5] On 31 August 2006, the day before the statute came into force, the front page of the *Daily Mail* warned families that they 'can't stop doctors taking organs', and predicted that the change in law is 'likely to cause considerable anguish to relatives who do not wish to see their loved one go under the knife after death'.[6]

'The value attached by many people to the physical remains of recently dead loved ones has been underestimated,' concluded the Retained Organs Commission. 'Death does not abruptly sever loving relationships.'[7] However, the retained organs scandal had not dismayed the presumed consent lobby, led on this occasion by the British Medical Association, and it had seized on the Human Tissue Bill as an opportunity to renew its campaign.[8] Parliament did not acquiesce, but strengthened the authority of explanters by allowing them to use techniques of preserving organs of potentially bioavailable people attached to life-support machinery whilst the Organ Donor Register is checked, and/or agreement of surrogates sought, medical bioavailability investigated and practical arrangements made.

The statute carried forward offences under the Anatomy Act 1984 and the Human Organ Transplant Act 1989. It also provided several new offences, including retaining body/cadaver stuff without appropriate consent, or repurposing it without specific agreement. Penalties can be a fine or up to three years' imprisonment, or both.

Minimizing Risk of Harm

Safety had not been forgotten.[9] The statute anticipated the United Kingdom's obligation to transpose into law the European Union's Tissues and Cells Directive (2004/23/EC), which requires the adoption of precautionary measures broadly in line with those in the Food and Drug Administration's *Proposed Approach*, but only in relation to the extraction and creation of artefacts; products are covered by the Medical Devices Directive (2001/83/EC), which obliges manufacturers to ensure raw materials are safe and have been lawfully extracted. Unsurprisingly, the borderline between the two directives is ambiguous: when body/cadaver stuff/artefact becomes product/device is no clearer in Europe than in the United States.

Sources, stuff/artefacts and recipients move freely within the European Union, but a majority of member states were without adequate legally enforceable rules on minimizing harm to recipients. Member states are required to adopt comparable measures in order to facilitate cross-border trade, in much the same way as American states are required to allow free movement of goods and services throughout the United States.

However, as happens in the United States, where each state adopts its own version of model acts drawn up by the Uniform Law Commission, how legislatures of member states interpret and implement directives varies.

Thousands of entities were extracting, processing, storing and distributing body/cadaver stuff within the European Union, in facilities varying from: a refrigerator or freezer in the corridor leading to an operating theatre in which surgeons kept their stash; 'just-in-time' enucleation of cadaver eyes at the deathbed or in a funeral home; 'garage entities' sometimes run by pathologists or mortuary staff; and sophisticated stashes organized by and for surgeons, such as the European Homograft Bank, in the Queen Astrid Military Hospital, Brussels, Belgium. The directive speaks in favour of non-profit status and/or public ownership, whilst at the same time recognizing, firstly, that processors can legitimately profit from products placed on the European market either by European commercial entities or wholly owned European subsidiaries or strategic allies of American conglomerates, and, secondly, that imports of American cadaver artefact/products cannot be blocked.

The directive requires that entities operate under a licence granted by a 'competent authority', which in the United Kingdom is the Human Tissue Authority, which the statute created, and which began work in April 2005. In order to be licensed, facilities must meet exacting standards, mostly concerning safety, and staff must abide by the authority's codes of practice, which are regularly updated, complete quantities of paperwork, and admit its inspectors. A licence is relatively expensive: the maximum initial annual fee was £4,500.

Structural Consolidation

Facilities of many British entities fell below the standard demanded for accreditation and licensing and were closed. But the regulations also strengthened the business case for investment that would allow concentration on one site of the National Blood Service's North London and East Anglia facilities, which had merged, and the recently enlisted Yorkshire Regional Tissue Bank and North Wales & Oswestry Tissue Bank, both full-service entities. The latter had been opened in 1992 by the International Atomic Energy Agency as a demonstration project in its programme of encouraging developing countries to use gamma radiation for terminal sterilization, which had been started in 1971, in order to promote peaceful applications of atomic energy.[10]

In July 2005, the National Blood Service opened a state-of-the-art facility on an industrial estate in Speke, Liverpool. One of the largest in Europe, with fourteen clean rooms, and more than fifty -80°C deep freezers, it is an updated version of the Navy Tissue Bank: skilled artisans fashion cadaver bone, skin, heart valves, tendons and blood vessels into fresh, frozen or freeze-dried artefacts. The Bethesda bank had its own operating theatre; the one in Speke has a 'Dedicated Retrieval Suite' in which corpses delivered by contracted funeral directors from local hospital mortuaries are disassembled in aseptic conditions by teams of technicians who are not medically trained but whom the Human Tissue Act 2004 permits to extract cadaver stuff. Situated within the catchment area of the Alder Hey Children's Hospital, a public relations exercise was undertaken to gain the support of local people, coroners and hospital staff, which included publicizing how and why cadaver stuff is repurposed, and providing assurances that following cosmetic restoration corpses would be returned to the mortuary within twenty-four hours.[11]

In October 2005, the National Blood Service merged with UK Transplant to form NHS Blood and Transplant. The Special Health Authority has a national monopoly of explanted cadaver organs, but not of other cadaver stuff, save for cadaver skin following closure of the Stephen Kirby Skin Bank in 2006. Almost everything else in what is now known as NHS Blood and Transplant Tissue Services' inventory is available from another domestic entity, or as an import. Indeed, in September 2006 it was revealed that in twenty-five British hospitals, products manufactured in the United States out of cadaver bone that had been extracted by Mastromarino's Biomedical Tissue Service had been implanted in eighty British bodies.[12] NHS Blood and Transplant Tissue Services only provisions medically indicated procedures. Customer service staff assist in scheduling surgery by regularly issuing current stock levels, and occasionally arranging a bespoke cadaver artefact.

Although compatibility, the Corneal Transplant Service's original justification, had been discredited, it had remained piggybacked on the organ explanters' infrastructure, despite differences in collecting sites, sources, pace of operation and inventory method. Organ explantation is organized around unpredictability and compatibility; a 24-hour telephone line is staffed so that whenever a bioavailable corpse is identified its freshly explanted organs can be rushed from explanting site to the transplant team responsible for the patients who are the source's closest immunological match. Cadaver eyes enucleated in collecting sites, mostly mortuaries, across the nation were delivered to the Corneal Transplant Service's facilities either in Bristol or

Manchester, where corneas were excised and stored in organ culture for up to twenty-eight days. Orders had to be placed eight weeks in advance of keratoplasty, but there was no guarantee they would be met.

The Corneal Transplant Service was operating an inventory method that brings to mind that of the original blood bank at Cook County Hospital: its users held an account, which was credited when cadaver eyes were deposited, debited when corneas were withdrawn, and a year end balance calculated. In the year of the merger, Moorfields hospital had deposited seventy-two, withdrawn 128 and ended the year with a debit of fifty-six; the Queen Victoria Hospital, East Grinstead, had deposited 549 cadaver eyes, withdrawn seventy-four, and ended the year with a credit balance of 475.[13] The enucleator responsible for this outstanding record was a full-time member of staff who made round trips in her own car, sometimes clocking up more than 500 miles in a day, enucleating eyes of corpses resting in hospital wards, mortuaries, hospices, funeral parlours and private homes in the south-east corner of England.[14] However, the ethos of reciprocity that had been the hallmark of the NHS had been extinguished by the transformation of hospitals into independent cost-conscious entities called 'Trusts', which are held accountable and penalized for a bottom line written in red ink. Her managers refused to allow her to support an outside organization, especially when each year many of its users were contributing fewer than fifty cadaver eyes.

NHS Blood and Transplant decided to subsidize the Corneal Transplant Service; eight of the most productive collecting sites were each appointed as an 'Eye Retrieval Centre', and each one awarded £70,000 to support a 'Cornea Retrieval Nurse', who identifies potentially suitable corpses, seeks agreement of recently bereaved kinfolk, reviews medical bioavailability, enucleates cadaver eyes, secures the globes in moist chambers, and sends them packed in an ocular tissue transplant box via courier to either Bristol or Manchester.[15]

The small charge that hitherto had been levied to cover the cost of transport was increased to recoup some of the cost of the eye retrieval scheme.[16] But 'major grafters' at the Queen Victoria and Moorfields hospitals calculated that an in-house entity would be cheaper and more convenient than the Corneal Transplant Service. Independence was declared. Although the facility at Moorfields had been refurbished in the 1990s at a cost of £80,000, it did not meet current standards.[17] In 2007, it was upgraded and renamed the Moorfields Lions Eye Bank in recognition of the civil society's financial support. Cadaver corneas are seldom wasted because each year around 600 people undergo keratoplasty at Moorfields, and a recipient can usually be found at short notice.[18] The Queen Victoria Hospital's eye bank provisions its own surgeons and any surplus to their requirement is distributed at a price to another hospital. Both independent entities meet the occasional shortage with expensive American imports.

Without Cooperation

'Mortuaries are a vital part of the service the NHS gives to patients who die in hospital and to their bereaved families and friends', as the Minister of State for NHS Reform Norman Warner (Labour) acknowledged in 2006 in his introduction to the first official guidance on mortuary etiquette. 'The services they provide are often overlooked. But good care after death is not an optional extra. If things go wrong in a hospital mortuary, the impact on bereaved families can be devastating.'[19] Yet the guidance, which was another response to the retained organs scandal, is without reference to conduct in the mortuary no-place, although it is where NHS Blood and Transplant Tissue Services' regional donation teams disassemble corpses, and a majority of cadaver eyes are enucleated.

The following year, without fanfare, unofficial no-places were effectively prohibited by the Human Tissue (Quality and Safety for Human Application) Regulations 2007 requiring formal 'service-level agreements' drawn up between extractors and collecting sites, but only those that Human Tissue Authority inspectors have accredited as fit for purpose, which effectively prohibited extraction in private homes, hospital wards, nursing homes and makeshift funeral parlour premises.

Accredited mortuaries are overcrowded. Yet their productivity as collecting sites is diminishing. Firstly, fewer corpses are medically bioavailable. Life expectancy is lengthening, but death typically is preceded by years of declining health, and repeated admission to hospital where patients are exposed to bacteria that accelerate the speed of putrefaction, which is easily detected by a poke and sniff test. Additionally, dementia, which increases with age, raises suspicion of contamination by prions. Moreover, medical criteria of bioavailability, which are increasingly exacting, are more demanding in relation to potential sources of cadaver stuff, including eyes, than potential sources of cadaver organs. This is because different considerations are applied in weighing risk of harm against potential benefit: one contaminated corpse can infect more than fifty recipients of cadaver stuff/ artefacts/products but, save for victims of massive burns, their condition is not life threatening; one contaminated corpse can extend the life of several people facing imminent death, thus the risk of harm from contaminated cadaver organs is sometimes worth taking

Secondly, stuff can be extracted up to twenty-four hours following death but fewer corpses meet the deadline. Hospital mortuaries are unattended by staff overnight, and in some parts of the country funeral directors hold a contract to keep the corpses of people who die outside hospital in the refrigerator on their premises, and deliver corpses scheduled for a post-mortem examination to the nearest accredited hospital mortuary when it opens.[20] Thirdly, the quality of some cadaver stuff – veins, arteries and tendons – diminishes with age. And, although some anatomical pathology technicians – the current job title of British mortuary staff – are trained enucleators, for their part, ophthalmologists are increasingly reluctant to graft an eighty-year-old cadaver cornea on to a twenty-year-old eye.[21]

All extractors value cadaver stuff of relatively young, healthy people admitted to hospital following an untimely and unanticipated death. Transplant coordinators are given first call. However, whereas American organ procurement organizations have a legal and financial incentive to 'maximize the gift', NHS Blood and Transplant treat extractors of homovital and homostatic stuff as if they belong to different, even hostile tribes. Indeed, in 2015, NHS Blood and Transplant Tissue Services and the Corneal Transplant Service were merged to form a separate 'homostatic division' within NHS Blood and Transplant.

In the year ending March 2016, eyes were enucleated out of 270 of the 1,364 corpses from which organs were explanted. Despite having their authority strengthened by the Human Tissue Act, many transplant coordinators faced with an unlimited pledge suffer 'list fatigue', especially in relation to enucleation of cadaver eyes, for fear of putting kinfolk off.[22] In the same year, 2,611 dead people were recorded as 'cornea only donors'. Cornea retrieval nurses consult the record of admissions to their hospital's mortuary in order to identify potentially bioavailable corpses. Other extractors, including the independent eye banks, rely on notification by the staff of hospital accident and emergency departments, bereavement departments and hospices, coroner's officers, police and sometimes next of kin. The majority of kinfolk are 'cold-called' by one of the trained nurses who operate the telephone lines of the Tissue Services National Referral Centre, in Speke. Around 400 corpses are disassembled annually out of around 6,000 notifications.[23] In 2012, the American industry's throughput was at least 30,000.[24]

CHAPTER 37

From Mortuary to Shopping Cart?

In 2007, forty-eight members of the American Association of Tissue Banks distributed 2,110,200 cadaver artefacts/products.[1] Some were labelled 'donated gift' in order to comply with a recommendation of a working group of industry representatives that had been convened two years earlier to consider how to repair the reputational damage sustained following the *Orange County Register* and *Chicago Tribune*'s exposés. The label's message might provide an accurate description of the source's motivation, but entities responsible for extraction, processing, distribution and retailing continued to expect reimbursement, and anticipate a comfortable margin allowed in standard acquisition costs.[2]

First-person consent (or 'first-person authorization' or 'donor designation' as it is increasingly called in the United States, following recognition of the category error) was being adopted around this time; its capacity to increase supplies of cadaver stuff depends on actual and potential pledgers believing that beneficiaries of corpse philanthropy are neither shareholders nor purchasers of the lunch-hour facelift but patients undergoing medically indicated procedures. Entities began formulating codes of corporate ethics, which acknowledge that their business decisions have social consequences, and providing reassurance that patient welfare is not being sacrificed on the altar of profitability.[3] Burn victims, who hitherto had provided poor lobby fodder, provided a narrative. 'For many years there has been a void in a comprehensive program dedicated to serving the needs of burn patients and the Institute will help meet the needs of these critically ill patients,' is how, in 2007, LifeNet Health's chief executive officer Rony Thomas explained why the Skin & Wound Allograft Institute had been opened in Baltimore, Maryland. 'Living Legacy is excited to expand our tissue processing services on a national basis', added Charlie Alexander, chief executive officer of Living Legacy Foundation, an organ procurement organization and LifeNet's partner in the institute. 'We have historically had a strong dedication to burn victims and we are pleased to help this effort now on a national basis.'[4] Community Tissue Services created the Bubba the Bear Program for children with burn wounds. The non-profit conglomerate, which claims to be the largest provider of cadaver skin for severely burned patients in the United States, donates plush bears to burns centres that staff can bandage to match those of their young patients.[5]

LifeNet and Community Tissue Services are two of ten American entities that rescued the mantle of responsibility for burn victims from skin banks as they were disappearing. No one can predict when split-thickness cadaver skin might be needed to dress burn wounds. Contracts agreed with the 160 or so American burns centres promise them priority over customers for branded acellular cadaver skin products, even though square inch for square inch these command more than four times as many dollars.[6] Demand for cadaver skin is rocketing to meet these large, growing markets: for instance, in 2010, a single unnamed American entity processed skin extracted from 12,000 corpses, which was nearly twice as many as had been processed in the United States five years previously, and many more than the 7,943 corpses that had had their organs explanted that year.[7]

Health problems associated with ageing provide large and expanding markets for cadaver stuff repurposed as support for collapsing bladders, as dressings for chronic leg ulcers, in reconstruction of breasts removed during cancer surgery, to provide firm foundations for dental implants and as healthy replacements of diseased heart valves. American entities are not revolutionaries intent on changing therapeutic paradigms. Innovation focuses on enhancing convenience, sometimes in collaboration with a medical device manufacturer, to create cadaver products that are relatively easy to use, and which can encourage less-skilled physicians to offer them. Convenience cadaver products tend to be more expensive than the cadaver artefacts they replace, but some of the extra cost can be offset by a reduction in staff and theatre time, and length of hospital stay.

Little wonder then that these markets are growing. Some cadaver products can be ordered 'off the shelf' from a catalogue: Musculoskeletal Transplant Foundation lists over one thousand 'allograft choices', including acellular cadaver skin products, which are available in a range of strengths, thicknesses, widths and lengths.[8] Tissue Banks International, which in 2016 was rebranded KeraLink International, has a different business model: each year, it provides around 15,000 cadaver corneas that have been customized to a surgeon's specification.[9]

Facing American Competition

'There are striking parallels with the food sector,' observed the Belgians involved in various capacities in extracting and repurposing cadaver stuff. 'The rising liberalization of agro-industrial markets was also accompanied by technological advances and the introduction of an EU regulatory framework.'[10] Eight years after publication of the EU Tissue and Cells Directive, they envisaged a future in which first-generation European entities had suffered a fate similar to that of small-scale producers of 'tasty local foods', which had been driven out of business by manufacturers of mass-produced, standardized, branded convenience foods.

The Belgians blamed the regulatory framework that member states had had to transpose into national law for their pessimism; it comforts second-generation entities operating complex supply chains linking multiple collecting sites, high-volume processing facilities and recipients at home and abroad, which sometimes wind through several different countries, even continents. European entities typically disassemble corpses and process cadaver stuff for repurposing locally or nationally; their throughput is relatively small, and compliance with the regulations expensive and a bureaucratic rigmarole. Moreover, the Belgians were sceptical of the directive's effectiveness: the measures it prescribed originated in the American food-manufacturing industry and were apparently unable to prevent outbreaks of serious food poisoning or scandals about mislabelling and adulteration.

The Belgians were also reacting to American convenience cadaver artefact/products that are retailed in Europe through subsidiaries, sales offices, alliances agreed with European entities or in response to orders placed over the phone or via email. The traffic in dollars is two way: American processors' insatiable appetite for cadaver stuff feeds an income stream in collecting sites, particularly in low-income member states that had been part of the Soviet Union. The United Kingdom is denied these dollars. For whereas during the 1970s the collection of cadaver pituitary glands had testified to the productive potential of British mortuaries, which had attracted foreign bioprospectors, the Food and Drug Administration is fearful of British prion stowaways and refuses entry to imports of anything extracted out of British corpses.

What, how much and by whom American cadaver artefacts/products are imported into the United Kingdom is not recorded. However, the Speke facility was facing robust competition both from within and outside the NHS.

Sales were falling, its inventory was sometimes full, and occasionally a bio-available corpse was refused.[11] In 2010, private-sector management consultants were brought in to advise on how its future might be ensured. Direct competition with imports was ruled out. Two strategies were advised: provide niche products for needs unmet within the NHS; and offer spare processing capacity to start-ups organized around technical innovation.

In 2012, NHS Blood and Transplant Tissue Services added dCELL® to its product range.[12] Launched nearly two decades after Alloderm®, it is the first decellularized wound dressing manufactured out of British cadaver skin. dCELL®'s proprietary process was modified from one developed by the Tissue Regenix Group, which had been spun off from the University of Leeds in 2006. Its motto is 'Commercializing academic research in regenerative medicine'.[13] dCELL® claims to offer two advantages over equivalent commercial cadaver products: a longer shelf life and lower price. However, its reception has not been recorded.

Statistics might suggest that British eye banking's viability is not threatened. Around 3,700 corneas were grafted in 2015/16, compared to just over 2,000 at the turn of the millennium.[14] However, the subsidy, which had protected the Corneal Transplant Service, is no longer politically justifiable. In 2012, the Corneal Transplant Service, which provisions around 90 per cent of operations, charged £450 for a cadaver cornea, when the actual cost was around £1,150. On paper the price was attractive compared to that of an American import, which cost between $2,500 and $3,000, but imports are dispatched immediately and arrive pre-cut, whereas orders for a Corneal Transplant Service cornea had to be placed eight weeks in advance of scheduled surgery, and surgeons had to spend around forty minutes trimming it prior to grafting.

The Corneal Transplant Service was unprepared for competition; it was out-dated and inefficient. Time between death and enucleation was on average 17.2 hours compared to 9.25 hours in the United States; time between death and distribution from either the Bristol or Manchester eye banks was 19.1 days compared to 2.7 days.[15] Its supply chains were far longer and more convoluted compared with those of the 'major grafter's' at Moorfields and Queen Victoria hospitals. Forty years after it was introduced in the United States, adoption of *in situ* excision of cadaver corneas began to be considered, and trials of pre-cut cadaver corneas begun.

Business Models

On both sides of the Atlantic, demand for cadaver stuff/artefacts/products is growing. On both, first-person authorization has been adopted as a device that provides pledges of cadaver stuff with legal authority without taking corpses into a controversial property regime. Both American and British sources of cadaver stuff are rewarded in the currency of regard, but the entities they provision have very different ambitions, ethos and business models.

In its report on the *Orange County Register* and *Chicago Tribune* exposés, the inspector general of the Department of Health and Human Services had agreed with industry associations, and patient support groups, that market forces and financial incentives are an essential part of the industry (as they are for the entire American health-care system), and could and should not be eliminated.[16] But whether or not American corpse philanthropy should provision exports was not addressed. Corporate ethics have not been extended overseas. For its part, NHS Blood and Transplant acknowledged that marketing of its 'tissue business' had been neglected. It decided to include a tick box for 'tissue' on the Organ Donor Register menu when it was relaunched in July 2015. The Department of Health supported the NHS 'tissue businesses' adopting a more commercial approach, but warned that commercial was not to be understood as profit making; profits in this business, it pontificated, are inappropriate.[17] What the NHS requires is cost-effective procedures to treat the ageing British population.[18] However, what is cost-effective is undecided: should it be achieved through relatively cheaply produced cadaver artefacts or by convenience cadaver products? If the latter is the case, then without significant investment required to climb up the value chain, imports will increase, and British entities either be forced to close, or follow in the footsteps of domestic food manufacturers that concentrate on producing niche artisan products that retail at prices only the wealthy can afford.

NOTES

Introduction

1. Elizabeth Haigh, 'The Roots of Vitalism of Xavier Bichat', *Bulletin of the History of Medicine*, 49 (1975), pp. 72–86; John V. Pickstone, 'Bureaucracy, Liberalism and the Body in Post-revolutionary France: Bichat's Physiology and the Paris School of Medicine', *History of Science*, xix (1981), pp. 115–42.
2. Jon Turney, *Frankenstein's Footsteps: Science, Genetics and Popular Culture* (New Haven and London: Yale University Press), 1998, p. 22.
3. Sociologist Kieran Healy coined 'cultural work'. 'Corpse philanthropy' has been appropriated from medical historian Susan Lederer's 'body philanthropy'. Kieran Healy, *Last Best Gifts: Altruism and the Market for Human Blood and Organs* (Chicago and London: Chicago University Press, 2006), p. 24; Susan E. Lederer, *Flesh and Blood: Organ Transplantation and Blood Transfusion in Twentieth-century America* (Oxford: Oxford University Press, 2008), p. xii.
4. Bernard Fantus, 'The Therapy of the Cook County Hospital', *Journal of the American Medical Association*,109 (1937), pp. 128–31.
5. Emile J. Farge, 'Eye Banking: 1944 to the Present', *Survey of Ophthalmology*, 30 (4) (1989), p. 261.
6. Richard Titmuss, 'The Gift Relationship: From Human Blood to Social Policy', in Ann Oakley and John Ashton (eds), *The Gift Relationship: From Human Blood to Social Policy, Expanded and Updated Edition* (New York: The New Press, 1997 [1970]), pp. 57–315.

1 Skin Donors-on-the-Hoof

1. Susan E. Lederer, *Flesh and Blood: Organ Transplantation and Blood Transfusion in Twentieth-century America* (Oxford: Oxford University Press, 2008), p. 12.
2. Steven Connor, *The Book of Skin* (London: Reaktion Books, 2004), p. 29.
3. James Barrett Brown, Minot P. Fryer, Peter Randall and Milton Lu, 'Postmortem Homografts as "Biological Dressings" for Extensive Burns and Denuded Areas', *Annals of Surgery*, 138 (4) (1953), pp. 618–30.
4. Suzanne Dechillo, 'At the Burn Center, Painstaking Victories', *New York Times* (18 December 1983).
5. Henk J. Klasen, *History of Free Skin Grafting: Knowledge or Empiricism* (Berlin and Heidelberg: Springer-Verlag, 1981); P. Santoni-Rugiu and P. J. Sykes, *A History of Plastic Surgery* (Berlin and Heidelberg: Springer-Verlag, 2007), pp. 126–39; Leland R. Chick, 'Brief History and Biology of Skin Grafting', *Annals of Plastic Surgery*, 21 (4) (1988), pp. 358–65.

6. Sander L. Gilman, *Making the Body Beautiful: A Cultural History of Aesthetic Surgery* (Princeton: Princeton University Press), 1999.
7. M. Felix Freshwater and Thomas J. Krizek, 'Skin Grafting of Burns: A Centennial', *Journal of Trauma*, 11(10) (1971), pp. 862–5.
8. Blair O. Rogers, 'Guide and Bibliography for Research into the Skin Homograft Problem', *Plastic and Reconstructive Surgery*, 7 (1957), 169–201; Lederer, *Flesh and Blood*, pp. 166–70.
9. James Elkins, *Pictures of the Body: Pain and Metamorphosis* (Stanford: Stanford University Press, 1999), p. 43.
10. Lederer, *Flesh and Blood*, pp. 5–20.
11. Ibid., p. 17.
12. Ibid., p. 10.
13. Ibid., p. 12.
14. Ibid., p. 13.
15. Thomas Gibson, 'Zoografting: A Curious Chapter in the History of Plastic Surgery', *British Journal of Plastic Surgery*, 8 (1955), pp. 234–41.
16. Klasen, *History of Free Skin Grafting*, p. 133.
17. Joseph E. Murray, 'Organ Transplantation: The Practical Possibilities', in G.E.W. Wolstenholme and Maeve O'Connor (eds), *Ethics in Medical Progress: A Ciba Foundation Symposium* (London: J. & A. Churchill, 1966), p. 60.
18. E. R. Mayhew, *The Reconstruction of Warriors: Archibald McIndoe, the Royal Air Force and the Guinea Pig Club* (London and Pennsylvania: Greenhill Books, 2004), pp. 58–61.
19. Frank P. Underhill, 'Changes in Blood Concentration with Special Reference to the Treatment of Extensive Superficial Burns', *Annals of Surgery*, 86 (6) (1927), pp. 840–49.
20. Vilray P. Blair, James Barrett Brown and William G. Hamm, 'The Early Care of Burns and the Repair of their Defects', *Journal of the American Medical Association*, 98 (16) (1932), pp. 1355–9.
21. Eric J. Stelnicki, V. Leroy Young, Tom Francel and Peter Randall, 'Vilray P. Blair, his Surgical Descendants, and their Roles in Plastic Surgical Development', *Plastic, and Reconstructive Surgery*, 103 (7) (1992), pp. 1990–99.
22. Frank McDowell, 'James Barrett Brown, M.D., 1899–1971: Obituary', *Plastic and Reconstructive Surgery*, 48 (1) (1971), pp. 101–4.
23. Lynn D. Ketchum, 'An Historical Account of the Development of the Calibrated Dermatome', *Annals of Plastic Surgery*, 1 (6) (1978), pp. 608–11; Kathryn Lyle Stephenson, 'Earl C. Padgett: As I Remember', *Annals of Plastic Surgery*, 6 (2) (1981), pp. 142–53.
24. James E. Bennett and Stephen R. Miller, 'Evolution of the Electro-dermatome', *Plastic and Reconstructive Surgery*, 45 (2) (1970), pp. 131–3.

2 Pioneers of 'Eye Banking'

1. Vladimir Filatov, 'Ophthalmic Surgery in the Russian Army', *Journal of the American Medical Association*, 120 (1) (1942), pp. 54–5.
2. George O. D. Rosenwasser and Miriam Rosenwasser, 'Vladimir Filatov 1875–1956', in Mark J. Mannis and Avi M. Mannis (eds), *Corneal Transplantation: A History in Profiles* (Ostend: Hirschberg History of Ophthalmology, The Monographs, volume 6, 1999), pp. 170–91.
3. Vladimir Filatoff, 'The Grafting of Cornea', *American Braille Press* (1933), pp. 18–22.
4. Vladimir Filatov, *My Path in Science* (Moscow: Foreign Languages Publishing House, 1957), p. 27.
5. S. S. Yudin, 'Transfusion of Cadaver Blood', *Journal of the American Medical Association*, 106 (12) (1936), pp. 997–8.
6. Donald J. Doughman, 'Tissue Storage', in Jay H. Krachmer, Mark J. Mannis and Edward J. Holland (eds), *Cornea: Fundamentals of Cornea and External Disease*, volume 1 (St Louis: Mosby, 1997), pp. 509–17.
7. Filatov, *My Path in Science*, p. 28.

8. Mark G. Field, 'Soviet Medicine', in Roger Cooter and John Pickstone (eds), *Medicine in the 20th Century* (London: Harwood Academic Publishers, 2000), pp. 51–66.
9. David Paton, 'The Founder of the First Eye Bank: R. Townley Paton, M.D.', *Refractive and Corneal Surgery*, 7 (1991), p. 193.
10. R. Townley Paton, 'Eye-Bank Program', *American Journal of Ophthalmology*, 41 (3) (1956), pp. 419–24.
11. Katherine B. Rhoads, 'Meeting the Challenge', *EBAA Newsight* (November 1965), pp. 10–14.
12. R. Townley Paton, 'Sight Restoration through Corneal Grafting', *Sight-Saving Review*, 15 (1) (1945), pp. 3–11.
13. William H. Schneider, 'Blood Transfusion between the Wars', *Journal of the History of Medicine*, 58 (2003), pp. 187–224.
14. Bernard Fantus, 'The Therapy of the Cook County Hospital', *Journal of the American Medical Association*, 109 (1937), pp. 128–31.
15. Susan E. Lederer, *Flesh and Blood Organ Transplantation and Blood Transfusion in Twentieth-century America* (Oxford: Oxford University Press, 2008), pp. 107–42; Kara W. Swanson, *Banking on the Body: The Market in Blood, Milk and Sperm in Modern America* (Cambridge, MA and London: Harvard University Press, 2014), pp. 49–66.
16. Emile J. Farge, 'Eye Banking: 1944 to the Present', *Survey of Ophthalmology*, 30 (4) (1989), p. 261.
17. 'First Year of the Eye-Bank for Sight Restoration', unpublished manuscript (May 1946), Eye-Bank for Sight Restoration, New York.
18. 'Yorkville Club Fosters Eye Bank', *The Lion* (March 1946), p. 15.
19. '"Eye Bank" Offers Sight to Many Blind Who Had Lost Hope', *The Lion* (September 1945), p. 17.
20. Phil Hirsch, 'Society for Sight', *The Lion* (February 1954), pp. 24–5, 29.
21. 'Cornea Bank on the Way', *The Lion* (February 1946), p. 37.
22. Filatov, *My Path in Science*, p. 35.
23. Alexis de Tocqueville, *Democracy in America* (Ware: Wordsworth, 1998), p. 199.
24. Susan Rose-Ackerman, 'Altruism, Nonprofits, and Economic Theory', *Journal of Economic Literature*, 34 (1996), pp. 701–28.
25. Olivier Zunz, *Philanthropy in America* (Princeton and Oxford: Princeton University Press, 2012), p. 2.
26. Paul Martin, *We Serve: A History of the Lions Clubs* (Washington DC: Regnery Gateway, 1991).
27. Robert Casey and W.A.S. Douglas, *The World's Biggest Doers: The Story of the Lions* (Chicago: Wilcox & Follett Co., 1949), p. 46.
28. Martin, *We Serve*, p. 59.
29. Rosemary Stevens, *In Sickness and in Wealth: American Hospitals in the Twentieth Century* (New York: Basic Books, 1989), pp. 140–41.

3 'Doctor, I see you!': Marketing Corpse Philanthropy

1. David Paton, 'The Founder of the First Eye Bank: R. Townley Paton, M.D.', *Refractive and Corneal Surgery*, 7 (1991), p. 193.
2. Kieran Healy, *Last Best Gifts: Altruism and the Market for Human Blood and Organs* (Chicago and London: Chicago University Press, 2006), p. 24; Susan E. Lederer, *Flesh and Blood: Organ Transplantation and Blood Transfusion in Twentieth-century America* (Oxford: Oxford University Press, 2008), p. xii
3. Eckert Goodman, 'How Anyone Can Leave the Legacy of Sight', *Daily News* (9 September 1958), Eye-Bank for Sight Restoration newspaper cuttings file.
4. Arlie Hochschild, 'Emotion Work, Feeling Rules, and Social Structure', *American Journal of Sociology*, 83 (3) (1979), pp. 551–75.
5. Thomas Lynch, *The Undertaking: Life Studies from the Dismal Trade* (London: Jonathan Cape, 1997), pp. 23–4.
6. Robert Pogue Harrison, *The Dominion of the Dead* (Chicago and London: University of Chicago Press, 2003), p. 92.

7. First draft of a report dictated to Mrs Putnam at the Cosmopolitan Club, unpublished manuscript (1955), Eye-Bank for Sight Restoration
8. The Eye-Bank for Sight Restoration, *A Gift Like the Gifts of God* (1945).
9. Avner Offer, 'Between the Gift and the Market: The Economy of Regard', *Economic History Review*, 50 (3) (1997), pp. 450–76.
10. David Burk, 'Doctor, I See You!' *Sunday News New York* (18 August 1957), The Eye-Bank for Sight Restoration newspaper cuttings file.
11. 'Woman Who Had Corneal Transplant Can Now See, First Sight is Crucifix', *Paterson, N.J. News* (22 June 1956), Eye-Bank for Sight Restoration newspaper cuttings file.
12. Paton, 'The Founder of the First Eye Bank', p. 193.
13. Editorial, *American Journal of Ophthalmology*, 30 (1947), p. 921.
14. 'Fete 20,000th Donor', *The Lion*, (February 1963), p. 63.
15. 'Coffee "Break" is Boon to Eye Foundation', *The Lion* (August 1961), p. 50.
16. *The Lion* (October 1961), p. 19.
17. 'Legacies of Light', *The Lion* (May 1963), pp. 28–30.
18. 'Peer-less Campaign', *The Lion* (April 1971), p. 47.

4 Bioprospecting in Mortuaries

1. Percival E. Jackson, *The Law of Cadavers and of Burial and Burial Places*, 2nd edn (New York: Prentice-Hall, 1950), p. 127.
2. Cori Hayden, *When Nature Goes Public: The Making and Unmaking of Bioprospecting in Mexico* (Princeton and Oxford: Princeton University Press, 2003).
3. R. Townley Paton, 'Donor Material and the Law', in Benjamin Rycroft (ed.), *Corneal Grafts* (London: Butterworth, 1955), p. 227.
4. R. Townley Paton, Unpublished Report, The Eye-Bank for Sight Restoration, n.d.
5. Emile J. Farge, 'Eye Banking: 1944 to the Present', *Survey of Ophthalmology*, 30 (4) (1989), pp. 260–61.
6. 'Autopsies Inquiry is Made City-wide', *New York Times* (3 November 1953); William Michelfelder, 'Autopsy Probers Score Dr. Werne of Queens', *New York Times* (19 March 1954).
7. Randy Hazlick and Debra Combs, 'Medical Examiner and Coroner Systems: History and Trends', *Journal of the American Medical Association*, 279 (11) (1998), pp. 870–74.
8. Michael R. Haines, 'The Population of the United States, 1790–1920', in Stanley L. Engerman and Robert E. Gallman (eds), *The Cambridge Economic History of the United States, Volume II: The Long Nineteenth Century* (Cambridge: Cambridge University Press, 2000), pp. 143–205.
9. Julie Rugg, 'From Reason to Regulation: 1760–1850', in Peter C. Jupp and Clare Gittings (eds), *Death in England: An Illustrated History* (Manchester: Manchester University Press, 1999), pp. 202–9.
10. Jackson, *The Law of Cadavers*, p. 176.
11. Gary Laderman, *A Cultural History of Death and the Funeral Home in Twentieth-century America* (Oxford: Oxford University Press, 2003).

5 The Doctrinal Tyranny of Skin

1. Medical Research Council, Consideration of problems arising out of the use of weapons productive of burns, FD 1/6300, n.d., National Archives, hereafter NA.
2. James Barrett Brown and Frank McDowell, 'Massive Repairs of Burns with Thick Split-skin Grafts', *Annals of Surgery*, 115 (1942), pp. 658–74.
3. W. C. Noble, *Coli: Great Healer of Men* (London: William Heinemann Medical Books, 1974); C. L. Oakley, 'Leonard Colebrook, 1883–1967', *Biographical Memoirs of Fellows of the Royal Society*, 17 (1971), pp. 91–138.
4. Noble, *Coli*, p. 36.
5. S. R. Douglas, A. Fleming and L. Colebrook, 'On Skin Grafting – A Plea for its More Extensive Application', *The Lancet*, 2 (1917), p. 5.

6. Noble, *Coli*, p. 61.
7. Ibid., p. 72.
8. N. A. Mitchison, 'Peter Brian Medawar', *Biographical Memoirs of Fellows of the Royal Society*, 35 (1990), pp. 283–301.
9. Peter Medawar, *Memoirs of a Thinking Radish* (Oxford: Oxford University Press, 1991), p. 77.
10. T. Gibson and P. B. Medawar, 'The Fate of Skin Homografts in Man', *Journal of Anatomy*, 77 (4) (1943), pp. 299–310.
11. Brown and McDowell, 'Massive Repairs of Burns', pp. 658–74.
12. Leo Loeb, *The Biological Basis of Individuality* (Chicago: University of Chicago Press, 1945).
13. P. B. Medawar, 'General Problems of Immunity', in G.E.W. Wolstenholme and Margaret P. Cameron (eds), *Preservation and Transplantation of Normal Tissues: A CIBA Foundation Symposium* (London: J. &. A. Churchill, 1954), pp. 1–20.
14. W. P. Longmire Jr, J. A. Cannon and R. A. Weber, 'General Surgical Problems of Tissue Transplantation', in G.E.W. Wolstenholme and Margaret P. Cameron (eds), *Preservation and Transplantation of Normal Tissues: A CIBA Foundation Symposium* (London: J. & A. Churchill, 1954), pp. 28–43.

6 Growth Hormone Soup

1. Herbert M. Evans and J. A. Long, 'The Effect of the Anterior Lobe of the Hypophysis Administered Intraperitoneally on Growth, Maturity and the Oestrus Cycles of the Rat', *Anatomical Record*, 21 (1921), p. 61.
2. Development of Vascular System File, Box 38, The Herbert McLean Evans' Papers, (hereafter HME Papers).
3. E. C. Amoroso and G. W. Corner, 'Herbert McLean Evans 1882–1971', *Biographical Memoirs of Fellows of the Royal Society*, 18 (1972), pp. 83–186; Leslie L. Bennett, 'Herbert McLean Evans: A Rare Spirit', *Perspectives in Biology and Medicine* (1978), pp. 90–103.
4. Michael Bliss, *The Discovery of Insulin* (Boston, MA: Faber and Faber, 1988).
5. James H. Madison, *Eli Lilly: A Life, 1885–1977* (Indianapolis: Indiana Historical Society, 1989), p. 63.
6. A. S. Parkes, *Sex, Science and Society* (Newcastle-upon-Tyne: Oriel Press, 1966), p. 43.
7. Roy O. Greep, 'Reproductive Endocrinology: Concepts and Perspectives, An Overview', *Recent Progress in Hormone Research*, 34 (1978), pp. 1–23.
8. Philip E. Smith, 'General Physiology of the Anterior Hypophysis', *Journal of the American Medical Association*, 104 (7) (1935), pp. 548–53.
9. A. S. Parkes, 'Herbert McLean Evans: An Interview', *Journal of Reproduction and Fertility*, 19 (1969), pp. 1–49.
10. Jimmy M. Skaggs, *Prime Cut: Livestock Raising and Meatpacking in the United States, 1607–1983* (College Station, TX: A & M University Press, 1986).
11. J. Russell Ives, *The Livestock and Meat Economy of the United States* (Chicago: AMI Center for Continuing Education American Meat Industry, 1966), p. 213.
12. Susan Strasser, *Waste and Want: A Social History of Trash* (New York: Owl Books, 1999), p. 29.
13. I confess to stealing and repurposing 'biotrash' from medical historian Sarah Hodges, University of Warwick, who coined it to describe clinical waste such as spent syringes, blood bags, bandages and such like, whereas I confine it to biological stuff susceptible to putrefaction.
14. Gay Hawkins and Stephen Muecke, 'Introduction', in Gay Hawkins and Stephen Muecke (eds), *Culture and Waste: The Creation and Destruction of Value* (Lanham and Boulder: Rowman & Littlefield, 2003), p. xi.
15. Martin O'Brien, 'Rubbish Values: Reflections on the Political Economy of Waste', *Science as Culture*, 8(3) (1999), pp. 269–95; Martin O'Brien, *A Crisis of Waste? Understanding the Rubbish Society* (London: Routledge, 2007).
16. William Cronon, *Nature's Metropolis: Chicago and the Great West* (New York and London: W. W. Norton, 1991), pp. 207–59.

17. Robert M. Aduddell and Louis P. Cain, 'The Consent Decree in the Meatpacking Industry, 1920-1956', *Business History Review*, 55 (3) (1981), pp. 359-78.
18. Roger Horowitz, *Putting Meat on the American Table: Taste Technology, Transformation* (Baltimore: The Johns Hopkins University Press, 2006), p. 89.
19. Roger Horowitz, '"Where Men Will Not Work": Gender, Power, Space and the Sexual Division of Labor in America's Meatpacking Industry, 1890-1990', *Technology and Culture*, 38 (1) (1997), pp. 187-213.
20. George Corner, *The Seven Ages of a Medical Scientist: An Autobiography* (Philadelphia: University of Pennsylvania Press, 1981), p. 247.
21. Applications 1922-3 Files, Carton 13, HME Papers.
22. E. M. Tansy, 'Ergot to Ergometrine: An Obstetric Renaissance', in Anne Hardy and Lawrence Conrad (eds), *Women and Modern Medicine* (Amsterdam and New York: Editions Rodopi B.V., 2001), pp. 195-216.
23. William Engelbach, 'The Growth Hormone: Report of a Case of Juvenile Hypopituitarism Treated with Evans' Growth Hormone', *Endocrinology*, 81 (1932), pp. 1-19.
24. Dr W. A. Reilly, Department of Pediatrics, University of California at Los Angeles Medical School to Herbert Evans, May 1935, Apr.-Sep.1935 File, Box 1, HME Papers.

7 The American Market for a Growth-promoting Substance

1. Robert W. Fogel, Stanley L. Engerman, Roderick Floud, Gerald Friedman, Robert A. Margo, Kenneth Sokoloff, Richard H. Steeckel, T. James Trussell, Georgia Villaflor and Kenneth W. Eachter, 'Secular Changes in American and British Stature and Nutrition', *Journal of Interdisciplinary History*, 14 (2) (1983), p. 467.
2. The experience of living with short stature in the United States is captured in Andrew Solomon, *Far From the Tree: Parents, Children, and the Search for Identity* (New York: Scribner, 2012), pp. 115-68. For an American man's experience of short stature, see Stephen S. Hall, *Size Matters: How Height Affects the Health, Happiness and Success of Boys – and the Men they Become* (Boston: Houghton Mifflin Company, 2006).
3. Robert Bogdan, *Freak Show: Presenting Human Oddities for Amusement and Profit* (Chicago and London: Chicago University Press, 1988).
4. Elizabeth Haiken, *Venus Envy: A History of Cosmetic Surgery* (Baltimore and London: The Johns Hopkins University Press, 1997).
5. For reasons of confidentiality potentially identifying information has been omitted. File: Requests for help from Patients File, Carton 14, HME Papers.
6. S. Charles Freed, 'Glandular Physiology and Therapy', *Journal of the American Medical Association*, 117 (1941), pp. 175-1182.
7. Merriley Borell, 'Brown-Sequard's Organotherapy and its Appearance in America at the End of the Nineteenth Century', *Bulletin of the History of Medicine*, 50 (1976), p. 312.
8. C. H. Li and Herbert McLean Evans, 'The Isolation of Pituitary Growth Hormone', *Science*, 99 (1944), pp. 183-4.
9. R. D. Cole, 'Choh Hao Li: April 21, 1913 – November 28, 1987', *Biographical Memoirs. National Academy of Sciences*, 70 (1996), pp. 221-39.
10. Edward Edelson, 'The Remarkable Dr Li: Master of the Master Gland', *Family Health* (August 1971), pp. 15-18.
11. 'Making the Hormone for Growth', *Medical World News* (21 February 1969), pp. 32-5.
12. 'Pure Hormone', *Time* magazine (20 March 1944), p. 48.
13. Editor of the *Jersey Journal*, to Herbert McLean Evans, 7 August 1946, Fl File, HME Papers.
14. Naomi Pfeffer, 'How Abattoir "Biotrash" Connected the Social Worlds of the University Laboratory and the Disassembly Line', in David Cantor, Christian Bonah and Matthias Dörries (eds), *Meat, Medicine and Human Health in the Twentieth Century* (London: Pickering & Chatto, 2010), pp. 63-76.
15. *New York Times* (29 September 1946), p. 1.
16. Armour Laboratories to Herbert McLean Evans, Bra-Bri File, Carton 1, HME Papers.
17. United Packing Company 1945-46 File, Carton 22, HME Papers.

18. For Leslie Bennet's recollection of the clinical trial, see Nancy M. Rockafellar, *Conversations with Dr Leslie Batty Bennett: The Research Tradition at UCSF* (UCSF Oral History Program, Department of History and Health Sciences, University of California, San Francisco, 1992); Roberto F. Escamilla and Leslie Bennett, 'Pituitary Infantilism Treated with Purified Growth Hormone, Thyroid and Sublingual Methyltestosterone: A Case Report', *Journal of Clinical Endocrinology*, 11 (1951), pp. 221–8.
19. M. Raben 'Treatment of a Pituitary Dwarf with Human Growth Hormone', *Journal of Clinical Endocrinology Metabolism*, 18 (1958), pp. 901–3; M. Raben, 'Clinical Use of hGH', *New England Journal of Medicine*, 266 (1962), pp. 82–6; Clark T. Sawin, 'Maurice S. Raben and the Treatment of Growth Hormone Deficiency', *Endocrinologist*, 12 (2) (2002), pp. 73–6.
20. ' "Growth Hormone" Reported to Work', *New York Times* (24 August 1958).
21. Choh Hao Li and Harold Papkoff, 'Preparation and Properties of Growth Hormone from Human and Monkey Pituitary Glands', *Science* (28 December 1956), pp. 1293–4.
22. D. Cantor, 'Cortisone and the Politics of Drama, 1949–55', in J. Pickstone (ed.), *Medical Innovations in Historical Perspective* (Basingstoke: Macmillan, 1992), pp. 165–84; D. Cantor, 'Cortisone and the Politics of Empire: Imperialism and British Medicine, 1918–1955', *Bulletin of the History of Medicine*, 67 (1993), pp. 463–93; Nicolas Rasmussen, 'Steroids in Arms: Science, Government, Industry, and the Hormones of the Adrenal Cortex in the United States, 1930–1950', *Medical History*, 46 (2002), pp. 299–324; Nicolas Rasmussen, 'Of "Small Men", Big Science and Bigger Business: The Second World War and Biomedical Research in the United States', *Minerva*, 40 (2002), pp. 115–46.
23. Edelson, 'The Remarkable Dr Li'.
24. Rolf Luft to C. H. Li, 1 September 1954, Rolf Luft file, Carton 3, Choh Hao Li Papers (hereafter CHL Papers).

8 Civilian Burns: Prevention and Treatment

1. 'Death in the Fireplace', *The Lancet*, 248 (1946), pp. 833–4.
2. J. C. Lawrence, 'Some Aspects of Burns and Burns Research at Birmingham Accident Hospital 1944–93: A.B. Wallace Memorial Lecture, 1994', *Burns*, 21 (1995), pp. 403–13; J. F. North, 'The Development of Plastic Surgery in the West Midlands Region', *British Journal of Plastic Surgery*, 40 (1987), pp. 317–22.
3. Noble, *Coli: Great Healer of Men* (London: William Heinemann Medical books, 1974), p. 96.
4. Ibid., p. 101.
5. L. Colebrook and V. Colebrook, 'The Prevention of Burns and Scalds: Review of 1000 Cases', *The Lancet*, 254 (1949), pp. 181–8.
6. Colebrook published these data in at least twenty-seven articles in peer-reviewed journals, broadsheet and tabloid newspapers, magazines, journals and newsletters of civil society organizations such as the Women's Institute. On his retirement, his successors at the Birmingham Burns Unit continued to collect, analyse and publish data. See, for instance, L. Colebrook, V. Colebrook, J. P. Bull and D. M. Jackson, 'The Prevention of Burning Accidents: A Survey of the Present Position', *British Medical Journal*, 1 (1956), pp. 1379–86; J. P. Bull, D. M. Jackson and C. Walton, 'Causes and Prevention of Domestic Burning Accidents', *British Medical Journal*, 2 (1964), pp. 1421–7.
7. Ministry of Housing and Local Government, *Accidents in the Home: Fireguards*, Circular No. 13/57, 1957, MH 78/370, NA.
8. Lawrence, 'Some Aspects of Burns and Burns Research'.
9. D.A.K. Black, 'Treatment of Burn Shock with Plasma and Serum', *British Medical Journal*, 2 (1940), pp. 693–7.
10. R. B. Bourdillon and L. Colebrook, 'Air Hygiene in Dressing-rooms for Burns or Major Wounds', *The Lancet*, 1 (6399) (1946), pp. 561–5.
11. L. Colebrook, 'The Control of Infection in Burns', *The Lancet*, 1 (6511) (1948), pp. 893–9.
12. L. Colebrook, *A New Approach to the Treatment of Burns and Scalds* (London: Fine Technical Publications, 1950).

13. L. Colebrook and V. Colebrook, 'A Suggested National Plan to Reduce Burning Accidents Based on a Study of 200 Cases', *The Lancet*, 258 (1951), pp. 579–84.
14. 'The Medical Services in the Korean War', *The Lancet*, 261 (1953), pp. 134–5.
15. Patrick Clarkson, 'A-bomb Casualties and Arrangements for their Care', *British Journal of Plastic Surgery*, 3 (1951), pp. 188–93; Air Commodore G. H. Morley, Royal Air Force Consultant in Plastic Surgery, 'Thoughts on Burns from Thermo-nuclear Weapons', n.d., AIR 20/9485, NA.
16. Ministry of Health, *Treatment of Burns*, 1952, MH 160/635, NA.

9 Extending Shelf Life After Death

1. B. G. Sparkes, 'Treating Mass Burns in Warfare, Disaster or Terrorist Strikes', *Burns*, 23 (1997), pp. 238–47.
2. Susan Lindee, *Suffering Made Real: American Science and the Survivors at Hiroshima* (Chicago and London: Chicago University Press, 1997).
3. Eileen Welsome, *The Plutonium Files: America's Secret Medical Experiments in the Cold War* (New York: Delta Trade Paperbacks, 1999); Jonathan D. Moreno, *Undue Risk: Secret State Experiments on Humans* (New York and London: Routledge, 2001).
4. Leonard F. Bush and C. Zent Garber, 'The Bone Bank', *Journal of the American Medical Association*, 137 (7) (1948), pp. 588–94.
5. G. W. Hyatt, *A Report of the Bone Bank, March 1949*, File NM7 01 001, Box 75, RG52, National Archives, College Park, Maryland, hereafter USNA.
6. G. W. Hyatt, 'The Founding of the US Navy Tissue Bank', *Transplantation Proceedings*, 8 (2) Supplement 1 (1976), p. 17.
7. G. W. Hyatt, 'The Navy's Wonderful Tissue Bank', *Hospital Management* (May 1956), pp. 41–4.
8. A. S. Parkes, *Sex, Science and Society* (Newcastle-upon-Tyne: Oriel Press, 1966), pp. 248–75; D. E. Pegg, 'The History and Principles of Cryopreservation', *Seminars in Reproductive Medicine*, 20 (1) (2002), pp. 5–13.
9. G.E.W. Wolstenholme and Margaret P. Cameron (eds), *Preservation and Transplantation of Normal Tissues: A CIBA Foundation Symposium* (London: J. & A. Churchill, 1954).
10. G. W. Hyatt, *Research Proposal: Novel Approaches to the Treatment of Naval Casualties, 1950*, File NM 7 01 001, Box 75, RG52, USNA.
11. Earl W. Flosdorf, 'Drying Penicillin by Sublimation in the United States and Canada', *British Medical Journal* (17 February 1945), pp. 216–18; Earl F. Flosdorf, *Freeze-drying: Drying By Sublimation* (New York: Rheinhold Publishing Company, 1949).
12. Hyatt, 'The Founding of the US Navy Tissue Bank', p. 17.
13. 'Life From Death', *Time* magazine (21 May 1956), p. 81.
14. W. Ronald Strong, 'The Tissue Bank, Its Operation and Management', in Wolstenholme and Cameron (eds), *Preservation and Transplantation of Normal Tissues*, p. 222.
15. Rollin W. Franks, United States Navy Tissue Bank, interview with author (Bethesda, Maryland, December 1998).
16. William M. Abbott and Everett Dupree, 'The Procurement, Storage, and Transplantation of Lyophilized Human Cadaver Dura Mater', *Surgery, Gynecology & Obstetrics*, 130 (1970), pp. 112–18.
17. Rollin W. Franks, United States Navy Tissue Bank, interview with author (Bethesda, Maryland, December 1998).
18. Strong, 'The Tissue Bank, Its Operation and Management', p. 224.
19. G. W. Hyatt, T. C. Turner, C.A.L. Bassett, J. W. Pate and P. N. Sawyer, 'New Methods for Preserving Bone, Skin and Blood Vessels', *Postgraduate Medicine*, 12 (1952), pp. 239–54.
20. Hyatt, 'The Founding of the US Navy Tissue Bank', p. 19.
21. Jim Ostrander, United States Navy Tissue Bank, interview with author (Bethesda, Maryland, December 1998).
22. R. E. Billingham, 'Concerning the Origins and Prospects of Cryobiology and Tissue Banks', *Transplant Proceedings*, Supplement, 8 (1976), p. 8.
23. 'Truce is Signed, Ending the Fighting in Korea', *New York Times* (27 July 1953).

24. Donald M. Goldstein and Harry J. Maihafer, *The Korean War: The Story and Photographs* (Washington DC: Basseys, 2000), p. 129.
25. George Hyatt, The Tissue Bank and its clinical program, Minutes, Naval Medical Research No. 14 – 27 September 1957, Appendix F.MED: Committee on Naval Medical Research, 1954–1957 Meetings: Eighth – Fourteenth. Washington DC, National Research Council (hereafter NRC).
26. James F. Dunnigan and Albert A. Nofi, *Dirty Little Secrets of the Vietnam War* (New York: St. Martin's Press, 1999), p. 241.
27. Howard A. Rusk, 'Medicine at War – 1: Low Mortality of Vietnam Wounded is Tribute to Quality of Health Aids', *New York Times* (22 May 1966).
28. William C. Trier and Kenneth W. Sell, 'United States Navy Skin Bank', *Plastic and Reconstructive Surgery*, 41 (6), (1968), pp. 543–8.
29. J. K. Herman, 'The Navy Tissue Bank: The World's First Still Sets the Standards for Others to Follow', *U.S. Navy Medicine*, 71 (5) (1980), p. 7.

10 Cadaver Eyes, Death Denial and the National Health Service

1. E. R. Mayhew, *The Reconstruction of Warriors: Archibald McIndoe, the Royal Air Force and the Guinea Pig Club* (London and Pennsylvania: Greenhill Books, 2004).
2. J. P. Bennett, 'A History of the Queen Victoria Hospital, East Grinstead', *British Journal of Plastic Surgery*, 41 (1988), pp. 422–40.
3. J. W. Tudor Thomas, 'The Technique of Corneal Transplantation as Applied in a Series of Cases', *Ophthalmological Society's Transactions*, 55 (1935), p. 55; J. W. Tudor Thomas, 'Corneal Transplantation: Results of a Series of 56 Operations on 48 Eyes', *British Medical Journal* (8 October 1938), pp. 740–42.
4. Cuttings file, R3, Royal National Institute for the Blind Archive, hereafter RNIB.
5. Benjamin Rycroft, 'The Organization of a Regional Eye Bank', *The Lancet*, 279 (1962), 147–51.
6. Russell M. Davies, 'Archibald McIndoe, His Times, Society, and Hospital', *Annals of the Royal College of Surgeons*, 59 (1977), pp. 359–67.
7. H. H. Skeoch to the National Institute of the Blind Information Officer, 28 October 1946, R3, RNIB.
8. F. W. Law, the Faculty of Ophthalmologists, to Eggar, National Institute for the Blind, 17 December 1947, Folder 735, RNIB.
9. Stewart Duke-Elder to Wilson Jameson, Chief Medical Officer, 19 May 1949, MH 58/500, NA.
10. Memo on Corneal Grafting, Shaw to Dr Banks, 15 February 1948, MH 58/500, NA.
11. Elaine Blond with Barry Turner, *Marks of Distinction: The Memoirs of Elaine Blond* (London: Valentine, Mitchell, 1988), pp. 112–13.
12. *Kent and Sussex Courier* (17 October 1947), p. 5.
13. Hugh McLeave, *McIndoe: Plastic Surgeon* (London: Frederick Muller, 1961), p. 157.
14. *Kent and Sussex Courier* (19 October 1951), p. 7.
15. Raymond Gelding, Cabinet Office, to Michael Reed, Ministry of Health, 6 December 1951, MH 58/500, NA.
16. Philippe Ariès, *The Hour of Our Death* (Harmondsworth: Peregrine, 1983), p. 560.
17. Charles Webster, *The National Health Service: A Political History* (Oxford: Oxford University Press, 1998), p. 113.
18. Jonathan Sawday, *The Body Emblazoned: Dissection and the Human Body in Renaissance Culture* (London: Routledge, 1995), p. 55.
19. *Kent and Sussex Courier* (9 November 1951), p. 5.

11 Whose Corpse Is It?

1. 'Passage of a Bill', *Kent and Sussex Courier* (20 June 1952), p. 5.
2. Elaine Scarry, *The Body in Pain: The Making and Unmaking of the World* (Oxford: Oxford University Press, 1985), p. 81.

3. 'Through Churchill's Eyes?', *Daily Sketch* (22 May 1956).
4. Copy of note from Miss A. M. Palmer, n.d., R3/3, RNIB.
5. Eye bequests. Note on the Corneal Grafting Act 1952, n.d., R3/3, RNIB.
6. John Kord Lagemann and Don Everitt, 'They Pass on the Gift of Sight', *Reader's Digest* (January 1965).
7. B. W. Rycroft, 'Second-hand Sight', *Medicine Illustrated* (1953), pp. 582-4.
8. Ruth Richardson, *Death, Dissection, and the Destitute* (Harmondsworth: Pelican, 1988), p. 293.
9. Ministry of Health, Corneal Grafting Act, 26 June 1952, MH 150/80, NA.
10. Rudolf Klein, *The Politics of the National Health Service*, 2nd edn (London: Longman, 1989), p. 42.
11. Ibid., p. 44.
12. Rosemary Stevens, *Medical Practice in Modern England: The Impact of Specialization and State Medicine* (New Haven and London: Yale University Press, 1966), p. 190.

12 Collecting British Cadaver Pituitary Glands

1. Philip Randle, 'Frank George Young', *Biographical Memoirs of Fellows of the Royal Society*, 36 (1990), pp. 583-99.
2. Marion Nestle, *Safe Food: The Politics of Food Safety* (Berkeley, Los Angeles and London: University of California Press, 2010), p. 52.
3. 'The Utilisation of Slaughterhouse By-products', *Meat Trades Journal and Cattle Salesman's Gazette* (9 January 1941), p. 41.
4. 'Gland Products', *Meat Trades Journal and Cattle Salesman's Gazette* (3 July 1941), p. 9.
5. S. L. Simpson, 'Dwarfism, Hormones, and Dieting', *British Medical Journal*, 2 (1947), p. 977.
6. Meeting of clinicians about growth hormones, 14 June 1957, FD 1/8367, NA.
7. A panel appointed by the Clinical Endocrinology Committee of the Medical Research Council, 'The Effectiveness in Man of Human Growth Hormone', *The Lancet* (1959), pp. 7-12.
8. 'Children of Short Stature', *British Medical Journal*, 3 (1967), pp. 187-8; J. M. Tanner and R. H. Whitehouse, 'Growth Response of 26 Children with Short Stature Given Growth Hormones', *British Medical Journal*, 2 (1967), pp. 69-75; J. M. Tanner, 'Towards Complete Success in the Treatment of Growth Hormone Deficiency: A Plea for Earlier Ascertainment', *Health Trends*, 7 (1975), pp. 61-5; R.D.G. Milner, T. Russell-Fraser, C.G.D. Brook, P. M. Cotes, J. W. Farquhar, J. M. Parkin, M. A. Preece, G.J.A.I. Snodgrass, A. Stuart Mason, J. M. Tanner and F. P. Vince, 'Experience with Human Growth Hormone in Great Britain: The Report of the MRC Working Party', *Clinical Endocrinology*, 11 (1979), pp. 15-38.
9. Russell Fraser, Hon. Secretary, MRC Clinical Endocrinology Committee, 'Directions for Collection of Human Pituitaries for Later Growth Hormone Extraction' (n.d.), FD 7/3542, NA.
10. Charles Webster, *The Health Services since the War, Volume I. Problems of Health Care: The National Health Service before 1957* (London: Her Majesty's Stationery Office, 1988), pp. 1-9.
11. Tom Cobbleton, 'LVII-De Mortus', *The Lancet*, 274 (1959), pp. 1081-2.
12. Noëlie Vialles, *Animal to Edible* (Cambridge: Cambridge University Press, 1994), p. 51.
13. Independent Review Group on Retention of Organs at Post-Mortem, *Report on Strontium-90 Research* (Edinburgh: Scottish Executive Health Department, 2002), p. 19.
14. Raymond Gelding, Cabinet Office, to Ministry of Health, 6 December 1951, MH 58/500, NA.
15. Grafting of human tissues: notes for minister (n.d.) MH 58/497, NA.
16. Cabinet Home Affairs Committee, Human Tissues, etc., 22 November 1960, FD 9/4607, NA.

17. HC Deb 29 June 1961, vol. 643 cc 844 (Edith Pitt).
18. National Health Service, Human Tissue Act 1961, H.M.(61) 98.
19. C.D. to Dr A. L. Winner, 16 August 1963, MH 150/70, NA.
20. H. P. Hope, Office of the Parliamentary Counsel, to S. D. Musson, Ministry of Health, 10 October 1960, MH 58/497, NA.
21. W. L. Mancey to Enoch Powell, 31 January 1961, MH 58/497, NA.
22. Independent Review Group, *Report on Strontium-90 Research*, p. 19.
23. Ibid., p. 25.
24. Medical Research Council, 'Note on the preparation and distribution of hgh', 8 February 1960, FD 1/8371, NA.
25. Work study of the post of Coroner's Officer, September 1967, HO 375/19; Post-mortem room technicians pay claim 1969, MH58/331; Pay Claim – Post mortem technicians, 1970-72, MH 58/332, NA.
26. Viviana A. Zelizer, *The Social Meaning of Money: Pin Money, Pay Checks, Poor Relief, and Other Currencies* (New York: Basic Books, 1994), pp. 94-9.
27. Russell Fraser, Hon. Secretary, MRC Clinical Endocrinology Committee, Directions for collection of human pituitaries for later growth hormone extraction, 1958, FD 7/3542, NA.

13 Lionizing American Eye Banks

1. 'Eye Bank Committee', *American Journal of Ophthalmology*, 42 (1957), pp. 309-10.
2. Emile J. Farge, 'The Eye-Bank Association of America', in F. B. Brightbill (ed.), *Corneal Surgery: Theory, Technique and Tissue* (St Louis: Mosby, 1999), pp. 674-82.
3. *American Journal of Ophthalmology*, 54 (1962), pp. 309-10.
4. Paul Starr, *The Social Transformation of American Medicine* (New York: Basic Books, 1984), pp. 348-51; Rosemary Stevens, *In Sickness and in Wealth: American Hospitals in the Twentieth Century* (New York: Basic Books, 1989), pp. 216-19.
5. Richard Dunlop, 'Missouri's Brick and Mortar Monument', *The Lion* (November 1975), pp. 14-17.
6. Paul Martin, 'New Lions Eye Institute Fills a Multi-faceted Role', *The Lion* (October 1975), pp. 16-18.
7. Ted Blankenship, 'A Prudent Use of Funds', *The Lion* (June 1983), pp. 10-11, 35.
8. 'Iowa's Eye Bank', *The Lion* (November 1957), pp. 18-20.
9. 'Hospital Gets Eye Bank', *The Lion* (August 1961), p. 4.
10. Dunlop, 'Missouri's Brick and Mortar Monument', p. 16.
11. Gary Laderman, *Rest in Peace: a Cultural History of Death and the Funeral Home in Twentieth-Century America* (New York: Oxford University Press, 2003), p. 97.
12. Anatomical Gift Act, hearing before the Committee on the District of Columbia United States Senate, Ninety-First Congress, Second Session (Washington DC: US Government Printing Office, 1970), p. 80.
13. Paul P. Lee, 'Cornea Donation Laws in the United States', *Archives of Ophthalmology*, 107 (1989), pp. 1585-9.
14. 'Eye Enucleation: First Step on the Road to Sight', *The Lion* (December 1973), pp. 34-5.
15. Ted Blankenship, 'A New Eye Bank for Kansans', *The Lion* (June 1982), p. 25.
16. 'New Scope for Eye Bank', *The Lion* (May 1954), p. 19; 'Salute to the Pioneers', *The Lion* (January 1971), pp. 16-19.
17. Emile J. Farge, 'Ethics and Eye Banking', in Jay H. Krachmer, Mark J. Mannis and Edward J. Holland (eds), *Cornea: Fundamentals of Cornea and External Disease*, volume 1 (St Louis: Mosby, 1997), pp. 531-55.
18. 'Eyeball Network Helps Blind Regain Sight', *The Lion* (April 1963), p. 48.
19. 'Eyes by Wireless', *The Lion* (May 1965), pp. 22-4.
20. John W. Payne, 'New Directions in Eye Banking', *Transactions of the American Ophthalmological Society*, 78 (1980), pp. 982-1026.
21. John Harry King Jr, 'American Diplomacy Eye to Eye', *EBAA Newsight* (February 1965), pp. 29-33.

22. Bill Cordes, 'The Eye Bank Moves East', *The Lion* (May 1965), pp. 20–21.
23. 'Medico Expanding', *Sight-Saving Review*, 31 (4) (1961), p. 239.
24. Robert M. Janes, 'CARE and MEDICO of Canada', *Canadian Medical Association Journal*, 90 (13) (1964), p. 795.
25. Dr. Thomas A. Dooley, Biography, www.umsl.edu (accessed December 2012).
26. 'Gift of Vision', *The Lion* (May 1967), p. 43.
27. Paul Martin, *We Serve: A History of the Lions Clubs* (Washington DC: Regnery Gateway, 1991), p. 22.

14 A Gland Lost is a Gland Wasted

1. S. Douglas Frasier, 'The Not-so-good old days: Working with Pituitary Growth Hormone in North America, 1956–1985', *Journal of Pediatrics*, 131 (1), part 2 (1997), p. S4.
2. 'Pituitary Bank Set Up on the Coast', *New York Times* (21 July 1960).
3. 'Pituitary Collection Program', *CAP Bulletin*, 15 (11) (1961), p. 8.
4. Robert M. Blizzard, 'History of Growth Hormone Therapy', *Indian Journal of Pediatrics*, 79 (1) (January 2012), pp. 87–91.
5. Robert Brill, 'Pituitary Collection Project: Progress Report', *CAP Bulletin*, 16 (2) (1962), pp. 22–3.
6. Correspondence between Li and Croxatto is in 'C' file, Carton 4, CHL Papers.
7. Correspondence and minutes of meetings relating to the origins of the National Pituitary Agency are in National Pituitary Agency 1963 file, and National Pituitary Agency 1965 file, Carton 10, CHL Papers. See also, Philip Henneman, Paul Steinlauf, John Tullis and Robert Brill, 'Summary of the Final Report of the Ad Hoc Committee on Pituitary Collection', *CAP Bulletin* (July 1963), pp. 85–6; Robert A. Tolman, 'The NIADDK Hormone Distribution Program', in Salvatore Raiti and Robert A. Tolman (eds), *Human Growth Hormone* (New York: Plenum Medical Book Company, 1986), pp. 13–18.
8. Minutes of Meeting, Special Committee on Collection of Human Pituitary Glands, Bethesda, Maryland, 12 and 13 March 1962, Nat. Pit. Agency 1963 file, Carton 10, CHL Papers, p. 6.
9. Draft proposal for the support and coordination of collection of human pituitaries, p. 4, Nat. Pit. Agency 1963 file, Carton 10, CHL Papers.
10. Minutes of Meeting, Special Committee on Collection of Human Pituitary Glands, Carton 10, CHL Papers, p. 7.
11. 'Pituitary Gland Removal and Storage', n.d. Nat. Pit. Agency 1963 file, Carton 10, CHL Papers.
12. Fred G. Smith to Choh Hao Li, 24 July 1964. Peter H. Forsham file, Carton 4, CHL Papers.
13. 'Pituitary Appeal Made', *CAP Bulletin*, 19 (10) (1965), p. 259.
14. Susan Cohen and Christine Cosgrove, *Normal at Any Cost: Tall Girls, Short Boys, and the Medical Industry's Quest to Manipulate Height* (New York and London: Jeremy P. Tarcher/Penguin, 2009), p.77.
15. Howard C. Cartwright, 'Pituitary Collection, Research Get New Aid', *CAP Bulletin*, 22 (4) (1968), pp. 133–5.
16. William J. Reals, 'The National Pituitary Agency and Human Growth', *Pathologist*, 25 (8) (1972), pp. 191–2.
17. *Pathologist*, 26 (11) (1973), p. 456.
18. Salvatore Raiti, 'An Appreciative Word', *The Pathologist*, 30 (7) (1976), p. 255.
19. Final Report, Growth Hormone Committee, Pituitary Bank, n.d., 'E' 9 file, Carton 4, CHL Papers.
20. Minutes of Meeting, Special Committee on Collection of Human Pituitary Glands, Bethesda, Maryland, 12 and 13 March 1962, p. 5, Nat. Pit. Agency file, Carton 10, CHL Papers.
21. Minutes of the Growth Hormone Committee Meeting, 2 February 1960, 'E' file, Carton 9, CHL Papers.
22. Ibid.

23. Lloyd Shearer, 'We Can End Dwarfism!', *Parade* (22 August 1965), pp. 4–6.
24. Jennifer Cooke, *Cannibals, Cows and the CJD Catastrophe* (London: Random House, 1998), p. 49.
25. Final Report, Growth Hormone Committee, Pituitary Bank, n.d., 'E' 9 file, Carton 4, CHL Papers.
26. Harold H. Schmeck Jr, 'Progress on Hormone', *New York Times* (15 May 1966).
27. Office of Public Information, released 5 May 1966, in Nancy M. Rockafellar, *Conversations with Dr Leslie Batty Bennett: The Research Tradition at UCSF* (San Francisco: UCSF Oral History Program, Department of History and Health Sciences, University of California, San Francisco, 1992), pp. 146–8.
28. 'Human Growth Hormone is Synthesized for the First Time by California Scientists', *Wall Street Journal* (7 January 1971), p. 4.
29. News from University of California, San Francisco, Office of Public Information, 7 January 1971, in Rockafellar, *Conversations with Dr Leslie Batty Bennett*, pp. 149–51.
30. Hugh D. Niall, 'Revised Primary Structure for Human Growth Hormones', *Nature New Biology*, 230 (17 March 1971), pp. 90–91; 'Erroneous, Dr. Li . . . I Know, Dr. Niall', *Medical World News*, 2 April 1971, in Rockafellar, *Conversations with Dr Leslie Batty Bennett*, pp.154–5.

15 Who's in the Mortuary?

1. National End of Life Care Intelligence Network, Deprivation and Death: variation in place and cause of death, 2012, www.endoflifecare-intelligence.org.uk (accessed January 2014).
2. Philippe Ariès, *The Hour of Our Death* (Harmondsworth: Penguin, 1983), pp. 159–601.
3. Committee on Death Certification and the Investigation of Death by Coroners, *Report*, Cmnd. 4810 (London: Her Majesty's Stationery Office, 1971).
4. The Shipman Inquiry, Third Report: Death Certification and the Investigation of Deaths by Coroners, CM 5854 (London: The Stationery Office, 2003).
5. D. E. B. Powell to Dr P. Alwyn-Smith, 28 September 1965, BD 18/983, NA.
6. Lawrence Cohen, 'The Other Kidney: Biopolitics Beyond Recognition', in N. Scheper-Hughes and L. Wacquant (eds), *Commodifying Bodies* (London: Sage Publications, 2002), pp. 9–30; Lawrence Cohen, 'Operability, Bioavailability, and Exception', in A. Ong and S. J. Collier (eds), *Global Assemblages: Technology, Politics, and Ethics as Anthropological Problems* (Oxford: Blackwell Publishing, 2005), pp. 79–90.
7. Ruth Richardson, *Death, Dissection and the Destitute* (Harmondsworth: Pelican, 1988), pp. 121–9.
8. Michael Mulkay and John Ernst, 'The Changing Profile of Social Death', *European Journal of Sociology*, 22 (1991), pp. 172–96; Jana Králova, 'What is Social Death?', *Contemporary Social Science*, 10 (3) (2015), pp. 235–48.
9. Helen Sweeting and Mary Gilhooly, 'Doctor, Am I Dead? A Review of Social Death in Modern Societies', *Omega (Westport)*, 24 (4) (1991–2), pp. 251–69; Helen Sweeting and Mary Gilhooly, 'Dementia and the Social Phenomenon of Social Death', *Sociology of Health and Illness*, 19 (1) (1997), pp. 93–117.
10. Ariès, *The Hour of Our Death*, p. 571. The survey of British attitudes to and experience of death was conducted by British anthropologist Geoffrey Gorer (1905–1985), who formulated the concept of 'death denial'. Geoffrey Gorer, *Death, Grief and Mourning in Contemporary Britain* (New York: Doubleday, 1965).
11. The Creutzfeldt-Jakob Disease Litigation, *Plaintiffs v United Kingdom Medical Research Council* (19 July 1996) QBD, p. 31.
12. Tim Kelsey, 'Families Seek Damages over CJD Deaths', *The Independent* (16 August 1993), p. 3.
13. Clare Dyer, 'Growth Hormone Victims Seek Compensation', *British Medical Journal* (6 March 1993), p. 607.
14. Wilson Smith, Department of Bacteriology, University College London Hospital, to Professor Fraser, 31 December 1958, FD1/8370, NA.

16 Representational Dilemmas in Marketing Eye Pledges

1. Helen MacDonald, 'Conscripting organs: "Routine Salvaging" or Bequest? The Historical Debate in Britain, 1961–75', *Journal of the History of Medicine and Allied Sciences*, 70(3) (2014), p. 427.
2. HC Deb 20 December 1960, vol. 632 cc1253 (J. Enoch Powell).
3. HC Deb 29 June 1961, vol. 643 cc 844 (Edith Pitt).
4. Rycroft to Colligan, 14 January 1961, R3/3, RNIB.
5. Corneal Grafting – replies from hospital authorities, January 1961, MH 58/505, NA.
6. B. Rycroft, 'The Organisation of a Regional Eye Bank', *The Lancet*, 279 (1962), pp. 147–51; B. Rycroft, 'Contemporary Views on the Surgery and Biology of the Corneal Graft', *Annals of the Royal College of Surgeons of England*, 36 (1965), pp. 152–71.
7. John Lagemann and Don Everitt, 'They Pass on the Gift of Sight', *Reader's Digest* (1965), p. 3.
8. J. C. Colligan, Secretary General of the RNIB, to Rycroft, 10 January 1961, R3/3, RNIB.
9. Virginia Berridge and Alex Mold, 'Professionalisation, New Social Movements and Voluntary Action in the 1960s and 1970s', in Matthew Hilton and James McKay (eds), *The Ages of Voluntarism: How we got to the BIG SOCIETY* (Oxford: Published for the British Academy by Oxford University Press, 2011), pp. 114–34.
10. Anne Karpf, *Doctoring the Media: The Reporting of Health and Medicine* (London: Routledge, 1988), pp. 49–56.
11. Colligan to Rycroft. 12 August 1966, R3/3, RNIB.
12. Rycroft to Colligan, 9 August 1966, R3/3, RNIB.
13. Ann Reedy, art editor, *Birmingham Post*, to Ministry of Health Press Office, 29 December 1966, R3/3, RNIB.
14. John Agar, 'Modern Horrors: British Identity and Identity Cards', in Jane Caplan and John Torpey (eds), *Documenting Individual Identity: The Development of State Practices in the Modern World* (Princeton and Oxford: Princeton University Press, 2001), pp. 101–20.
15. Draft Memo to Mr Fish, September 1960, MH 150/79, NA.
16. Colligan to A. L. Winner, 5 September 1963, R3/3, RNIB.
17. National Health Service, Human Tissue Act 1961, H.M. (61) 98.
18. Mr Smith to Commander T. S. Jackson R.N. (Retd), 23 April 1963, N.A. Folder MH 151/35, NA.
19. Note for file, 7 December 1962, R3/3, RNIB.
20. Dr G. C. Milner, Ministry of Health to Colligan, 11 May 1962, R3/3, RNIB.
21. '150,000 Future Donors Sought for Regional Eye Bank', *The Times* (2 March 1965).
22. Morris M. Agnew, Chairman, The United Manchester Hospitals to Sir Bruce Fraser, Permanent Secretary, Ministry of Health, 6 August 1983, MH 150/79, NA.
23. Edward to Barber, 11 April 1964, R3/3A, RNIB.
24. J. Fish, Ministry of Health, to Secretaries of Regional Hospital Boards, 17 September 1964, MH 150/79, NA.
25. P. D. Trevor-Roper, 'Westminster Hospital: Suggestions for the Promotion of British Eye-banks' (n.d.), MH 150/79, NA.
26. Dr Milner to Colligan, 11 May 1962. R3/3 RNIB.
27. Rudolph Klein, *The Politics of the National Health Service*, 2nd edn (London: Longman, 1989) p. 51.
28. Bob Marchant, Blond McIndoe Centre, Queen Victoria Hospital, interview with author (East Grinstead, August 2005).
29. David Pegg, University of York/British Association of Tissue Banks, interview with author (York, November 2007).
30. Michael Jeffries, 'New Eyes for Old from the Deep-freeze', *Evening News* (23 February 1965).
31. P. V. Rycroft, Report of the East Grinstead Eye Bank, MH 150/79, NA.
32. T. A. Casey, 'Editorial: Corneal Storage', *British Journal of Ophthalmology*, 73 (1989), p. 774.

33. Elizabeth Shaw, Memo, 22 December 1964. MH 150/79, NA.
34. R. E. Billingham and B. W. Rycroft, 'The Preservation of the Donor Graft', in B. W. Rycroft (ed.), *Corneal Grafts* (London: Butterworth, 1955), pp. 195–207.
35. Rycroft, *Report on East Grinstead Eye Bank*, MH 150/79, NA.
36. 'Eye Banks', *British Medical Journal 1* (1964), p. 388.
37. J. Fish to Mr Williamson, 6 October 1965. N.A. Folder, MH 150/79, NA.
38. Birmingham Region: Eye Bank, 17 February 1965, MH 150/79, NA.
39. Central Health Services Council, Standing Medical Advisory Committee, *Corneal Grafting: Report of the Sub-Committee*, SAC(M)(CG)65, 1965, R3/3A, RNIB.
40. D. B. Archer and P. D. Trevor-Roper, 'Organization and Administration of Westminster-Moorfields Eye-Bank', *British Journal of Ophthalmology*, 51 (1) (1967), pp. 1–12.

17 Banking British Cadaver Skin

1. Charles Webster, *The Health Services since the War. Volume I: Problems of Health Care, the National Health Service before 1957* (London: Her Majesty's Stationery Office, 1988), p. 209.
2. T. L. Barclay, 'New Burns Unit in Wakefield, Yorkshire', in A. B. Wallace and A. W. Wilkinson (eds), *Research in Burns* (Edinburgh: E & S Livingstone, 1966), pp. 547–9.
3. J. Ellsworth Laing, 'Design of a New Burns Unit for the Wessex Regional Hospital Board', in Wallace and Wilkinson, *Research in Burns*, pp. 542–6.
4. John Watson, 'Experience of Burns Unit Design – Queen Victoria Hospital, East Grinstead, England', in Wallace and Wilkinson, *Research in Burns*, pp. 534–41.
5. Dulyn Thomas, Report on South East Metropolitan Regional Hospital Board Plastic Surgery Service, 1970, MH 160/483, NA.
6. Elaine Blond with Barry Turner, *Marks of Distinction: The Memoirs of Elaine Blond* (London: Valentine, Mitchell, 1988), pp. 145–54.
7. Peter Medawar, *Memoirs of a Thinking Radish* (Oxford: Oxford University Press, 1991), p. 135.
8. Murray, in G.E.W. Wolstenholme and Margaret P. Cameron (eds), *Preservation and Transplantation of Normal Tissues: A CIBA Foundation Symposium* (London: J. & A. Churchill, 1954), pp.19–23.
9. Ilana Löwy, 'Tissue Groups and Cadaver Kidney Sharing: Socio-cultural Aspects of a Medical Controversy', *International Journal of Technology Assessment in Health Care* (1986), pp. 195–218; Ilana Löwy, 'The Impact of Medical Practice on Biomedical Research: The Case of Human Leucocyte Antigens Studies', *Minerva*, 25 (1–2) (1987), pp. 171–200.
10. Tom Cochrane, Blond McIndoe Centre, Queen Victoria Hospital, interview with author (East Grinstead, August 2005).
11. Douglas Jackson, 'A Clinical Study of the Use of Skin Homografts for Burns', *British Journal of Plastic Surgery*, 7 (1) (7 April 1954), pp. 26–43.
12. Frank Dexter, 'Tissue Banking in England', *Transplantation Proceedings* (June 1976), Supplement, pp. 43–8; John N. Kearney, 'Yorkshire Regional Tissue Bank – Circa 50 Years of Tissue Banking', *Cell and Tissue Banking*, 7 (2006), pp. 259–64.
13. R. E. Billingham and P. Medawar, 'The Freezing, Drying and Storage of Mammalian Skin', *Journal of Experimental Biology*, 29 (1952), pp. 454–68.
14. Tom Cochrane, Blond McIndoe Centre, Queen Victoria Hospital, interview with author (East Grinstead, August 2005).
15. Tom Cochrane, Interim Report of the First Rayne Fellow in Burns Research, 1966, Blond McIndoe Research Unit Archives.
16. Sara Robinson, 'World of Hospital Unit is More than Skin Deep', *East Grinstead Observer* (17 November 1976).
17. T. Cochrane, 'The Low Temperature Storage of Skin: A Preliminary Report', *British Journal of Plastic Surgery*, 21 (2) (1968), pp. 118–25.
18. M. E. Hackett, 'Cadaver Homografts', *Proceedings of the Royal Society of Medicine*, 64 (1971), pp. 1292–4.

19. Kenneth Chambler, Report to the Trustees of the Rayne Foundation, n.d., Blond McIndoe Research Unit Archives.
20. Kenneth Chambler to the editor of the *British Medical Journal*, 16 March 1968, Blond McIndoe Research Unit Archives.
21. The Tissue Bank, East Grinstead (n.d.), Blond McIndoe Research Unit Archives.
22. Bob Marchant, Blond McIndoe Centre, Queen Victoria Hospital, interview with author (East Grinstead, August 2005).
23. John Watson to Russell M. Davies, Queen Victoria Hospital, 15 September 1967, Blond McIndoe Research Unit Archives.
24. Tom Cochrane, Blond McIndoe Centre, Queen Victoria Hospital, interview with author (East Grinstead, August 2005).
25. J. B. Batchelor and M. Hackett, 'HL-A Matching in Treatment of Burned Patients with Skin Allografts', *The Lancet* (19 September 1970), pp. 581–3; Mohini Roberts, 'The Role of the Skin Bank', *Annals of the Royal College of Surgeons*, 58 (1976), pp. 70–74.
26. Thomas E. Starz, *The Puzzle People: Memoirs of a Transplant Surgeon* (Pittsburgh and London: University of Pittsburgh Press, 1992), p. 112.
27. 'Transplantation and Burns', *The Lancet*, 305 (1975), pp. 1017–18.
28. Skin Bank Records, Blond McIndoe Research Unit Archives.
29. Andrew Burd, 'Once Upon a Time and the Timing of Surgery in Burns', *Journal of Plastic, Reconstructive and Aesthetic Surgery*, 61 (3) (2008), pp. 237–9.
30. Zora Janžekovič, 'A New Concept in the Early Excision and Immediate Grafting of Burns', *Journal of Trauma*, 10 (12) (1970), pp. 1103–8; Zora Janžekovič, 'The Burn Wound from the Surgical Point of View', *Journal of Trauma*, 15 (1) (1975), pp. 42–62; Zora Janžekovič, 'Once Upon a Time . . . How West Discovered East', *Journal of Plastic, Reconstructive and Aesthetic Surgery*, 61 (2008), pp. 240–44.
31. Reports of the visits to various burns units conducted by civil servants in 1970, MH 160/484, NA.
32. 'Facilities for Burns Treatment in the United Kingdom', *Burns*, 4 (4) (1987), pp. 297–300.
33. HC Deb 27 May 1977, vol. 932 cc 691 (Michael Ward).
34. Sixth Planning Cycle – 1978/79, Group Planning Statement 4, Acute Specialist Services, November 1978, MH 160/635, NA.
35. Antony F. Wallace, 'The First Decade of the North East Thames Regional Plastic and Surgery and Burns Unit', *British Journal of Plastic Surgery*, 38 (1985), pp. 422–5.

18 The Burn-prone Society

1. Leonard Colebrook, 'The Prevention of Burning Accidents in England and America', *Bulletin of the New York Academy of Medicine*, 27 (7) (1951), pp. 425–38.
2. 'Burned Girl, 9, Saved by 10 Prisoners' Skin', *New York Times* (9 January 1951).
3. 'Fire Victim for Whom 15 Donated Skin Recovers after Nearly a Year in Bellevue', *New York Times* (29 March 1951).
4. James Barrett Brown, Minot P. Fryer, Peter Randall and Milton Lu, 'Postmortem Homografts as "Biological Dressings" for Extensive Burns and Denuded Areas', *Annals of Surgery*, 138 (4) (1953), pp. 618–30; James Barrett Brown and Minot P. Fryer, 'Postmortem Homografts to Reduce Mortality in Extensive Burns', *Journal of the American Medical Association*, 156 (12) (1954), pp. 1163–5; James Barrett Brown and Minot P. Fryer, 'Skin Homografts from Postmortem Sources', *American Journal of Surgery*, 97 (1959), pp. 418–20.
5. R. J. Kagan, 'Human Skin Banking: Past, Present and Future', in D. M. Strong (ed.), *Advances in Tissue Banking: Volume 2* (Singapore: World Scientific, 1998), pp. 297–321.
6. James Barrett Brown, Minot P. Fryer and Thomas J. Zaydon, 'Establishing a Skin Bank: Use and Various Methods of Preservation of Postmortem Homografts', *Plastic and Reconstructive Surgery*, 16 (5) (1955), pp. 337–51.
7. Harold H. Schmeck Jr., 'Navy Plans Bank for Living Tissue', *New York Times* (26 May 1963).
8. Thomas R. Layton, 'U.S. Fire Catastrophes of the 20th Century', *Journal of Burn Care and Research*, 3 (1) (1982), pp. 21–8; David J. Barillo and Steven Wolf, 'Planning for

Burn Disasters: Lessons Learned from One Hundred Years of History', *Journal of Burn Care and Research*, 27 (5) (2006), pp. 622–34.
9. National Commission on Fire Prevention and Control, *America Burning* (Washington DC: US Government Printing Office, 1973).
10. Howard A. Rusk, 'Medicine at War – 1: Low Mortality of Vietnam Wounded is Tribute to Quality of Health Aids', *New York Times* (22 May 1966).
11. Jerry M. Shuck, Basil A. Pruitt and John A. Moncrief, 'Homograft Skin for Wound Coverage: A Study in Versatility', *Archives of Surgery*, 98 (4) (1969), pp. 472–8.
12. Kenneth Sell, 'Recent Progress in the Use of Tissues in Surgery', Minutes, Naval Medical Research, No. 31, 12–13 June 1967, Appendix F, MED: Committee on Naval Medical Research, Meetings: Thirtieth to Thirty-Second file, National Research Council, National Academies of Sciences, hereafter NRC.
13. Alan R. Dimick, '1978 Presidential Address: American Burn Association', *Journal of Trauma*, 19 (7) (1979), pp. 520–24.
14. Irving Feller and Keith H. Crane, 'Classification of Burn-care Facilities in the United States', *Journal of the American Medical Association*, 215 (3) (1971), pp. 463–6.
15. 'The Burn-prone Society', *Journal of the American Medical Association*, 231 (3) (1973), pp. 281–2.
16. Office of Planning, Evaluation and Legislation Health Service Administration Burn Care, *A Report to the Senate*, 1976. File ABA-Crozer-Chester Center, Box 19, American Burn Association Archives.
17. P. W. Eggers, 'Trends in Medicare Reimbursement for End-stage Renal Disease, 1974–1978', *Health Care Financial Review*, 6 (1) (1984), pp. 31–8.
18. Richard A. Rettig, 'The Politics of Organ Transplantation: A Parable of our Time', in James F. Blumstein and Frank A. Sloan (eds), *Organ Transplantation Policy: Issues and Prospects* (Durham, NC and London: Duke University Press, 1989), pp. 191–228; R. A. Rettig, 'Special Treatment – The Story of Medicare's ESRD Entitlement', *New England Journal of Medicine*, 364 (2011), pp. 596–8.
19. Barbara Ravage, *Burn Unit: Saving Lives After the Flames* (Cambridge, MA: Da Capo Press, 2004), p. 71.
20. E. Richard Brown, 'Public Hospitals on the Brink: Their Problems and their Options', *Journal of Health Politics, Policy and Law*, 7 (4) (1983), pp. 927–44.
21. Curtis P. Artiz, 'Planning a Burn Institute', in A. B. Wallace and A. W. Wilkinson (eds), *Research in Burns* (Edinburgh: E & S Livingstone, 1966), pp. 522–33.
22. Kurt Melstrom, 'The New York Firefighters Skin Bank', unpublished MPS II: Public Health Paper, 2001.
23. Lawrence K. Altman, 'A Burn Center is Opening in New York City – It's a First', *New York Times* (9 December 1976).
24. Douglas Hand, 'Saving Burn Victims', *New York Times* (15 September 1985).
25. R. J. Kagan, 'Human Skin Banking: Past, Present and Future', in Strong, *Advances in Tissue Banking*, pp. 297–321.
26. C. C. Bondoc and J. F. Burke, 'Clinical Experience with Viable Frozen Human Skin and a Frozen Skin Bank', *Annals of Surgery*, 174 (3) (1971), pp. 371–81; J. F. Burke, W. C. Quinby, C. C. Bondoc, A. B. Cosimi, P. S. Russell and S. K. Szyfelbein, 'Immunosuppression and Temporary Skin Transplantation in the Treatment of Massive Third Degree Burns', *Annals of Surgery* (September 1975), pp. 183–97.
27. 'A Skin Bank for Burn Victims', *New York Times* (3 October 1979).
28. 'Skin Bank', *Texas Monthly* (July 1977).
29. Pauline Neff, 'A Supermarket of Transplants', *The Lion* (April 1982), pp. 10–13; Ellen L. Heck, 'Retrospective of a Skin Bank', *Journal of Burn Care and Rehabilitation*, 20 (1999), pp. 103–7.
30. Melstron, 'The New York Firefighters Skin Bank'.
31. Ronald Sullivan, 'Burn Center Sets Up First Skin Bank in New York', *New York Times* (2 April 1978).
32. Heather Jacobs, New York Firefighters Skin Bank, interview with author (New York, November 2006).

19 Harvesting the Dead

1. Willard Gaylin, 'Harvesting the Dead', *Harper's Magazine* (September 1974), pp. 23–30.
2. Lesley A. Sharp, 'Organ Transplantation as a Transformative Experience: Anthropological Insights into Restructuring of the Self', *Medical Anthropology Quarterly*, 9 (3) (1995), pp. 357–89; Mita Giacomini, 'A Change of Heart and a Change of Mind? Technology and the Redefinition of Death in 1968', *Social Science and Medicine*, 44 (1) (1997), pp. 1465–82; Margaret Lock, *Twice Dead: Organ Transplants and the Reinvention of Death* (Berkeley: University of California Press, 2002); Gary S. Belkin, 'Brain Death and the Historical Understanding of Bioethics', *Journal of the History of Medicine*, 58 (2003), pp. 325–61.
3. P. J. Hauptman and K. J. O'Connor, 'Procurement and Allocation of Solid Organs for Transplantation', *New England Journal of Medicine*, 336 (6) (1997), pp. 422–31.
4. The Presidential Commission for the Study of Bioethical Issues, *Controversies in the Determination of Death* (2008), www.bioethics.gov (accessed March 2011).
5. David J. Rothman, *Strangers at the Bedside: A History of How Law and Bioethics Transformed Medical Decision-making* (New York: Basic Books, 1991).
6. Michael Sappol, *A Traffic of Dead Bodies* (Princeton: Princeton University Press, 2002), pp. 98–135.
7. D. Gareth Jones, *Speaking for the Dead: Cadaver in Biology and Medicine* (Aldershot: Ashgate, 2000), pp. 47–50.
8. Allan D. Vestal, Rodman E. Taber and W. J. Shoemaker, 'Medico-legal Aspects of Tissue Homotransplantations', *Journal of the American Medical Association*, 159 (5) (1955), pp. 487–92.
9. 'Iowa's Eye Bank', *The Lion* (November 1957), pp. 18–20.
10. 'New Law will Aid Eye Bank Program', *The Lion* (October 1960), p. 45.
11. Alfred M. Sadler Jr, Blair L. Sadler and E. Blyth Stason, 'The Uniform Anatomical Gift Act', *Journal of the American Medical Association*, 206 (11) (1968), pp. 2501–6.
12. Francis D. Moore, *Give and Take: The Development of Tissue Transplantation* (Philadelphia: W.B. Saunders, 1964); Nicholas L. Tilney, *Transplant: From Myth to Reality* (New Haven: Yale University Press, 2003); Renée C. Fox and Judith P. Swazey, *The Courage to Fail: A Social View of Organ Transplants and Dialysis* (Chicago: University of Chicago Press, 1974); Renée C. Fox and Judith P. Swazey, *Spare Parts: Organ Replacement in American Society* (New York: Oxford University Press, 1992); Tony Stark, *Knife to the Heart: The Story of Transplant Surgery* (Basingstoke: Macmillan, 1996).
13. Alfred M. Sadler Jr, Blair L. Sadler and George E. Schreiner, 'A Uniform Card for Organ and Tissue Donation', *Modern Medicine* (29 December 1969), pp. 71–4.
14. Alfred M. Sadler Jr, Blair L. Sadler and E. Blyth Stason, 'Transplantation and the Law: Progress Towards Uniformity', *New England Journal of Medicine*, 282 (13) (1970), pp. 717–23.
15. Sadler et al., 'A Uniform Card for Organ and Tissue Donation', pp. 71–5.
16. 'Enlist Youngsters' Aid in Eye Bank Donor Pledges', *The Lion* (April 1976), p. 48.
17. Thomas D. Overcast, Roger W. Evans, Lisa E. Bowen, Marilyn M. Hoe and Cynthia L. Livak, 'Problems in Identification of Potential Organ Donors', *Journal of the American Medical Association*, 251 (12) (1984), pp. 1559–62.
18. The National Transplant Information Center, Records of the American National Red Cross, Box 89, File 500.002, RG200, United States National Archives.
19. F. P. Stuart, 'Need, Supply, and Legal Issues Related to Organ Transplantation in the United States', *Transplantation Proceedings*, 16 (1) (1984), pp. 87–94.
20. Overcast et al., 'Problems in Identification of Potential Organ Donors'.
21. Jeffrey Prottas, *The Most Useful Gift: Altruism and the Public Policy of Organ Transplants* (San Francisco: Jossey-Bass Publishers, 1994), p. 9.

20 Horse-trading in the Mortuary

1. The Creutzfeldt-Jakob Disease Litigation, *Plaintiffs v. United Kingdom Medical Research Council* (19 July 1996), QBD, pp 8–78 p. 30.
2. Directions for collection of human pituitaries, November 1974. FD 7/3547, NA.
3. A. Stockell Hartree, 'Separation and Partial Purification of the Protein Hormones from Human Pituitary Glands', *Biochemical Journal*, 100 (1966), pp. 76–1.
4. Visit to A. S. Hartree, Cambridge, 9 February 1971, FD 9/4609, NA.
5. Emily Green, 'A Wonder Drug that Carried the Seeds of Death', *Los Angeles Times* (21 May 2000).
6. The Creutzfeldt-Jakob Disease Litigation, p. 42.
7. J. R. Sharp, 'Background to the New "Orange Guide"', *Pharmaceutical Journal* (25 June 1983), pp. 718–21; Michael Murray, 'Good Pharmaceutical Manufacturing Practice', in Frank Wells (ed.), *Medicines: Good Practice Guidelines* (Belfast: The Queen's University of Belfast, 1990), pp. 17–27.
8. J. A. Holgate, Medicines Division, DHSS, Production of human growth hormone: note of a visit of 2nd May 1974, FD 7/3517, NA.
9. D.A.J. Tyrrell, Medical Research Council Clinical Research Centre, to Barbara Rashbass, Medical Research Council, 7 January 1974, FD 7/3545, NA.
10. Minutes of 5th meeting of the steering committee of the human pituitary collection, 7 June 1974, FD 7/3546, NA.
11. Anne Stockell Hartree to Barbara Rashbass, Medical Research Council, 5 July 1973, FD 7/3544, NA.
12. J. F. Heggie, 'Training of Post-mortem Room Technicians', *Journal of Clinical Pathology*, 20 (5) (1967), pp. 793–4.
13. J. F. Heggie to J. T. Woodlock, Ministry of Health, 13 January 1969, MH 158/331, NA.
14. Whitley Councils for the Health Services (Great Britain), PTB Circular 243, 1969, MH 158/331, NA.
15. James Wilkinson, 'Doctors' Tangle Over "Spare Parts" Cash Offer', *Daily Express* (18 July 1970), p. 3.
16. W. G. Robertson to Mrs Poole, Department of Health and Social Security, 20 July 1970, MH 150/405, NA.
17. B.O.B. Gidden to Mr Watt, 24 June 1975, MH 166/1185, NA.
18. Minutes of the 4th Meeting of the Steering Committee for the Human Pituitary Collection held on 8 February 1974, FD 7/3546, NA.
19. Frank Young, Chairman of the Medical Research Council Steering Committee for the Human Pituitary Gland Collection, January 1976, *Collection of Human Pituitary Glands*', MH 166/1185, NA.
20. R.D.G. Milner, The National Pituitary Collection: Report for the period 9 March to 28 September 1983, FD 7/3522, NA.
21. 'Body Robbers Scandal', *Sunday People* (29 August 1976), p. 1.
22. 'Taking Glands from Bodies Without Consent Condemned', *The Times* (30 August 1976), p. 2.
23. David Loshak, 'Stricter Controls on Removal of Human Organs Planned', *Daily Telegraph* (29 September 1976).
24. Cameron to Dr [illegible], 'Charges for human tissue of cadaveric origin', 26 May 1977, FD 7/3516, NA.
25. Human pituitary gland collection – discussion paper on arrangements for collection, MRC/DHSS meeting, 17 January 1978, FD 7/3516, NA.
26. R. M. Oliver, Department of Health and Social Security to Dear Pathologist, 25 April 1980, FD 7/3517, NA.
27. Malcolm Stuart, 'Pay Deal "Could Produce Midgets"', *The Guardian* (18 February 1981).
28. Milner, The National Pituitary Collection.

29. R. M. Oliver, Chairman of the Joint DHSS / MRC Advisory Committee for the National Pituitary Collection, 12 February 1981, HO 299/55, NA.
30. Robert Walgate, 'Pituitary Slump', *Nature*, 290, 5801 (1981), pp. 6–7.
31. Ibid., p. 7.

21 Value for Money in American Mortuaries

1. 'Code that controls growth', *LIFE* (October 14 1966), pp. 93–6.
2. Robert M. Blizzard, 'Statement', *Congressional Record – Senate* (25 January 1971), S28–S30.
3. National Institutes of Health, National Institute of Arthritis, Metabolism, and Digestive Diseases, Proposal for an evaluation of the effect of the National Pituitary Agency (NPA), December 1972, National Academy of Sciences, Division of Medical Science, Medical Sciences Committee on Evaluation of National Pituitary Agency. NAS-NRC Executive Offices Medical Sciences 1973, Com. On Evaluation of National Pituitary Agency, NRC.
4. Tom Gable, 'Calbiochem Drug Needs Final Tests, Shareholders Told', *Evening Tribune, San Diego* (18 June 1973).
5. Maureen McKelvey, *Evolutionary Innovation: Early Industrial Uses of Genetic Engineering* (Linköping, Sweden: Department of Technology and Social Change, Linköping University, 1984).
6. Medical Committee for the Evaluation of the National Pituitary Agency, *An Evaluation of the National Pituitary Agency*, 1974, pp. 79–84, Medical Committee for the Evaluation of the National Pituitary Agency, Division of Medical Sciences, NRC.
7. Ibid., pp. 117–26.
8. Ibid., pp. 91–2.
9. Susan Cohen and Christine Cosgrove, *Normal at Any Cost: Tall Girls, Short Boys, and the Medical Industry's Quest to Manipulate Height* (New York and London: Jeremy P. Tarcher/Penguin, 2009), p. 95.
10. Inquiry into the use of pituitary derived hormones in Australia and Creutzfeldt-Jakob Disease, *Report* (Canberra: Australian Government Publishing Service, 1994), pp. 734–42.
11. Committee for the Evaluation of the National Pituitary Agency, *An Evaluation of the National Pituitary Agency*, pp. 47–9.
12. Richard Titmuss, 'The Gift Relationship: From Human Blood to Social Policy', in Ann Oakley and John Ashton (eds), *The Gift Relationship: From Human Blood to Social Policy. Expanded and Updated Edition* (New York: The New Press, 1997 [1970]).
13. M. H. Cooper and A. J. Culyer, *The Price of Blood*, Hobart Paper No. 41 (London: Institute of Economic Affairs, 1968); Philippe Fontaine, 'Blood, Politics, and Social Science: Richard Titmuss and the Institute of Economic Affairs, 1957–1973', *Isis*, 93 (2002), pp. 401–34.
14. Committee for the Evaluation of the National Pituitary Agency, Minutes of First Meeting, May 11, 1973, p. 5. MED:COM for Evaluation of the National Pituitary Agency, Meetings 1973–1974, NRC.
15. John I. Coe, 'Collection of Pituitary Glands in Hennepin County', *Journal of Forensic Sciences*, 15 (1) (1970), pp. 14–17.
16. Robert M. Blizzard, 'History of Growth Hormone Therapy', *Indian Journal of Pediatrics*, 79 (1) (January 2012), p. 89.
17. 'Gland Theft Charge Jails Sylmar Man', *Los Angeles Times* (9 November 1966), p. 24.
18. Salvatore Raiti, 'The National Hormone and Pituitary Program: Achievements and Current Goals', in Salvatore Raiti and Robert A. Tolman (eds), *Human Growth Hormone* (New York: Plenum Medical Book Company, 1986), p. 5.
19. Donna L. Hoyert, 'The Changing Profile of Autopsied Deaths in the United States, 1972–2007', US Department of Health and Human Services (2011), www.cdc.gov/nchs/data/databasebriefs/db67.pdf (accessed May 2012).
20. Robert E. Anderson, 'The Autopsy – Benefits to Society', *American Journal of Clinical Pathology* (February 1978), p. 239.
21. Robert E. Anderson, James T. Weston, John E. Craighead, Paul E. Lacy, Robert W. Wissler and Rolla B. Hill, 'The Autopsy: Past, Present, and Future', *Journal of the American Medical Association*, 242 (10) (1979), pp. 1056–9; Rolla B. Hill and Robert E. Anderson, 'The

Autopsy Crisis Re-examined: The Case for a National Autopsy Policy', *Milbank Quarterly*, 69 (1) (1991), pp. 51–77.
22. R. D. Paegle and J. V. Klavins, 'Special Technologists for Autopsies', *Pathologist*, 23 (9) (1970), pp. 297–301.
23. Jane Levitt, 'The Growth of Technology and Corporate Profit-making in the Clinical Laboratories', *Journal of Health Politics, Policy and Law*, 8 (4) (1984), pp. 732–42.
24. Hoyert, 'The Changing Profile of Autopsied Deaths'.
25. Edwin E. Pontius, 'Financing Mechanisms for Autopsy', *Pathologist*, 69 (7) (1978), Supplement, pp. 245–7; Lou A. Orsini, 'Health Insurance and the Autopsy', *Pathologist*, 69 (7) (1978), Supplement, pp. 248–9; Raymond Yesner, 'Medical Center Autopsy Costs', *American Journal of Clinical Pathology*, 69 (1978), pp. 242–4; Michael A. Clark, 'The Value of the Hospital Autopsy: Is it Worth the Cost?', *American Journal of Forensic Medicine and Pathology*, 2 (1981), pp. 231–7; Jack Hadley, 'Medicaid Reimbursement of Teaching Hospitals', *Journal of Health Politics, Policy and Law*, 74 (4) (1983), pp. 911–26.
26. Matthew J. Hickman, Kristen A. Hughes, Kevin J. Strom and Jeri D. Ropero-Miller, *Medical Examiners and Coroners' Offices, 2004* (Washington DC: Department of Justice, Office of Justice Programs, 2007), http://bjs.ojp.usdoj.gov (accessed December 2011).
27. Randy Hazlick and Debra Combs, 'Medical Examiner and Coroner Systems: History and Trends', *Journal of the American Medical Association*, 279 (11) (1988), pp. 870–74.
28. Coe, 'Collection of Pituitary Glands in Hennepin County'.
29. Blizzard, 'History of Growth Hormone Therapy', p. 92.
30. Erik S. Jaffe, 'She's Got Bette Davis['s] Eyes': Assessing the Nonconsensual Removal of Cadaver Organs Under the Takings and Due Process Clauses, *Columbia Law Review* (1990), pp. 528–74.
31. Paul A. Lombardo, 'Consent and "Donations" from the Dead', *Hasting Center Report* (December 1981), pp. 9–11.
32. R. Hunter Manson, *Statutory Regulation of Organ Donation in the United States* (Richmond, VA: The South-Eastern Organ Procurement Foundation, 1986).

22 Financing High-volume Eye Banks

1. Ernest F. Imhoff, 'No Eye Doctor, But Vision Helps Blind see Testimonial: Colleagues, Friends Set Up Foundation to Continue the Work of Man Who Organized Corneal Transplant System Worldwide', *Baltimore Sun* (1 November 1997).
2. David Sudnow, *Passing On: The Social Organization of Dying* (Englewood Cliffs, NJ: Prentice-Hall, 1967).
3. Lawrence E. Holder, 'How Much Artifice in Asking for an Autopsy?', *Hospital Physician*, 7 (1971), pp. 87–91.
4. Sudnow, *Passing On*, p. 59.
5. Samuel Shem, *The House of God* (New York: Doubleday Dell, 1978), p. 152.
6. Frederick N. Griffith, Tissue Banks International, interview with author (Baltimore, April 2005).
7. C. R. Graham Jr, 'Eye Banking: A Growth Story', *Transplantation Proceedings*, 17 (6), Supplement 4 (1985), p. 108.
8. Frederick N. Griffith, Tissue Banks International, interview with author (Baltimore, April 2005).
9. Homicide victimization, 1950–2005, Bureau of Justice Statistics, http://bjs.ojp.usdoj.gov (accessed January 2010).
10. John W. Payne, 'New Directions in Eye Banking', *Transactions of the American Ophthalmological Society*, 119 (1) (1980), p. 989.
11. Wing Chu, 'The Past Twenty-five Years in Eye Banking', *Cornea*, 19 (5) (2000), p. 757.
12. Daniel M. Berman, *Death on the Job: Occupational Health and Safety Struggles in the United States* (New York: Monthly Review Press), 1978.
13. Herbert E. Kaufman, 'Tissue Storage Systems: Short and Medium Term', in Frederick S. Brightbill (ed.), *Corneal Surgery: Theory, Technique and Tissue*, 3rd edn (St Louis: Mosby, 2000), pp. 892–7.
14. Chu, 'The Past Twenty-five Years in Eye Banking', pp. 755–6.

15. Emile J. Farge, 'Eye Bank Association of America: 1982 Activities', *Cornea*, 2 (1) (1983), p. 677.
16. Viviana A. Zelizer, *The Social Meaning of Money: Pin Money, Pay Checks, Poor Relief, and Other Currencies* (New York: Basic Books, 1994), pp. 21–30.
17. Gilbert M. Gaul and Neill A. Borowski, *Free Ride: The Tax-exempt Economy* (Kansas City: Andrews and McMeel, 1993), p. 2.
18. Robert D. Putnam, *Bowling Alone: The Collapse and Revival of American Community* (New York: Simon and Schuster, 2000), p. 49; Dan Frost, 'Farewell to the Lodge', *American Demographics* (January 1996), pp. 40–45.
19. Lisa Champell, 'Gallup Survey Identifies Why Former Lions Quit their Clubs', *The Lion* (February 1994), pp. 22–3.
20. Burton A. Weisbrod, 'The Nonprofit Mission and its Financing: Growing Links Between Nonprofits and the Rest of the Economy', in Burton A. Weisbrod (ed.), *To Profit or Not to Profit: The Commercial Transformation of the Nonprofit Sector* (Cambridge: Cambridge University Press, 1998), pp. 1–22.
21. Graham, 'Eye Banking: A Growth Story', p. 109.
22. Frederick N. Griffith, Tissue Banks International, interview with author (Baltimore, April 2005).
23. Richard L. Fuller, 'Medical Legal Issues', in J.H. Krachmer, M. J. Mannis and E. J. Holland (eds), *Cornea* (Philadelphia: Elsevier Mosby, 1997), pp. 509–17.
24. Albert S. Leveille, Janie Benson, H. Dwight Cavanagh, Bruce I. Bodner and Robert H. Byers, 'Cost-Effectiveness in Eyebanking', *American Academy of Ophthalmology*, 89 (6) (1982), pp. 51A–53A.
25. Farge, 'Eye Bank Association of America', pp. 83–5; Graham, 'Eye Banking: A Growth Story', p. 108.
26. Frederick N. Griffith, 'The Promise of International Eye Banking', *International Ophthalmology*, 14 (1990), pp. 205–10.
27. Susan V. Lawrence, 'Year of the Lions in China: Club's Recruitment of Chinese May Trigger Societal Change', *The Lion* (November 2002), p. 8.

23 Regulation is Necessary, but How?

1. John R. Kateley, 'Establishing a Tissue Bank', in K. J. Fawcett and A.R. Barr (eds), *Tissue Banking* (Arlington, VA: American Association of Tissue Banks, 1987), pp. 17–27.
2. T. I. Malinin, 'University of Miami Tissue Bank: Collection of Post-mortem Tissue for Clinical Use and Laboratory Investigation', *Transplant Proceedings*, 8(2 Suppl. 1) (1976), pp. 53–8.
3. Robert E. Stevenson, United States Navy Tissue Bank staff, interview with author (Savannah, Georgia, April 2008).
4. *Ravenis v. Detroit General Hospital*, 234 N W 2d, 411 (1975).
5. Susan Bartlett Foote, *Managing the Medical Arms Race: Innovation and Public Policy in the Medical Device Industry* (Berkeley: University of California Press, 1992).
6. Kara W. Swanson, *Banking on the Body: The Market in Blood, Milk and Sperm in Modern America* (Cambridge, MA and London: Harvard University Press, 2014), pp. 125–30.
7. Julia D. Mahoney, 'The Market for Human Tissue', *Virginia Law Review*, 86 (2) (2000), pp. 164–223.
8. James T. Quirk III, Executive Vice President Lions Eye Bank of Delaware, personal communication, April 2005.
9. Michael J. Joyce, 'American Association of Tissue Banks: A Historical Reflection Upon Entering the 21st Century', *Cell and Tissue Banking*, 1 (2000), pp. 5–8.

24 The Blind Eye Act

1. 'Eye Surgeon Admits Secret Operations', *Sunday People* (16 January 1966).
2. Benjamin Rycroft, 'Eyes for Corneal Grafting', *British Medical Journal*, 2 (1966), p. 1452.
3. D. P. Choyce, 'Eyes for Corneal Grafting', *British Medical Journal* (24 December 1966), p. 1593.

4. Alex F. Cross, Honorary Consultant to the Royal National Institute for the Blind, to the Ministry of Health, 6 March 1968, MH 150/80, NA.
5. Douglas Gibbs, East Grinstead Eye Bank, July 1972, MH 150/80, NA.
6. John Armitage, Corneal Transplant Service, interview with author (Bristol, March 2005).
7. A. J. Kember to G. H. Weston, 2 December 1971; G. H. Weston, North West Metropolitan Regional Hospital Board, to W. G. Robertson, Department of Health and Social Security, 13 January 1972, MH 150/80, NA.
8. T. A. Casey, 15 May 1972, MH 150/80, NA.
9. Note of meeting to discuss the supply of eyes for corneal grafting, 17 October 1972, MH 150/80, NA.
10. Interview with Mr T. A. Casey by S.A.M.O. and Secretary of R.H.B., 22 December 1971, MH 150/80, NA.
11. HC Deb 23 May 1972, vol. 837 cc 322 (L. Pavitt).
12. W. G. Robertson to Mrs Poole, 8 February 1972, MH 150/80, NA.
13. Maurice Pappworth, 'Letter to the Editor', *The Lancet*, 291 (1968), p. 419.
14. Lorna Thomson, 'Renal Transplants – A Sin of Omission', *World Medicine* (29 February 1972), pp. 17–22.
15. Helen MacDonald, Conscripting Organs: "Routine Salvaging" or Bequest? The Historical Debate in Britain, 1961–1975', *Journal of the History of Medicine and Allied Science*, 70 (3) (2014), pp. 425–6.
16. Advisory Group on Transplantation Problems on the question of amending the Human Tissue Act 1961, *Advice*, Cmnd. 4106, (London: Her Majesty's Stationery Office, 1969).
17. MacDonald, 'Conscripting Organs', pp. 448–9.
18. Department of Health and Social Security Guidance Circular to NHS authorities: Human Tissue Act, 1961, 1975, HSC(IS)156, MH 150/405, NA.
19. Alison Smithies to Mr Garlick, Department of Health and Social Security, 'Post mortem consent forms', 16 August 1976, MH 150/405, NA.
20. 'Body Robbers Scandal', *Sunday People* (29 August 1976).
21. G. Machin, Autopsy Department, North Staffs. Royal Infirmary, to Autopsy Technicians, 4 February 1977, FD 9/4607, NA.
22. M. V. Wakefield-Richmond, Home Office, to Dr J. D. K. Burton, Coroner's Court, 16 February 1977, HO 299/55, NA.
23. Bernard Knight to Professor V.H.T. James, Medical Research Council, 16 March 1977, FD 9/4607, NA.
24. Department of Health and Social Security, Removal of human tissue at post mortem examination – Human Tissue Act 1961, 1977, HC(77)28, MH 150/405, NA.
25. 'Postmortem Tissue Problems', *British Medical Journal* (5 August 1978), p. 382.
26. R. M. Oliver to Dear Pathologist, Collection of Pituitary Glands, 14 July 1980, FD 7/3521, NA.
27. M. V. Wakefield-Richmond, Note of a meeting at DHSS on 27 November 1980, MH 150/405, NA.
28. Minutes of the 5th meeting of the National Pituitary Collection, 16 December 1980, MH 150/405, NA.
29. The National Pituitary Collection: Report for the period 9 March to 28 September 1983, FD 7/3522, NA.
30. J. Harley to Dear Administrator, 11 May 1982, DA(82)9, FD 7/3522, NA.
31. The Creutzfeldt-Jakob Disease Litigation, *Plaintiffs v. United Kingdom Medical Research Council* (19 July 1996), QBD, p. 49.
32. M. E. Abrams, Chairman of the Advisory Committee for the National Pituitary Collection, to Dear Doctor, February 1984, MH 150/405, NA.
33. D. R. Bangham, National Institute for Biological Standards and Control, to Jenifer Gunning, MRC, 27 May 1986, FD 7/3523, NA.

25 Creating American Hybrid Extractors of Cadaver Stuff

1. Jeffrey M. Prottas, 'Obtaining Replacements: The Organizational Framework of Organ Procurement', *Journal of Health Politics, Policy and Law*, 8 (2) (1983), pp. 241–2.
2. Peter H. Schuck, 'Government Funding for Organ Transplants', in James F. Blumstein and

Frank A. Sloan (eds), *Organ Transplantation Policy: Issues and Prospects* (Durham, NC and London: Duke University Press, 1989), p. 171.
3. Richard. A. Rettig, 'The Politics of Organ Transplantation: A Parable of our Time', in Blumstein and Sloan (eds), *Organ Transplantation Policy*, pp. 207–9.
4. Janice Zeigler Cuzzell, 'Fighting City Hall: The Politics of Burn Care', *American Journal of Nursing*, 86 (2) (1986), pp. 194–5; A. R. Dimick, L. H. Potts, E.D. Charles Jr, J. Wayne and I. M. Reed, 'The Cost of Burn Care and Implications for the Future of Quality of Care', *Journal of Trauma*, 26 (3) (1986), pp. 260–66.
5. S. H. Denise, 'Regulating the Sale of Human Organs', *Virginia Law Review*, 71 (6) (1985), p. 1022.
6. Prottas, 'Obtaining Replacements', p. 246.
7. F. P. Stuart, 'Need, Supply, and Legal Issues Related to Organ Transplantation in the United States', *Transplantation Proceedings*, 16 (1) (1984), pp. 87–94.
8. Michele Goodwin, *Black Markets: The Supply and Demand of Body Parts* (New York: Cambridge University Press, 2006), p. 105.
9. Phil Gunby, 'Bill Introduced to Thwart Kidney Brokerage', *Journal of the American Medical Association*, 250 (17) (1983), pp. 2263–4.
10. *National Organ Transplant Act*, Hearing before the Subcommittee of the Committee on Ways and Means, House of Representatives, Ninety-Eighth Congress, Second Session on H.R. 4080, 9 February 1984, Serial 98-64, p. 138.
11. Ibid., pp. 130–31.
12. Ibid., pp. 131–6.
13. Emile J. Farge and Robert A. Fort, 'Corneal Preservation and Eye Banking', *Cornea*, 4 (1985/6), pp. 256–62.
14. Deni Bunis, 'Lions Clubs Thwarted Tissue Safety Standards Legislation: As a Congressman, Gore Proposed Legislation to Govern the Field', *Orange County Register* (19 April 2000), http://www.newslibrary.com (accessed 18 July 2000).
15. *National Organ Transplant Act*, p. 36.
16. James F. Blumstein, 'Government's Role', in Blumstein and Sloan (eds), *Organ Transplantation Policy*, p. 15.
17. Jeffrey M. Prottas, 'The Organization of Organ Procurement', *Journal of Health Politics, Policy and Law*, 14 (1) (1989), pp. 45–8.
18. Jeffrey M. Prottas, *The Most Useful Gift: Altruism and the Public Policy of Organ Transplants* (San Francisco: Jossey-Bass Publishers, 1994), p. 67.
19. Blumstein, 'Government's Role', p. 18.
20. Ibid., p. 25.
21. Prottas, *The Most Useful Gift*, pp. 44–8.

26 Sharing Pledges and Cadaver Stuff

1. Elizabeth D. Ward, *Timbo: A Struggle for Survival* (London: British Kidney Patients Association, 1996), p. 31.
2. W. G. Robertson to Mr Brandes, 5 October 1971, MH 150/438, NA.
3. L. H. Brandes to Mr Gelding, 20 April 1972, MH 150/439, NA.
4. Ibid.
5. Ward, *Timbo*, p. 34.
6. Prevention of Blindness Sub-Committee Minutes, 30 January 1973, R3/3A, RNIB.
7. Correspondence and papers relating to the discussion on how to revamp the blue card eye pledge scheme, R3/3A, RNIB.
8. 'New Donor Card', DHSS Press Release, 4 June 1981, R3/3, RNIB.
9. Alan Lewis and Martin Snell, 'Increasing Kidney Transplantation in Britain: The Importance of Donor Cards, Pubic Opinion and Medical Practice', *Social Science and Medicine*, 22 (10) (1986), pp. 1075–80.
10. Script of advertisement, 12 May 1981, R3/3, RNIB.
11. Notes on procedures connected with prevention of blindness services, January 1984, R3/3, RNIB.
12. Lucille Hall, RNIB. Press & Publicity Officer, to Vehicle Licensing Centre, Swansea, 18 May 1981, R3/3A, RNIB.

13. David Graves, 'Ask Everyone to be an Organ Donor, Says Surgeon', *Daily Telegraph* (23 February 1984).
14. Catherine Gillespie to RNIB, 20 May 1982, R3/3, RNIB.
15. Organ donation, DHSS meeting 28 January 1986, R3/3, RNIB.
16. S. D. Nelson and G. H. Tovey, 'National Organ Matching and Distribution Service', *British Medical Journal*, 1 (1974), pp. 622–4.
17. J. R. Batchelor, T. A. Casey, D. C. Gibbs, D. F. Lloyd, A. Werb, S. S. Prasad and A. James, 'HLA Matching and Corneal Grafting', *The Lancet* (13 March 1976), pp. 551–4.
18. David Easty, Corneal Transplant Service, interview with author (Bristol, March 2005).
19. Iris Fund Eye Information Centre: Analysis of Questionnaire, 17 June 1982, MH 150/650, NA.
20. Miss M. Heath to Mr Jewesbury, 2 December 1982, MH 150/650, NA.
21. T. A. Casey, 'Eye Banking', *British Medical Journal*, 288 (1984), p. 5.
22. D. L. Easty, 'Eye Banking', *Transplantation Proceedings*, 21 (1) (1989), pp. 3120–22.
23. HC Deb 18 April 1986, vol. 95 cc1162 (Jeremy Hanley).
24. W. J. Armitage, S. J. Moss, D. L. Easty and B. A. Bradley, 'Supply of Corneal Tissue in the United Kingdom', *British Journal of Ophthalmology*, 74 (1990), pp. 685–7; A.E.A. Ridgeway and B. Marcyniuk, 'The U.K. Cornea Transplant Service and the Establishment of the David Lucas Manchester Eye Bank', in P. Sourdille (ed.), *Evolution of Microsurgery*. Dev Opththalmol (Basel: Karrer, 1991), pp. 44–9.
25. B. Bradley and S. Burr, *An Introduction to a New National Service for Centres Engaged in Corneal Transplantation* (London: Iris Fund for the Prevention of Blindness, 1983).
26. *Cornea Transplant Service Eye Bank* (6 April 1987), Iris Fund for the Prevention of Blindness.
27. The Collaborative Corneal Transplantation Studies (CCTS), 'Effectiveness of Histocompatibility Matching in High-risk Corneal Transplantation', *Archives of Ophthalmology*, 110 (10) (1992), pp. 1392–403; A. Vail, S. M. Gore, B. A. Bradley, D. L. Easty, C. A. Rogers and W. J. Armitage, 'Conclusions of the Corneal Transplant Follow Up Study', *British Journal of Ophthalmology*, 81 (1997), pp. 631–6; W. J. Armitage, 'HLA Matching and Corneal Transplantation', *Eye*, 18 (2004), pp. 231–2.

27 Iatrogenesis: Disregarding Risk in Plain Sight

1. Warwick Anderson, *The Collectors of Lost Souls: Turning Kuru Scientists into Whitemen* (Baltimore, MD: Johns Hopkins University Press, 2008).
2. W. J. Hadlow, 'Scrapie and Kuru', *The Lancet* (5 September 1959), pp. 289–90; William J. Hadlow, 'Kuru Likened to Scrapie: The Story Remembered', *Philosophical Transactions of the Royal Society, Biological Sciences*, 363 (1510) (2008), p. 3644; David M. Asher, 'Kuru: Memories of the NIH Years', *Philosophical Transactions of the Royal Society, Biological Sciences*, 363 (1510) (2008), pp. 3618–25.
3. D. C. Gajdusek, C. Gibbs and M. Alpers, 'Experimental Transmission of a Kuru-like Syndrome to Chimpanzees', *Nature*, 209 (1966), pp. 794–6.
4. C. Gibb, D. Carleton Gajdusek and D. Asher, 'Creutzfeldt-Jakob Disease (Spongiform Encephalopathy); Transmission to the Chimpanzee', *Science*, 161 (1968), pp. 388–9.
5. World Health Organization Scientific Group, *Viral Hepatitis*, Technical Report Series No. 512 (Geneva: World Health Organization, 1973).
6. Marion Nestle, *Safe Food: The Politics of Food Safety* (Berkeley, Los Angeles, and London: University of California Press, 2010), p. 16.
7. Ian Burrell and John Furbisher, 'Warnings Over Brain Disease Ignored', *Sunday Times* (5 September 1983), p. 5.
8. P. Duffy, J. Wolf, G. Collins et al., 'Possible Person-to-person Transmission of Creutzfeldt-Jakob Disease', *New England Journal of Medicine*, 290 (1974), p. 692.
9. Inquiry into the use of pituitary derived hormones in Australia, *Report*, pp. 309–31.
10. The Creutzfeldt-Jakob Disease Litigation, *Plaintiffs v. United Kingdom Medical Research Council* (19 July 1996), QBD, p. 41.
11. Inquiry into the use of Pituitary Derived Hormones in Australia, *Report*, pp. 106–7.
12. Salvatore Raiti, 'The National Hormone and Pituitary Program: Achievements and Current Goals', in S. Raiti and R.A. Tolman (eds), *Human Growth Hormone*, (New York: Plenum Medical Book Company, 1986), p. 10.

13. Reuters in Paris, 'French Doctors Questioned about Tainted Gland Death', *The Guardian* (21 July 1993), p. 8; Alexander Dorozynski, 'French to Investigate Deaths from Growth Hormone', *British Medical Journal*, 307 (1993), p. 281.
14. Emily Green, 'A Wonder Drug that Carried the Seeds of Death', *Los Angeles Times* (21 May 2000).
15. C. Bernoulli, J. Siegfried, G. Baumgartner, F. Regli, T. Rabinowicz, D. C. Gajdusek and C. J. Gibbs, 'Danger of Accidental Person-to-person Transmission of Creutzfeldt-Jakob Disease By Surgery', *The Lancet*, 309 (1977), pp. 478–9.
16. D. Carleton Gajdusek, C. Gibbs, D. Asher, P. Brown, A. Diwan, P. Hoffman, G. Nemo, R. Rohwer and L. White, 'Precautions in Medical Care of, and in Handling Materials from Patients with Transmissible Virus Dementia (Creutzfeldt-Jakob Disease)', *New England Journal of Medicine*, 297 (23) (1977), pp. 1253–8.
17. The Creutzfeldt-Jakob Disease Litigation, pp. 56–7.
18. Ibid., p. 58.
19. S. Douglas Frasier, 'The Not-so-good old days: Working with Pituitary Growth Hormone in North America, 1956–1985', *Journal of Pediatrics*, 131 (1), part 2 (1997), p. S3.
20. Reuters in Paris, 'French Doctors Questioned'; Dorozynski, 'French to Investigate Deaths from Growth Hormone'.
21. Maureen McKelvey, 'Engineering the Biological and Political: Enabling Industrial Use of Genetic Engineering in Sweden', *Polhem: Tisskrift för Teknikhistoria*, 12 (1994), pp. 216–59.
22. R.D.G. Milner, 'Growth Hormone 1985', *British Medical Journal* (7 December 1985), pp. 1593–4; Kimberly Glasbrenner, 'Technology Spurt Resolves Growth Hormone Problem, Ends Shortage', *Journal of the American Medical Association*, 255 (5) (1985), pp. 581–7.
23. Michael J. Cronin, 'Pioneering Recombinant Growth Hormone Manufacturing: Pounds Produced Per Mile of Height', *Journal of Pediatrics*, 131 (1) (1997), pp. S5–S7.
24. Douglas S. Diekema, 'Is Taller Really Better? Growth Hormone Therapy in Short Children', *Perspectives in Biology and Medicine*, 34 (1) (1990), p. 116.
25. Paul Brown, 'Human Growth Hormone Therapy and Creutzfeldt-Jakob Disease: A Drama in Three Acts', *Pediatrics*, 81 (1) (1988), p. 87.
26. Paul Brown, D. Carleton Gajdusek, C. J. Gibbs and David M. Asher, 'Potential Epidemic of Creutzfeldt-Jakob Disease From Human Growth Hormone Therapy', *New England Journal of Medicine*, 313 (1985), pp. 728–31.
27. Paul Brown, Jean Philippe Brandel, Michael Preece and Takeshi Sato, 'Iatrogenic Creutzfeldt-Jakob Disease: The Waning of an Era', *Neurology*, 67 (2006), pp. 389–93.
28. David Body and Jonathan Glasson, 'Iatrogenic Creutzfeldt-Jakob Disease: Litigation in the UK and Elsewhere', *Clinical Risk*, 11 (2005), pp. 4–13; Andrea Boggio, 'The Compensation of the Victims of the Creutzfeldt-Jakob Disease in the United Kingdom', *Medical Law International*, 7 (2005), pp. 149–67.

28 Ask, or Don't Ask: Inconsistencies in Collecting Sites

1. Wing Chu, 'The Past Twenty-five Years in Eye Banking', *Cornea*, 19(5)(2000), p. 762.
2. Jaffe, 'She's Got Bette Davis['s] Eyes: Assessing the Nonconsensual Removal of Cadaver Organs under the Takings and Due Process Clauses', *Columbia Law Review* (1990), p. 538.
3. *Georgia Lions Eye Bank v. Lavant*, 255 Ga. 60, 335 S.E.2d 127 (1985).
4. *State of Florida v. Powell, et ux., et al.*, Appellees, 497 So.2d 1188 (Fla. 1986).
5. Kathleen S. Andersen and Daniel M. Fox, 'The Impact of Routine Inquiry Laws on Organ Donation', *Health Affairs* (Winter 1988), pp. 65–78.
6. Andrew H. Malcolm, 'Human Organ Transplants Gain With New State Laws', *New York Times* (1 June 1986).
7. Andersen and Fox, 'The Impact of Routine Inquiry Laws', p. 75.
8. Office of Evaluation and Inspections, Office of Inspector General, *Organ Procurement Organizations and Tissue Recovery* (Washington DC: The Department of Health and Human Services, 1994), p. 5.

9. Chad A. Thomson, 'Appendix A: Organ Transplantation in the United States: A Brief Legislative History', in Sally Satel (ed.), *When Altruism Isn't Enough* (Washington DC: American Enterprise Institute for Public Policy Research, 2008), p. 137.
10. J.M. Prottas, *The Most Useful Gift: Altruism and the Public Policy of Organ Transplants* (San Francisco: Jossey-Bass Publishers, 1994), pp. 12–13; Jennifer L. Mesich-Brant and Lawrence J. Grossback, 'Assisting Altruism: Evaluating Legally Binding Consent in Organ Donation Policy', *Journal of Health Politics, Policy and Law*, 30 (4) (2005), pp. 687–717.
11. J. S. Wolf, E. M. Servino and H. N. Nathan, 'National Strategy to Develop Public Acceptance of Organ and Tissue Donation', *Transplantation Proceedings*, 29 (1997), pp. 1477–8; E. M. Servino, H. Nathan and J. S. Wolf, 'Unified Strategy for Public Education in Organ and Tissue Donation', *Transplantation Proceedings*, 29 (1997), p. 3247.
12. Ralph Frammolino, 'Harvest of Corneas at Morgue Questioned', *Los Angeles Times* (2 November 1997); Ralph Frammolino, 'Objections to Cornea Policy Date Back to 1992', *Los Angeles Times* (22 November 1997).
13. Michele Goodwin, *Black Markets: The Supply and Demand of Body Parts* (New York: Cambridge University Press, 2006), pp. 120–23.
14. Ann W. O'Neill, 'Parents Tell of Grief in Lawsuit Over Harvest of Son's Cornea', *Los Angeles Times* (13 May 1998).
15. Donald Jason, 'The Role of the Medical Examiner/Coroner in Organ and Tissue Procurement for Transplantation', *American Journal of Forensic Medicine and Pathology*, 15 (3) (1994), pp. 192–202.
16. Michele Goodwin, 'Rethinking Legislative Consent Law', *De Paul Journal of Health Care Law*, 5 (2) (2002), p. 259.
17. Julia D. Mahoney, 'The Market for Human Tissue', *Virginia Law Review*, 86 (2) (2000), p. 185, fn. 85.
18. Ralph Frammolino, 'Reforms on Cornea Harvesting OKd', *Los Angeles Times* (1 October 1998).
19. David Orentlicher, 'Presumed Consent to Organ Donation: Its Rise and Fall in the United States', *Rutgers Law Review*, 61 (2009), pp. 295–331.
20. Michele Goodwin, 'Deconstructing Legislative Consent Law: Organ Taking, Racial Profiling and Distributive Justice', *Virginia Journal of Law and Technology* (2001), pp. 1522–687.
21. Gerald J. Cole, President, Tissue Banks International, Slide Presentation to the Food and Drug Administration Transmissible Spongiform Encephalopathies Advisory Committee, 18 January 2001, www.fda.gov/ohrms/dockets/ac/01/slides (accessed May 2016).
22. US Department of Health and Human Services, Health Care Financing Administration, Final Rule for Organ, tissue and eye donation (22 June 1998, 63 Fed. Reg. 33856).
23. Martha W. Anderson and Rene Schapiro, 'From Donor to Recipient: The Pathway and Business of Donated Tissues', in Stuart J. Youngner, Martha W. Anderson and Renie Schapiro (eds), *Transplanting Human Tissue: Ethics, Policy, and Practice* (New York and Oxford: Oxford University Press, 2004), p. 5.
24. *Tissue Banks: Is the Federal Government's Oversight Adequate?* Hearing *New York Times* before the Permanent Subcommittee on Investigations of the Committee on Governmental Affairs, 24 May 2001 (Washington DC: Government Printing Office), p. 258.
25. United Network for Organ Sharing, www.unos.org/data (accessed May 2008).
26. Patricia Aiken-O'Neill and Mark J. Mannis, 'Summary of Corneal Transplant Activity: Eye Bank Association of America', *Cornea*, 21 (1) (2002), pp. 1–3.

29 Climbing up the Value Chain

1. Elizabeth Haiken, *Venus Envy: A History of Cosmetic Surgery* (Baltimore and London: Johns Hopkins University Press, 1997), p. 161.
2. Stephen H. Miller, 'Competitive Forces and Academic Plastic Surgery', *Plastic and Reconstructive Surgery*, 101 (5) (1998), pp. 1389–99; K. T. Ta, John A. Persing, Henry Chauncey, H. Bradley and Stephen H. Miller, 'Effects of Managed Care on Teaching,

Research, and Clinical Practice in Academic Plastic Surgery', *Annals of Plastic Surgery*, 48 (4) (2002), pp. 348–54.
3. Donor Alliance, www.donoralliance.org (accessed May 2016).
4. Timothy F. Kirn, 'How Does Tissue Banking Work? Virginia Bank While Not Typical, May Offer Some Insights', *Journal of the American Medical Association*, 258 (3) (1987), pp. 304–5.
5. Andrew A. Skolnick, 'Tissue Bank Expands Facilities, Efforts', *Journal of the American Medical Association*, 266 (10) (1991), pp. 1329–31.
6. Michael Waldhole, 'Red Cross's Plan To Procure Organs Could Hurt Smaller Organizations', *Wall Street Journal* (8 August 1984), p. 33.
7. Community Tissue Services, www.communitytissue.org (accessed April 2016).
8. Community Tissue Services Annual Report 2013, www.communitytissue.org (accessed May 2016).
9. Kirn, 'How Does Tissue Banking Work?'
10. *Regulation of Human Tissue Banks*, Hearing before the Subcommittee on Regulation, Business Opportunities, and Technology, Washington DC, 15 October 1993, Serial No. 103-53 (Washington DC: US Government Printing Office, 1994), p. 198.
11. LifeCell Corporation – Company Profile, Information, Business Description, History, Background Information on LifeCell Corporation, http://www.referenceforbusiness.com (accessed May 2016).
12. Donor Alliance's 25 Years of Living History, www.donoralliance.org (accessed May 2016).
13. Anthony Clark, 'RTI Founder Spin Off Thriving Biotech Firms in Gainesville Area', *Gainesville Sun* (19 February 2011), http://www.gainesville.com/article (accessed 22 May 2015).
14. Martha W. Anderson and Scott Bottenfield, 'Tissue Banking – Past, Present, and Future', in Stuart J. Youngner, Martha W. Anderson and Renie Schapiro (eds), *Transplanting Human Tissue: Ethics, Policy, and Practice* (New York and Oxford: Oxford University Press, 2004), p. 26.
15. Christopher Truitt, *The Dark Side of Tissue Donation* (Publisher: Author, 2009), p. 55.
16. Stephen J. Hedges and William Gaines, 'Donor Bodies Milled into Growing Profits; Little-regulated Industry Thrives on Unsuspecting Families; Cadavers' Average Worth: $80,000', *Chicago Tribune* (21 May 2000), p. 1.
17. *Regulation of Human Tissue Banks*, p. 28.
18. Mark Katches, William Heisel and Ronald Campbell, 'The Body Brokers', *Orange County Register* (16 April 2000), http://www.newslibrary.com (accessed July 2000).
19. *Organ Donation: Assessing Performance of Organ Procurement Organizations,* Statement for the Record by Bernice Steinhardt, Director Health Services Quality and Public Health Issues, Health, Education, and Human Services Division (Washington DC: United States General Accounting Office, 1998), p. 6.
20. Jeffrey M. Prottas 'Competition for Altruism: Bone and Organ Procurement in the United States', *Milbank Quarterly*, 70 (2) (1992), pp. 299–317.
21. Office of Evaluation and Inspections, Office of Inspector General, *Organ Procurement Organizations and Tissue Recovery* (Washington DC: Department of Health and Human Services, 1994), pp. 5–6.
22. Jeffrey Prottas, *The Most Useful Gift: Altruism and the Public Policy of Organ Transplants* (San Francisco: Jossey-Bass Publishers, 1994), p. 39.
23. Office of Evaluation and Inspections, Office of Inspector General, *Organ Procurement Organizations and Tissue Recovery*, pp. 5–6.
24. Stephen J. Hedges and William Gaines, 'Donor Bodies Milled into Growing Profits' …, *Chicago Tribune* (21 May 2000), p. 1.
25. Anderson and Bottenfield, 'Tissue Banking – Past, Present, and Future', p. 22.
26. Midwest Transplant Network, www.mwtn.org (accessed June 2016).
27. James Brian Quinn, Frederick G. Hilmer, 'Strategic Outsourcing', *Sloan Management Review* (Summer 1994), pp. 43–55.
28. Hedges and Gaines, 'Donor Bodies Milled into Growing Profits'.
29. Robert A. Katz, 'The Re-gift of Life: Can Charity Law Prevent Non-profit Firms from

Exploiting Donated Tissue and Non-profit Tissue Banks?', *De Paul Law Review*, 55 (3) (2006), p. 969.
30. *Tissue Banks: is the Federal Government's Oversight Adequate? Hearing before the Permanent Subcommittee on Investigations of the Committee on Governmental Affairs*, United States Senate, One Hundred Seventh Congress, First Session, 24 May 2001 (Washington DC: US Government Printing Office), p. 1.

30 Contagious Corpses

1. Lawrence K. Altman, '3 Die of AIDS After Getting Organs from Man Infected with the Virus', *New York Times* (18 May 1991).
2. Report of speech delivered by David M. Link, Director of the Food and Drug Administration, 27 November 1978, *American Association of Tissue Banks Newsletter*, 3 (2) (April 1979), p. 12.
3. Gilbert M. Gaul, *The Blood Brokers – How Blood, the 'Gift of Life', Became a Billion Dollar Industry* (1989), www.bloodbook.com (accessed January 2016).
4. J. Strobos, Food and Drug Administration, 'Safety of Tissue Allografts, American Association of Tissue Banks Newsletter', 16 (1) (16 February 1993), pp. 3–4.
5. Paula A. Treichler, *How to Have a Theory in an Epidemic: Cultural Chronicles of AIDS* (Durham, NC and London: Duke University Press, 1999).
6. Stephen J. Hedges and William Gaines, 'Donor Bodies Milled into Growing Profits' …, *Chicago Tribune* (21 May 2000), p. 1.
7. *Tissue banks and the dangers of tainted tissues: the need for Federal Regulation*, Hearing before the Committee on Governmental Affairs, 14 May 2003 (Washington DC: US Government Printing Office, 2003), p. 6.
8. *Regulation of Human Tissue Banks, Hearing before the Subcommittee on Regulation, Business Opportunities, and Technology*, Washington DC, 15 October 1993, Serial No. 103-53 (Washington DC: US Government Printing Office, 1994) p. 378.
9. Michael Leechman, 'Regulation of the Human Tissue Industry: A Call For Fast-track Regulation', *Louisiana Law Review*, 65 (1) (2004), pp. 448–9.
10. Marion Nestle, *Safe Food: The Politics of Food Safety* (Berkeley, Los Angeles, and London: University of California Press, 2010), p. 34.
11. *Regulation of Human Tissue Banks*, p. 8.
12. Ibid., pp. 107–10.
13. Hedges and Gaines, 'Donor Bodies Milled into Growing Profits'.

31 British Prions

1. Frank A. Burnham, 'The Rendering Industry – A Historical Perspective', in Don A. Franco and Winfield Swanson (eds), *The Original Recyclers* (The Animal Protein Producers Industry, The Fats & Proteins Research Foundation, The National Renderers Association, 1996), pp. 1–16.
2. Richard Rhodes, *Deadly Feasts: Tracking the Secrets of a Terrifying New Plague* (New York: Simon and Schuster, 1997), p. 176.
3. Brian J. Ford, *BSE: The Facts* (London: Corgi Books, 1996), pp. 21–2.
4. Rudolph Klein, The Politics of the *National Health Service*, 2nd edn (London: Longman, 1989), p. 299.
5. Allyson M. Pollock, *NHS plc: The Privatisation of our Health Care* (London: Verso, 2004).
6. John Kearney, NHS Tissue Services, interview with author (Liverpool, July 2007).
7. British Association of Tissue Banks, 'Standards for tissue banking, June 1995', *Transplant Medicine*, 6 (1996), 155–8.
8. R. J. Michaud, K. J. Drabu, 'Bone Allograft Banking in the United Kingdom', *Bone and Joint Journal*, 76-B (1994), pp. 350–51.
9. Klaus Hoeyer, 'After Novelty: The Mundane Practices of Ensuring a Safe and Stable Supply of Bone', *Science as Culture*, 19 (2) (2010), pp. 123–50.
10. E. A. Caffrey, 'Lessons Learnt from Setting Up a Regional Tissue Bank' (abstract), *Transfusion Medicine*, 5 (supplement 1) (1995), p. 2.

11. David Pegg, University of York/British Association of Tissue Banks, interview with author (York, November 2007).
12. Deidre Fehily, NHS Blood Service, interview with author (London, January 1999).
13. A. M. Murphy, 'The Development of Tissue Banking in Scotland' (abstract), *Transfusion Medicine*, 5 (supplement 1) (1995), p. 2.
14. 'Early Excision of Thermal Burns – An International Round-table Discussion, Geneva, June 22, 1987', *Journal of Burns Care and Rehabilitation*, 9 (5) (1988), pp. 549–61.
15. Working Group of the British Burn Association and Hospital Infection Society, 'Report on the Principles of Design of Burns Unit,' *Journal of Hospital Infection*, 19 (1991), pp. 63–6.
16. National Burn Care Review Committee, *Standards and Strategy for Burn Care: A Review of Burn Care in the British Isles* (British Burn Association, 2001).
17. E. Freedlander, S. Boyce, M. Ghosh, D. R. Ralston and S. MacNeil, 'Skin Banking in the UK: The Need for Proper Organization', *Burns*, 24 (1998), pp. 19–24.
18. Nigel Hawkes, 'Friends Donate Skin in Effort to Save Fire Victim', *The Times* (26 June 1984).
19. *The Stephen Kirby Skin Bank and Research Unit* (Richmond, Twickenham and Roehampton Healthcare NHS Trust, n.d.).
20. A.C.J. de Backere, 'Euro Skin Bank: Large Scale Skin-banking in Europe Based on Glycerol-preservation of Donor Skin', *Burns*, 20 (1) (1994), pp. S4–S9.
21. Sarah Harris, 'Hospital's Skin Bank Tribute to Tragic Susan', *Western Daily Press – Bristol* (3 November 1994).
22. David R. Ralston, Sue G. Boyce, Sheila MacNeil and Eric Freedlander, 'Demand has Outstripped Supply in Sheffield's Skin Bank', *British Medical Journal*, 313 (1996), p. 1485.
23. S. R. Myers, M. R. Macheseny, R. M. Warwick and P. D. Cussons, 'Skin Storage', *British Medical Journal*, 313 (1996), p. 439.
24. HC Deb 8 April 1998, vol. 310 cc 353 (Frank Dobson).
25. Janet Morgan, 'Blood to be Screened for CJD', *British Medical Journal*, 313 (1996), p. 441.
26. Committee on Microbiological Safety of Blood and Tissues for Transplantation, *Guidance on the Microbiological Safety of Human Tissue and Organs Used in Transplantation* (London: NHS Executive, 1996).
27. Council of Europe, Committee of Ministers to Member States on Human Tissue Banks, Recommendation No. R (94) 1.
28. Maya L. Ponte, 'Insights into the Management of Emerging Infections: Regulating Creutzfeldt-Jakob Disease Transfusion Risk in the UK and the US', *PLoS Medicine*, 3 (10) (2006), pp. 1751–64.
29. John Warden, 'UK Blood Products are Banned', *British Medical Journal*, 316 (1998), p. 723.
30. Ric R. Grummer and Terry Klopfenstein, 'Utilizing Rendered Products: Ruminants', in Franco and Swanson (eds), *The Original Recyclers*, pp. 129–74.
31. Donald G. McNeil Jr, 'U.S. Reduces Testing for Mad Cow Disease Citing Few Infections', *New York Times* (21 July 2006).
32. David Derbyshire, "Patient's death from CJD is linked to blood transfusion', *Daily Telegraph* (18 December 2003).

32 Compassion and Commerce

1. M. Katches, W. Heisel and R. Campbell, 'The Body Brokers', *Orange County Register*, (16 April 2000).
2. *Tissue Banks: is the Federal Government's Oversight Adequate?*, Hearing before the Permanent Subcommittee on Investigations of the Committee on Governmental Affairs, United States Senate, One Hundred Seventh Congress, First Session, 24 May 2001 (Washington DC: US Government Printing Office), p. 7.
3. Diane Wilson and Glenn Greenleaf, 'The Availability of Allograft Skin for Large Scale Medical Emergencies in the United States', *Cell Tissue Banking*, 15 (2014), pp. 35–40.
4. Glenn Greenleaf, 'Skin Transplantation: Clinical Applications and Current Issues', in Stuart J. Youngner, Martha W. Anderson and Renie Schapiro (eds), *Transplanting Human*

Tissue: Ethics, Policy, and Practice (New York and Oxford: Oxford University Press, 2004), p. 62.
5. Andrew A. Skolnick, 'Tissue Bank Expands Facilities, Efforts', *Journal of the American Medical Association*, 266 (10) (1991).
6. Stephen Williamson, Shriners Skin Bank, Galveston, Texas, Presentation, American Association of Tissue Banks Annual Meeting, Savannah, 2008.
7. Nancy Gallo, New York Firefighters Skin Bank, interview with author (Savannah, Georgia, April 2008).
8. Peter G. Shakespeare, 'The Role of Skin Substitutes in the Treatment of Burn Injuries', *Clinics in Dermatology*, 23 (2005), p. 415–18.
9. Katches, Heisel and Campbell, 'The Body Brokers'.
10. Stephen J. Hedges and William Gaines, 'Donor Bodies Milled into Growing Profits'.
11. Stephen H. Miller, 'Competitive Forces and Academic Plastic Surgery', *Plastic and Reconstructive Surgery* (April 1998), pp. 1389–9.
12. Hedges and Gaines, 'Donor Bodies Milled into Growing Profits'.
13. Martha W. Anderson and Scott Bottenfield, 'Tissue Banking – Past, Present, and Future', in Stuart J. Youngner, Martha W. Anderson and Renie Schapiro (eds), *Transplanting Human Tissue: Ethics, Policy, and Practice* (New York and Oxford: Oxford University Press, 2004), p. 19.
14. William Heisel, 'Tissue-group Steps Closer to Full Disclosure', *Orange County Register*, 13 (September 2000), http://www.ocregister.com (accessed 15 October 2006).
15. *Tissue Banks: is the Federal Government's Oversight Adequate?*, p. 21.
16. Office of Evaluation and Inspections, Office of Inspector General, *Informed Consent in Tissue Donation: Expectations and Realities* (Washington DC: Department of Health and Human Services, 2001).
17. *Tissue Banks: is the Federal Government's Oversight Adequate?*, p. 16
18. Office of Evaluation and Inspections, *Informed Consent in Tissue Donation*, p. v.
19. Ibid., pp. ii–iv.
20. Ibid., pp. 7–9.
21. Ibid., pp. 19–21.
22. Karen Sokohl, 'First Person Consent: OPOs Across the Country are Adapting to the Change', *Update* (September–October 2003), pp. 2–7, www.unos.org (accessed June 2010).
23. Jennifer L. Mesich-Brant and Lawrence J. Grossback, 'Assisting Altruism: Evaluating Legally Binding Consent in Organ Donation Policy', *Journal of Health Politics, Policy and Law*, 30 (4) (2005), pp. 688–9.
24. David Orentlicher, 'Presumed Consent to Organ Donation: Its Rise and Fall in the United States', *Rutgers Law Review*, 61 (2009), p. 307, fn. 58; Sheldon E. Kurtz, Christina Woodward Strong and David Gerasimow, 'The 2006 Revised Uniform Anatomical Gift Act – A Law to Save Lives', *Health Lawyers News* (February 2007), pp. 44–9.
25. Robert D. Truog, 'Consent for Organ Donation – Balancing Conflicting Ethical Obligations', *New England Journal of Medicine* (20 March 2008), pp. 1209–11.
26. Ana S. Iltis, Michael A. Rie and Anji Wall, 'Organ Donation, Patients' Rights, and Medical Responsibilities at the End of Life', *Critical Care Medicine*, 37 (1) (2009), pp. 310–16.
27. Sheldon Zink and Stacey Wertlieb, 'A Study of the Presumptive Approach to Consent for Organ Donation: A New Solution to an Old Problem', *Critical Care Nurse*, 26 (2) (2006), pp. 129–36.
28. Laura A. Siminoff, Heather M. Traino and Nahida Gordon, 'Determinants of Family Consent to Tissue Donation', *Journal of Trauma, Injury, Infection, and Critical Care*, 69 (4) (2010), pp. 956–63.
29. Patricia Mulvania, 'Gift of Life Donor Program, Philadelphia, Pennsylvania', Presentation, American Association of Tissue Banks Annual Meeting, March 2008.

33 A Roadmap for the Future

1. *Tissue Banks and the Dangers of Tainted Tissues: the Need for Federal Regulation*, Hearing before the Committee on Governmental Affairs, 14 May 2003 (Washington DC: US Government Printing Office, 2003), p. 14.

2. Centers for Disease Control and Prevention, 'Update: Allograft-associated Bacterial Infections – United States, 2002,' *Journal of the American Medical Association*, 287 (13) (2002), pp. 1642–4.
3. Sandra Blakeslee, 'Recall is Ordered at Large Supplier of Implant Tissue', *New York Times* (15 August 2002).
4. Jason L. Williams, 'Patient Safety or Profit: What Incentives are Blood Shield Laws and FDA Regulations Creating for the Tissue Banking Industry?' *Indiana Health Law Review*, 2 (295) (2005), pp. 295–328.
5. Warren E. Leary, 'Rules to Cover Human Tissue in Products', *New York Times* (1 March 1997).
6. Robert Pear, 'F.D.A. Delays Regulation of Tissue Transplants', *New York Times* (16 May 2003).
7. Department of Health and Human Services, Food and Drug Administration, 'Human Cells, Tissue, and Cellular and Tissue-based Products: Establishment, Registration and Listing', *Federal Register*, 66 (13) (2001), p. 5454.
8. Department of Health and Human Services, Food and Drug Administration, 'Eligibility Determination for Donors of Human Cells, Tissue, and Cellular Tissue-based Products; Final Rule and Notice', *Federal Register*, 69 (101) (2004), p. 29817.
9. *Tissue Banks: is the Federal Government's Oversight Adequate?*, Hearing before the Permanent Subcommittee on Investigations of the Committee on Governmental Affairs, United States Senate, One Hundred Seventh Congress, First Session, 24 May 2001 (Washington DC: US Government Printing Office), p. 8.
10. Ibid., p. 45.
11. Adam Liptak, 'Inmate Count in U.S. Dwarfs Other Nations', *New York Times* (23 April 2008).
12. Bianca Miranda, Alessandro Nanni Costa, Jacinto Sánchez-Ibáñez and Eliana Porta, 'Gathering Donor History: Ensuring Safe Tissue for Transplant', in Ruth M. Warwick, Deirdre Fehily, Scott A. Brubaker and Ted Eastlund (eds), *Tissue and Cell Donation: An Essential Guide* (Oxford: Wiley-Blackwell, 2009), pp. 98–115.
13. Department of Health and Human Services, Food and Drug Administration, 'Eligibility Determination', pp. 29804–5.
14. Marion Nestle, *Safe Food: The Politics of Food Safety* (Berkeley, Los Angeles, and London: University of California Press, 2010), pp. 67–9.
15. Department of Health and Human Services, Food and Drug Administration, 'Current Good Tissue Practice for Manufacturers of Human Cellular and Tissue-Based Products; Inspection and Enforcement; Proposed Rule', *Federal Register*, 66 (5) (2001), p. 1545.

34 Repairing the Past

1. Public Inquiry into children's heart surgery at the Bristol Royal Infirmary, *Interim Report: Removal and Retention of Human Material* (2000), in Public Inquiry into children's heart surgery at the Bristol Royal Infirmary 1984–1995, *Report: Learning from Bristol*, CM 5207(1) (London: The Stationery Office, 2001), Annex C; The Royal Liverpool Children's Inquiry, *Report* HC12-11 (London: The Stationery Office, 2001); Public Inquiry into children's heart surgery, *Report: Learning from Bristol*; Department of Health, Department for Education and Employment, Home Office, *The Removal, Retention and Use of Human Organs and Tissue from Post-mortem Examination: Advice from the Chief Medical Officer* (London: The Stationery Office, January 2001); Lord Justice Clarke, *Public Inquiry into the Identification of Victims Following Major Transport Accidents*, volume 1, Cm 5012 (London: The Stationery Office, 2001); Independent Review Group on Retention of Organs at Post-Mortem, *Final Report* (Edinburgh: The Stationery Office, 2001); Independent Review Group on Retention of Organs at Post-Mortem, *Report on Strontium-90 Research*; Department of Health, *Isaacs Report: The Investigation of Events that Followed the Death of Cyril Mark Isaacs* (London: The Stationery Office, 2003).
2. Department of Health, Department for Education and Employment, Home Office, *The Removal, Retention and Use of Human Organs and Tissue from Post-mortem Examination*, p. 2.
3. Chief Medical Officer, *Report of a Census of Organs and Tissues Retained by Pathology Services in England* (London: The Stationery Office, 2001).

4. 'Eye Transplants "Hit by Alder Hey"', BBC News (6 August 2001), http://news.bbc.co.uk (accessed May 2013).
5. History of the Organ Donor Register, updated May 2011, https://wwworgandonation.nhs.uk/ukt/newsroom/fac (accessed November 2012).
6. Bill New, Michael Solomon, Robert Dingwall and Jean McHale, *A Question of Give and Take: Improving the Supply of Organs for Transplantation* (London: King's Fund Institute), p. 14.
7. HC Deb 29 July 1998, vol. 317 column 316 (B. Blizzard).

35 Globalizing the Gift

1. Philip J. Guyett Jr, *Heads, Shoulders, Knees and Bones* (Publisher: Author, 2011), p. 180.
2. *Report of the Grand Jury In Re County Investigating Grand Jury XXI*, In the Court of Common Pleas, First Judicial District of Pennsylvania, Criminal Trial Division, Misc. No. 05–011487, 2007, pp. 6–8.
3. *Kennedy-McInnis et al. v. Biomedical Tissue Services, Ltd. et al.*, No. 13–CV-05140.(2016)
4. Christopher Truitt, *The Dark Side of Tissue Donation* (Publisher: Author 2009), pp. 68–71.
5. Guyett Jr, *Heads, Shoulders, Knees and Bones*, p. 241.
6. *Regulation of Human Tissue Banks,* pp. 13–16.
7. 'Musculoskeletal Transplant Foundation Acquires Assets of American Red Cross Tissue Services', www.thefreelibrary.com (accessed January 2017).
8. Organ Procurement and Transplantation Network, National Data, https://optn.transplant.hrsa.gov/data (accessed August 2016).
9. Data for 2003 are not publicly available. However, in 2004, eyes/corneas were extracted from 41,131 American corpses: www. restoresight.org (accessed March 2013).
10. Scott A. Brubaker and Nancy Senst, 'The Gift of Tissue and Eye Donation', in Dianne La Pointe Rudow, Linda Ohler and Teresa Shafer (eds), *A Clinician's Guide to Donation and Transplantation* (Lenexa: NATCO, 2006), p. 373.
11. 'Osteotech Rejects MTF Approach' (1 October 2005), http: www.highbeam.com (accessed March 2013).
12. Musculoskeletal Transplant Foundation, 'LifePoint Forges New Tissue Donation Partnership, 2005', http://www.mtf.org/pdf/life_point.pdf (accessed May 2008).
13. LifeLine, www.lifelineofohio.org/about/index.aspx (accessed May 2008).
14. Iowa Donor Network, www.iowadonornetwork.org/hospitals.asp (accessed May 2008).
15. Tutogen Medical, Inc., History, www.fundinguniverse.com (accessed May 2016).
16. *Report of the Grand Jury In Re County Investigating Grand Jury XXI*, In the Court of Common Pleas, First Judicial District of Pennsylvania, Criminal Trial Division, Misc. No. 05–011487 (2007), p. 43.
17. Martha W. Anderson and Rene Schapiro, 'From Donor to Recipient: The Pathway and Business of Donated Tissues', in Stuart J. Youngner, Martha W. Anderson and Renie Schapiro (eds), *Transplanting Human Tissue: Ethics, Policy, and Practice* (New York and Oxford: Oxford University Press, 2004), p. 10.
18. Guyett Jr, *Heads, Shoulders, Knees and Bones*, p. 251.
19. Vlad Lavrov, Kate Willson and Thomas Maier, 'Five Charged Over Alleged Body Parts Theft' (21 December 2012), www.icij.org (accessed July 2016).
20. Thomas Maier, Kate Willson and Michael Hudson, 'Pentagon, Congress Probe Tissue Contracts' (5 September 2012), www.icij.org (accessed July 2016).
21. D. Michael Strong and Naoshi Shinozaki, 'Coding and Traceability for Cells, Tissue and Organs for Transplantation', *Cell and Tissue Banking*, 11 (2010), pp. 305–23.
22. Eye Bank Association of America, *Annual Report 2015*, www.restoresight.org (accessed November 2016).
23. Jennifer Y. Li and Mark J. Mannis, 'Eye Banking and the Changing Trends in Contemporary Corneal Surgery', *International Ophthalmology Clinics*, 50 (3) (2010), pp. 101–12.
24. Vision Share, www.vision share.org (accessed March 2013).
25. James H. Leimkuhler, Tissue Banks International, interview with author (Baltimore, April 2005).

36 Consolidation without Cooperation

1. Interim guidance includes: Royal College of Pathologists, *Guidelines for the Retention of Tissue and Organs at Post-Mortem Examination* (London: Royal College of Pathologists, 2000); Department of Health, *Organ Retention: Interim Guidance on Post-mortem Examination* (London: Department of Health, 2000); Department of Health, Department of Education and Employment, Home Office, *The Removal, Retention and Use Of Human Organs and Tissue from Post-mortem Examination: Advice from the Chief Medical Officer* (January 2001); Medical Research Council, *Human Tissue and Biological Samples for use in Research: Operational and Ethical Guidelines* (London: Medical Research Council, 2001); Department of Health, Welsh Assembly Government, NHS Central Office for Research Ethics Committees (COREC), *The Use of Human Organs and Tissue: An Interim Statement* (London: Department of Health, 2003).
2. *Human Bodies, Human Choices. The Law on Human Organs and Tissue in England and Wales: A Consultation Report* (London: Department of Health, July 2002).
3. *Human Bodies, Human Choices. Summary of Responses to the Consultation Report* (London: Department of Health, April 2003).
4. The Department of Culture, Media and Sports established a Working Party on Human Remains, which considered how to respond to claims for return of cadaver stuff/artefacts, more than 100 years old, held outside major museums and other institutions, such as universities. Department of Culture, Media and Sports, *Guidance for the Care of Human Remains in Museums* (London: Department of Culture, Media, and Sports, 2004).
5. Qualifying relationships are set out and ranked in the following descending pecking order: spouse or partner (meaning they have lived together in an enduring family relationship, whether of the same or different sexes); parent or child; brother or sister; grandparent or grandchild; niece or nephew; step-father or step-mother; half-brother or half-sister; friend of long standing. A lower-ranked person cannot overturn a decision.
6. Julia Wheldon, 'Families Can't Stop Doctors Taking Organs', *Daily Mail* (31 August 2006), p. 1.
7. Retained Organs Commission, *Remembering the Past, Looking to the Future. The Final Report of the Retained Organs Commission, Including the Summary Accountability Report for 2003/2004* (London: Department of Health, 2004), p. 10.
8. David Price, 'The Human Tissue Act 2004', *Modern Law Review*, 68 (5) (2005), pp. 798–821.
9. The British Association for Tissue Banking, the Department of Health, and the Council of Europe, had issued guidance on safety, which was unenforceable in law. Department of Health, *Transplantation, A Code of Practice for Tissue Banks Providing Tissue of Human Origin for Therapeutic Purposes* (London: Department of Health, 2001); Alina Tatarenko, 'European Regulations and their Impact on Tissue Banking', *Cell and Tissue Banking*, 7 (2006), pp. 231–5.
10. Glyn O. Philips and Jorge M. Pedraza, 'The International Atomic Energy Agency (IAEA) Programme in Radiation and Tissue Banking: Past, Present and Future', *Cell and Tissue Banking*, 4 (2003), pp. 69–76.
11. Emma Winstanley and Anthony J. Clarkson, 'The Dedicated Donation Facility. Routine Notification: A Novel Approach', *Blood Matters*, 22 (2007), pp. 13–14; Tracey Long-Sutehall, Emma Winstanley, Anthony J. Clarkson and Magi Sque, 'Evaluation of the Experience of Family Members Whose Deceased Relative Donated Tissue at the NHSBT Dedicated Donation Facility, at Speke, Liverpool', *Cell and Tissue Banking*, 13 (4) (2012), pp. 537–46.
12. Sarah-Kate Templeton, 'Stolen Body Parts were Implanted in NHS Patients', *Sunday Times* (10 September 2006).
13. NHS Blood and Transplant, *Transplant Activity in the UK* (2006), p. 38, www.nhsby.nhs.uk.
14. Andrea Rowe, Blond McIndoe Centre, Queen Victoria Hospital, interview with author (East Grinstead, August 2005).
15. Pamela Keely, East Anglian Eye Bank Coordinator, presentation at the British Association of Tissue Banks Annual Meeting, 2004.
16. Stephen Kaye, The Royal Liverpool University Hospital, interview with author (Liverpool, July 2007).

17. Peter K. Leaver, *The History of Moorfields Eye Hospital*, volume 3 (London: The Royal Society of Medicine Press, 2004), p. 112.
18. Frank Larkin, Moorfields Eye Hospital, interview with author (London, May 2013).
19. Department of Health, *Care and Respect in Death: Good Practice Guidance for Mortuary Staff* (London: Department of Health, 2006), http://www.dh.gov.uk (accessed May 2011).
20. Pamela Keely, East Anglian Eye Bank, interview with author (Norwich, May 2007).
21. The Royal College of Ophthalmologists, *Standards for Retrieval of Human Ocular Tissue Used in Transplantation, Research and Training* (2008), www.rcophjth.ac.uk (accessed December 2010).
22. John Armitage, Corneal Transplant Service, interview with author (Bristol, March 2005).
23. 'Tissue Donation', www.nhsbt.nhs.uk (accessed February 2017).
24. Laura A. Siminoff and Heather M. Traino, 'Consenting to Donation: An Examination of Current Practices in Informed Consent for Tissue Donation in the US', *Cell and Tissue Banking*, 14 (1) (2013), p. 85.

37 From Mortuary to Shopping Cart?

1. Jason E. LoVerdi, Accreditation Manager, American Association of Tissue Banks, personal communication, February 2017.
2. Andrew Siegel, Martha W. Anderson, Tracy C. Schmidt and Stuart J. Youngner, 'Informed Consent to Tissue Donation: Policies and Practice', *Cell and Tissue Banking*, 10 (2009), pp. 235–40.
3. Adrian Cadbury, 'Corporate Social Responsibility', *Twenty-first Century*, 1 (2006), pp. 5–21.
4. 'LifeNet Health establishes "The Skin and Wound Allograft Institute"' (17 September 2007), www.prnewswire.com (accessed August 2016).
5. Community Tissue Services, www.communitytissue.org (accessed August 2016).
6. Jean-Paul Pirnay, Alain Vanderkelen, Martin Zizi, Daniel De Vos, Thomas Rose, Geert Laire, Nadine Ectors and Gilbert Verbeken, 'Human Cells and Tissues: The Need for a Global Ethical Framework', *Bulletin of the World Health Organization*, 88 (2010), pp. 870–72.
7. Diane Wilson and Glenn Greenleaf, 'The Availability of Allograft Skin for Large-scale Medical Emergencies in the United States', *Cell and Tissue Banking*, 15 (2014), p. 36.
8. Musculoskeletal Transplant Foundation Catalog, www.mtf.org (accessed February 2017).
9. www.keralink.org (accessed February 2017).
10. Jean-Paul Pirnay, Alain Vanderkelen, Nadine Ectors, Christian Delloye, Denis Dufrane, Etienne Baudoux, Michel Van Brussel, Michael P. Casaer, Daniel De Vos, Jean-Pierre Draye, Thomas Rose, Serge Jennes, Pierre Neirinckx, G. Laire and M. Zizi, 'Beware of the Commercialization of Human Cells and Tissues: Situation in the European Union', *Cell and Tissue Banking*, 13 (2012), p. 493.
11. NHS Blood and Tissue, *Annual Report and Accounts, 2010/2011* (London: The Stationery Office), p. 4.
12. Penny Hogg, Paul Rooney, Eileen Ingham and John N. Kearney, 'Development of a Decellularised Dermis', *Cell and Tissue Banking*, 14 (3) (2013), pp. 465–74.
13. www.tissueregenix.com (accessed January 2017).
14. D. L. Tiarnan, D. L. Keenan, Mark N. A. Jones, Sally Rushton and Fiona M. Carley, 'Trends in the Indications for Corneal Graft Surgery in the United Kingdom, 1999 through 2009', *Archives of Ophthalmology*, 130 (5) (2012), pp. 621–8.
15. 'Ocular Tissue Pathway Service Review, Draft Report' (2012), p. 13, www.odt.nhs.uk/transplantation/advisory-group/ (accessed January 2017).
16. Andrew Siegel, Martha W. Anderson, Tracy C. Schmidt and Stuart J. Youngner, 'Informed Consent to Tissue Donation': Policies and Practice', *Cell and Tissue Banking*, 10 (2009), pp. 235–236.
17. Department of Health, Procurement, Investment and Commercial Division, NHS Blood and Transplant Commercial Review, 2011, p. 23, www.gov.uk (accessed February 2017).
18. NHS Blood and Tissue, *Annual Report and Accounts, 2012/2013* (London: The Stationery Office), p. 5.

BIBLIOGRAPHY

Primary Sources
Archives
United Kingdom

Blond McIndoe Research Unit Archives, Queen Victoria Hospital, East Grinstead
Colebrook, Leonard Papers, Wellcome Library for the History and Understanding of Medicine, London
Iris Fund for the Prevention of Blindness, London
The National Archives, London
Royal National Institute for the Blind, London
Young, Frank George Papers, University of Cambridge Library

United States of America

American Association of Tissue Banks, McLean, Virginia
American Burn Association Papers, National Institutes of Health Library, Bethesda, Maryland
College of American Pathologists, Northfield, Illinois
Evans, Herbert McLean Papers, University of California, Berkeley, California
Eye-Bank for Sight Restoration, New York
Li, Choh Hao Papers, University of California, San Francisco Medical Center, San Francisco, California
National Archives, College Park, Maryland
National Research Council, National Academies of Sciences, Washington DC

Government Reports
United Kingdom

Advisory Group on Transplantation Problems on the question of amending the Human Tissue Act 1961, *Advice*, Cmnd. 4106 (London: Her Majesty's Stationery Office, 1969)
Chief Medical Officer, *Report of a Census of Organs and Tissues retained by pathology services in England* (London: The Stationery Office, 2001)
Clarke, Lord Justice, *Public inquiry into the identification of victims following major transport accidents*, volume 1, CM 5012 (London: The Stationery Office, 2001)

Committee on Death Certification and Coroners, Report, Cmnd. 4810 (London: Her Majesty's Stationery Office, 1971)
Committee on Microbiological Safety of Blood and Tissues for Transplantation, *Guidance on the microbiological safety of human tissue and organs used in transplantation* (London: NHS Executive, 1996)
Council of Europe, *Committee of Ministers to Member States on Human Tissue Banks, Recommendation No. R (94) 1* (1994)
Department of Culture, Media and Sports, *Guidance for the Care of Human Remains in Museums* (London: Department of Culture, Media, and Sports, 2004)
Department of Health, *Organ Retention: Interim Guidance on post-mortem examination* (London: Department of Health, 2000)
—, *A code of practice for tissue banks providing tissue of human origin for therapeutic purposes* (London: Department of Health, 2001)
—, *Human bodies, human choices: the law on human organs and tissue in England and Wales: a consultation report* (London: Department of Health, 2002)
—, *Human bodies, human choices: summary of responses to the consultation report* (London: Department of Health, 2003)
—, *Isaacs Report: The investigation of events that followed the death of Cyril Mark Isaacs* (London: The Stationery Office, 2003)
Department of Health, Department for Education and Employment, Home Office, *The removal, retention and use of human organs and tissue from post-mortem examination: Advice from the Chief Medical Officer* (London: The Stationery Office, 2001)
Department of Health, Procurement, Investment and Commercial Division, *NHS Blood and Transplant Commercial Review* (London: Department of Health, 2011)
Department of Health, Welsh Assembly Government, NHS Central Office for Research Ethics Committees (COREC), *The use of human organs and tissue: an interim statement* (London: Department of Health, 2003)
Independent Review Group on Retention of Organs at Post-Mortem, *Final Report* (Edinburgh: The Stationery Office, 2001)
—, *Report on Strontium–90 Research* (Edinburgh: Scottish Executive Health Department, 2002)
Medical Research Council, *Human tissue and biological samples for use in research: operational and ethical guidelines* (London: Medical Research Council, 2001)
NHS Blood and Tissue, *Annual Report and Accounts, 2010/2011* (London: The Stationery Office, 2011)
—, *Annual Report and Accounts, 2012/2013* (London: The Stationery Office, 2013)
Public Inquiry into children's heart surgery at the Bristol Royal Infirmary 1984–1995, 'Interim report: Removal and retention of human material, 2000', in Public Inquiry into children's heart surgery at the Bristol Royal Infirmary 1984–1995, *Report: Learning from Bristol*, CM 5207(1), Annex C (London: The Stationery Office, 2001)
Public Inquiry into children's heart surgery at the Bristol Royal Infirmary 1984–1995, *Report: Learning from Bristol*, CM 5207(1) (London: The Stationery Office, 2001)
Retained Organs Commission, *Remembering the Past, Looking to the Future. The final report of the Retained Organs Commission, including the Summary Accountability Report for 2003/2004* (London: Department of Health, 2004)
Royal College of Pathologists, *Guidelines for the retention of tissue and organs at post-mortem examination* (London: Royal College of Pathologists, 2000)
The Royal Liverpool Children's Inquiry, *Report,* HC12–11 (London: The Stationery Office, 2001)
The Shipman Inquiry, *Third Report: Death Certification and the Investigation of Deaths by Coroners,* CM 5854 (London: The Stationery Office, 2003)

United States of America

Anatomical Gift Act, Hearing before the Committee of the District of Columbia United States Senate, Ninety-First Congress, Second Session (Washington DC: US Government Printing Office, 1970)

Department of Health and Human Services, Food and Drug Administration, 'Current Good Tissue Practice for Manufacturers of Human Cellular and Tissue-Based products; Inspection and Enforcement; Proposed Rule', *Federal Register*, 66(5), 1507 (2001)
—, 'Human cells, tissue, and cellular and tissue-based products: establishment, registration and listing', *Federal Register*, 66(13) 5447 (2001)
—, 'Eligibility determination for donors of human cells, tissues, and cellular and tissue-based products; Final Rule and Notice', *Federal Register*, 69 (101) 29785 (2004)
Department of Health and Human Services, Health Care Financing Administration, 'Medicare and Medicaid Programs; Hospital Conditions of Participation; Identification of Potential Organ, Tissue, and Eye Donors and Transplant Hospitals' Provision of Transplant-Related Data', *Federal Register*, 63, 33856 (1998)
National Commission on Fire Prevention and Control, *America Burning* (Washington DC: US Government Printing Office, 1973)
National Organ Transplant Act, Hearing before the Subcommittee of the Committee on Ways and Means, House of Representatives, Ninety-Eighth Congress, Second Session on H.R. 4080, 9 February 1984, Serial 98–64
Office of Evaluation and Inspections, Office of Inspector General, *Informed Consent in Tissue Donation: Expectations and Realities* (Washington DC: Department of Health and Human Services, 2001)
Office of Evaluation and Inspections, Office of Inspector General, *Organ Procurement Organizations and Tissue Recovery* (Washington DC: Department of Health and Human Services, 1994)
Organ Donation: Assessing Performance of Organ Procurement Organizations, Statement for the Record of Bernice Steinhardt, Director Health Services Quality and Public Health Issues, Health, Education, and Human Services Division (Washington DC: United States General Accounting Office, 1998)
Regulation of Human Tissue Banks, Hearing before the Subcommittee on Regulation, Business Opportunities, and Technology, Washington DC, 15 October 1993, Serial No. 103–53 (Washington DC: US Government Printing Office, 1994)
Tissue banks and the dangers of tainted tissues: the need for Federal Regulation, Hearing before the Committee on Governmental Affairs, 14 May 2003 (Washington DC: US Government Printing Office, 2003)
Tissue Banks: is the Federal Government's Oversight Adequate?, Hearing before the Permanent Subcommittee on Investigations of the Committee on Governmental Affairs, 24 May 2001 (Washington DC: US Government Printing Office, 2001)

Australia

Inquiry into the use of pituitary derived hormones in Australia and Creutzfeldt-Jakob Disease, *Report* (Canberra: Australian Government Publishing Service, 1994)

Law Reports and Cases

United Kingdom

The Creutzfeldt-Jakob Disease Litigation, *Plaintiffs v. United Kingdom Medical Research Council* (19 July 1996), QBD, pp. 8–78

United States of America

Georgia Lions Eye Bank v. Lavant, 255 Ga. 60, 335 S.E.2d 127 (1985)
Kennedy-McInnis et al. v. Biomedical Tissue Services, Ltd. et al., No. 13–CV–05140(2016)
Ravenis v. Detroit General Hospital, 234 N W 2d, 411 (1975).
Report of the Grand Jury In Re County Investigating Grand Jury XXI, In the Court of Common Pleas, First Judicial District of Pennsylvania, Criminal Trial Division, Misc. No. 05–011487 (2007)
State of Florida v. Powell, et ux., et al., Appellees, 497 So.2d 1188 (Fla. 1986)

Hansard

HC Deb 20 December 1960, vol. 632 cc1253 (J. Enoch Powell)
HC Deb 29 June 1961, vol. 643 cc 844 (Edith Pitt)
HC Deb 23 May 1972, vol. 837 cc 322 (L. Pavitt)
HC Deb 27 May 1977, vol. 932 cc 691 (Michael Ward)
HC Deb 18 April 1986, vol. 95 cc1162 (Jeremy Hanley)
HC Deb 8 April 1998, vol. 310 cc 353 (Frank Dobson)
HC Deb 29 July 1998, vol. 317 column 316 (B. Blizzard)

Newspaper and Magazine Articles

Altman, Lawrence K., '3 Die of AIDS after Getting Organs from Man Infected with the Virus', *New York Times* (18 May 1991)
—, 'A Burn Center is Opening in New York City – It's a First', *New York Times* (9 December 1976)
Aunt Agatha, 'Letter', *Kent and Sussex Courier* (17 October 1947), p. 5
Autopsies Inquiry is Made City-wide', *New York Times* (3 November 1953)
Blakeslee, Sandra, 'Recall is Ordered at Large Supplier of Implant Tissue', *New York Times* (15 August 2002)
Blankenship, Ted, 'A New Eye Bank for Kansans', *The Lion* (June 1982), p. 25
Bunis, Deni, 'Lions Clubs Thwarted Tissue Safety Standards Legislation: As a Congressman, Gore Proposed Legislation to Govern the Field', *Orange County Register* (19 April 2000)
Burrell, Ian and John Furbisher, 'Warnings Over Brain Disease Ignored', *Sunday Times* (5 September 1983), p. 5
Champell, Lisa, 'Gallup Survey Identifies Why Former Lions Quit Their Clubs', *The Lion* (February 1994), pp. 22–3
Clark, Anthony, 'RTI Founder Spin Off Thriving Biotech Firms in Gainesville Area', *Gainesville Sun* (19 February 2011) http://www.gainesville.com/article.
Cordes, Bill, 'The Eye Bank Moves East', *The Lion* (May 1965), pp. 20–21
'Through Churchill's eyes?', *Daily Sketch* (22 May 1956)
Dechillo, Suzanne, 'At Burn Center, Painstaking Victories', *New York Times* (18 December 1983)
Dunlop, Richard, 'Missouri's Brick and Mortar Monument', *The Lion* (November 1975), pp. 14–17
'Eye Transplant "Hit by Alder Hey"', *BBC News* (6 August 2001)
Frammolino, Ralph, 'Harvest of Corneas at Morgue Questioned', *Los Angeles Times* (2 November 1997)
Gable, Tom, 'Calbiochem Drug Needs Final Tests, Shareholders Told', *Evening Tribune, San Diego* (18 June 1973)
Gaul, Gilbert M., 'The Blood Brokers – How Blood, the "Gift of Life", Became a Billion Dollar Industry', *Philadelphia Inquirer* (24 September 1989)
Gaylin, Willard, 'Harvesting the Dead', *Harper's Magazine* (September 1974), pp. 23–30
Graves, David, 'Ask Everyone to be an Organ Donor, Says Surgeon', *Daily Telegraph* (23 February 1998)
Green, Emily, 'A Wonder Drug that Carried the Seeds of Death', *Los Angeles Times* (21 May 2000)
Hand, Douglas, 'Saving Burn Victims', *New York Times* (15 September 1985)
Harris, Sarah, 'Hospital's Skin Bank Tribute to Tragic Susan', *Western Daily Press – Bristol* (3 November 1994)
Hawkes, Nigel, 'Friends Donate Skin in Effort to Save Fire Victim', *The Times* (26 June 1984)
Hedges, Stephen J., and William Gaines, 'Donor Bodies Milled into Growing Profits; Little-regulated Industry Thrives on Unsuspecting Families; Cadavers' Average Worth: $80,000', *Chicago Tribune* (21 May 2000), p. 1
Heisel, William, 'Tissue-group Steps Closer to Full Disclosure', *Orange County Register* (13 September 2000)
Hirsch, Phil, 'Society for Sight', *The Lion* (February 1954), pp. 24–5, 29

Imhoff, Ernest F., 'No Eye Doctor, but Vision Helps Blind See. Testimonial: Colleagues, Friends Set Up Foundation to Continue the Work of Man who Organized Corneal Transplant System Worldwide', *Baltimore Sun* (1 November 1997)
Jeffries, Michael, 'New Eyes for Old from the Deep-freeze', *Evening News* (23 February 1965)
Katches, Mark, William Heisel and Ronald Campbell, 'The Body Brokers', *Orange County Register* (16 April 2000)
Kelsey, Tim, 'Families Seek Damages Over CJD Deaths', *The Independent* (16 August 1993), p. 3
—, 'Mortuary Man Tells of Work on Bodies', *The Independent* (4 October 1993), p. 2
Kent and Sussex Courier (17 October 1947), p. 5
Kent and Sussex Courier (19 October 1951), p. 7
Kent and Sussex Courier (9November 1951), p. 5
Kent and Sussex Courier, 'Passage of a Bill' (20 June 1952), p. 5
King Jr, John Harry, 'American Diplomacy Eye to Eye', *EBAA Newsight* (February 1965), pp. 29–33
Kurtz, Sheldon E., Christina Woodward Strong and David Gerasimow, 'The 2006 Revised Uniform Anatomical Gift Act – a Law to Save Lives', *Health Lawyers News* (February 2007), pp. 44–9
Lagemann, John Kord, and Don Everitt, 'They Pass on the Gift of Sight', *Reader's Digest* (January 1965)
Lawrence, Susan V., 'Year of the Lions in China: Club's Recruitment of Chinese May Trigger Societal Change', *The Lion* (November 2002), p. 8
Leary, Warren E., 'Rules to Cover Human Tissue in Products', *New York Times* (1 March 1997)
'"Eye Bank" Offers Sight to Many Blind Who Had Lost Hope', *The Lion* (September 1945), p. 17
'Cornea Bank on the Way', *The Lion* (February 1946), p. 37
'Yorkville Club Fosters Eye Bank', *The Lion* (March 1946), p. 15
'New Scope for Eye Bank', *The Lion* (May 1954), p. 19
'Iowa's Eye Bank', *The Lion* (November 1957), pp. 18–20
'New Law Will Aid Eye Bank Program', *The Lion* (October 1960), p. 45
'Hospital Gets Eye Bank', *The Lion* (August 1961), p. 4
'"Coffee Break" is Boon to Eye Foundation', *The Lion* (August 1961), p. 50
The Lion (October 1961), p. 19
'Fete 20,000th Donor', *The Lion* (February 1963), p. 6
'Eyeball Network Helps Blind Regain Sight', *The Lion* (April 1963), p. 48
'Legacies of Light', *The Lion* (May 1963), pp. 28–30
'Eyes By Wireless', *The Lion* (May 1965), pp. 22–4
'Gift of Vision', *The Lion* (May 1967), p. 43
'Salute to the pioneers', *The Lion* (January 1971), pp. 16–19
'Peer-less Campaign', *The Lion* (April 1971), p. 47
'Eye Enucleation: First Step on the Road to Sight', *The Lion* (December 1973), pp. 34–5
'Enlist Youngsters' Aid in Eye Bank Donor Pledges', *The Lion* (April 1976), p. 48
'A Prudent Use of Funds', *The Lion* (June 1983), pp. 10–11, 35
Liptak, Adam, 'Inmate Count in U.S. Dwarfs Other Nations', *New York Times* (23 April 2008)
'Gland Theft Charge Jails Sylmar Man', *Los Angeles Times* (9 November 1966), p. 24
'Objections to Cornea Policy Date Back to 1992', *Los Angeles Times* (22 November 1997)
'Reforms on Cornea Harvesting OKd', *Los Angeles Times* (1 October 1998)
Loshak, David, 'Stricter Controls on Removal of Human Organs Planned', *Daily Telegraph* (29 September 1976)
McNeil Jr, Donald G., 'U.S. Reduces Testing for Mad Cow Disease, Citing Few Infections', *New York Times* (21 July 2006)
Malcolm, Andrew H., 'Human-organ Transplants Gain with New State Laws', *New York Times* (1 June 1986)
Martin, Paul, 'New Lions Eye Institute Fills a Multi-faceted Role', *The Lion* (October 1975), pp. 16–18

'Gland Products', *Meat Trades Journal and Cattle Salesman's Gazette* (3 July 1941), p. 9
'The Utilisation of Slaughterhouse By-products', *Meat Trades Journal and Cattle Salesman's Gazette* (9 January 1941), p. 41
'Making the Hormone for Growth', *Medical World News* (21 February 1969), pp. 32–5
Michelfelder, William, 'Autopsy Probers Score Dr. Werne of Queens', *New York Times* (19 March 1954)
Neff, Pauline, 'A Supermarket of Transplants', *The Lion* (April 1982), pp. 10–13, 37
'Burned Girl, 9, Saved by 10 Prisoners' Skin', *New York Times* (9 January 1951)
'Fire Victim for Whom 15 Donated Skin Recovers After Nearly a Year in Bellevue', *New York Times* (29 March 1951)
'Truce is Signed, Ending the Fighting in Korea', *New York Times* (27 July 1953)
'Autopsies Inquiry is Made City-wide', *New York Times* (3 November 1953)
'"Growth Hormone" Reported to Work', *New York Times* (24 August 1958)
'Pituitary Bank Set Up on Coast', *New York Times* (21 July 1960)
'A Skin Bank for Burn Victims', *New York Times* (3 October 1979)
O'Neill, Ann W., 'Parents Tell of Grief in Lawsuit Over Harvest of Son's Cornea', *Los Angeles Times* (13 May 1998)
Pear, Robert, 'F.D.A. Delays Regulation of Tissue Transplants', *New York Times* (16 May 2003)
LifeNet Health Establishes 'The Skin and Wound Allograft Institute', *PR Newswire* (17 September 2007)
Report of speech delivered by David M. Link, Director of the Food and Drug Administration, 27 November 1978', *American Association of Tissue Banks Newsletter* (2) (3 April 1979), p. 12
Reuters in Paris, 'French Doctors Questioned About Tainted Gland Death', *The Guardian* (21 July 1993), p. 8
Rhoads, Katherine B., 'Meeting the Challenge', *EBAA Newsight* (November 1965), pp. 10–14
Rusk, Howard A., 'Medicine at War – 1: Low Mortality of Vietnam Wounded is Tribute to Quality of Health Aids', *New York Times* (22 May 1966)
Schmeck Jr, Harold H., 'Navy Plans Bank for Living Tissue', *New York Times* (26 May 1963)
—, 'Progress on Hormone', *New York Times* (15 May 1966)
Shearer, Lloyd, 'We Can End Dwarfism!', *Parade* (22 August 1965), pp. 4–6
Sullivan, Ronald, 'Burn Center Sets Up First Skin Bank in New York', *New York Times* (2 April 1978)
'Body Robbers Scandal', *Sunday People* (29 August 1976), p. 1
Templeton, Sarah-Kate, 'Stolen Body Parts were Implanted in NHS Patients', *Sunday Times* (10 September 2006)
'Skin Bank', *Texas Monthly* (July 1977)
'Pure Hormone', *Time* magazine (20 March 1944), p. 48
'Life from Death', *Time* magazine (21 May 1956), p. 81
'150,000 Future Donors Sought for Regional Eye Bank', *The Times* (2 March 1965)
'Taking Glands from Bodies Without Consent Condemned', *The Times* (30 August 1976), p. 2
Waldhole, Michael, 'Red Cross's Plan to Procure Organs Could Hurt Smaller Organizations', *Wall Street Journal* (8 August 1984), p. 33
'Human Growth Hormone is Synthesized for the First Time by California Scientists', *Wall Street Journal* (7 January 1971), p. 4
Wheldon, Julia, 'Families Can't Stop Doctors Taking Organs', *Daily Mail* (31 August 2006), p. 1
Wilkinson, James, 'Doctors' Tangle Over "Spare Parts" Cash Offer', *Daily Express* (18 July 1970), p. 3
Wilson, Kate, 'Abusing the "Gift" of Tissue Donation', *Huffington Post* (18 September 2012)

Published Works

Aiken-O'Neill, Patricia, and Mark J. Mannis, 'Summary of Corneal Transplant Activity: Eye Bank Association of America', *Cornea*, 21 (2002), pp. 1–3
American Journal of Ophthalmology, 54 (1962), pp. 309–10

Anderson, Martha W., and Rene Schapiro, 'From Donor to Recipient: The Pathway and Business of Donated Tissues', in Stuart J. Youngner, Martha W. Anderson and Renie Schapiro (eds), *Transplanting Human Tissue: Ethics, Policy, and Practice* (New York and Oxford: Oxford University Press, 2004), pp. 3-13

Anderson, Martha W., and Scott Bottenfield, 'Tissue Banking – Past, Present, and Future', in Stuart Youngner, Martha W. Anderson and Renie Schapiro (eds), *Transplanting Human Tissue: Ethics, Policy, and Practice* (New York and Oxford: Oxford University Press, 2004), pp. 14-35

Anderson, Robert E., 'The Autopsy – Benefits to Society', *American Journal of Clinical Pathology*, 69 (1978), pp. 239-41

Anderson, Robert E., James T. Weston, John E. Craighead, Paul E. Lacy, Robert W. Wissler and Rolla B. Hill, 'The Autopsy: Past, Present, and Future', *Journal of the American Medical Association*, 242 (10) (1979), pp. 1056-9

Armitage, W. J., 'HLA Matching and Corneal Transplantation', *Eye*, 18 (2004), pp. 231-2

Armitage, W. J., S. J. Moss, D. L. Easty, and B. A. Bradley, 'Supply of Corneal Tissue in the United Kingdom', *British Journal of Ophthalmology*, 74 (1990), pp. 685-7

Artiz, Curtis, P., 'Planning a Burn Institute', in A. B. Wallace and A. W. Wilkinson (eds), *Research in Burns* (Edinburgh: E & S Livingstone, 1966), pp. 522-33

Asher, David M., 'Kuru: Memories of the NIH Years', *Philosophical Transactions of the Royal Society, Biological Sciences*, 363 (1510) (2008), pp. 3618-25

Ashford, Paul, 'Traceability', *Cell and Tissue Banking*, 11 (2010), pp. 329-33

Backere, A.C.J. de, 'Euro Skin Bank: Large Scale Skin-banking in Europe Based on Glycerol-preservation of Donor Skin', *Burns* 20 (1) (1994), pp. S4-S9

Barclay, T. L., 'New Burns Unit in Wakefield, Yorkshire', in A. B. Wallace and A. W. Wilkinson (eds), *Research in Burns* (Edinburgh: E & S Livingstone, 1966), pp. 547-9

Barillo, David J. and Steven Wolf, 'Planning for Burn Disasters: Lessons Learned from One Hundred Years of History', *Journal of Burn Care & Research*, 27 (5) (2006), pp. 622-34

Batchelor, J. R. and M. Hackett, 'HL-A Matching in Treatment of Burned Patients with Skin Allografts', *The Lancet*, 296 (1970), pp. 581-3

Batchelor, J. R., T. A. Casey, D. C. Gibbs, D. F. Lloyd, A. Werb, S. S. Prasad and A. James, 'HLA Matching and Corneal Grafting', *The Lancet*, 307 (1976), pp. 551-4

Bennett, J. P., 'A History of the Queen Victoria Hospital, East Grinstead', *British Journal of Plastic Surgery*, 41 (1988), pp. 422-40

Bennett, James E. and Stephen R. Miller, 'Evolution of the Electro-dermatome', *Plastic and Reconstructive Surgery*, 45 (2) (1970), pp. 131-3

Bernoulli, C., J. Siegfried, G. Baumgartner, F. Regli, T. Rabinowicz, D. C. Gajdusek and C. J. Gibbs, 'Danger of Accidental Person-to-person Transmission of Creutzfeldt-Jakob Disease by Surgery', *The Lancet*, 309 (1977), pp. 478-9

Billingham, R. E., 'Concerning the Origins and Prospects of Cryobiology and Tissue Banks', *Transplant Proceedings*, Supplement, 8 (1976), pp. 7-13

Billingham, R. E. and B. W. Rycroft, 'The Preservation of the Donor Graft', in B.W. Rycroft (ed.), *Corneal Grafts* (London: Butterworth, 1955), pp. 195-207

Billingham, R. E. and P. Medawar, 'The Freezing, Drying and Storage of Mammalian Skin', *Journal of Experimental Biology*, 29 (1952), pp. 454-68

Black, D.A.K., 'Treatment of Burn Shock with Plasma and Serum', *British Medical Journal*, 2 (1940), pp. 693-7

Blair, Vilray P., James Barrett Brown and William G. Hamm, 'The Early Care of Burns and the Repair of their Defects', *Journal of the American Medical Association*, 98 (16) (1932), pp. 1355-9

Blizzard, Robert M., 'History of Growth Hormone Therapy', *Indian Journal of Pediatrics*, 79 (2012), pp. 87-91

Blond, Elaine, with Barry Turner, *Marks of Distinction: The Memoirs of Elaine Blond* (London: Valentine, Mitchell, 1988)

Body, David, and Jonathan Glasson, 'Iatrogenic Creutzfeldt-Jakob Disease: Litigation in the UK and Elsewhere', *Clinical Risk*, 11 (2005), pp. 4-13

Boggio, Andrea, 'The Compensation of the Victims of the Creutzfeldt-Jakob Disease in the United Kingdom', *Medical Law International*, 7 (2005), pp. 149–67

Bondoc, C. C., and J. F. Burke, 'Clinical Experience with Viable Frozen Human Skin and a Frozen Skin Bank', *Annals of Surgery*, 174 (3) (1971), pp. 371–81

Bourdillon, R. B. and L. Colebrook, 'Air Hygiene in Dressing-rooms for Burns or Major Wounds', *The Lancet*, 1 (6399) (1946), pp. 561–5

Brill, Robert, 'Pituitary Collection Project: Progress Report', *CAP Bulletin*, 16 (2) (1962), pp. 22–3

British Association of Tissue Banks, 'Standards for Tissue Banking, June 1995', *Transplant Medicine*, 6 (1996), pp. 155–8

'Eye Banks', *British Medical Journal*, 1 (1964), p. 388

'Postmortem Tissue Problems', *British Medical Journal*, 2 (1978), p. 382

Brown, E. Richard, 'Public Hospitals on the Brink: Their Problems and their Options', *Journal of Health Politics, Policy and Law*, 7 (4) (1983), pp. 927–44

Brown, James Barrett, and Frank McDowell, 'Massive Repairs of Burns with Thick Split-skin Grafts', *Annals of Surgery*, 115 (1942), pp. 658–74

Brown, James Barrett, and Minot P. Fryer, 'Postmortem Homografts to Reduce Mortality in Extensive Burns', *Journal of the American Medical Association*, 156 (12) (1954), pp. 1163–5

—, 'Skin Homografts from Postmortem Sources', *American Journal of Surgery*, 97 (1959), pp. 418–20

Brown, James Barrett, Minot P. Fryer, Peter Randall and Milton Lu, 'Postmortem Homografts as "Biological Dressings" for Extensive Burns and Denuded Areas', *Annals of Surgery*, 138 (4) (1953), pp. 618–30

Brown, James Barrett, Minot P. Fryer and Thomas J. Zaydon, 'Establishing a Skin Bank: Use and Various Methods of Preservation of Postmortem Homografts', *Plastic and Reconstructive Surgery* 16 (5) (1955), pp. 337–51

Brown, Paul, 'Human Growth Hormone Therapy and Creutzfeldt-Jakob Disease: A Drama in Three Acts', *Pediatrics*, 81 (1988), pp. 85–92

Brown, Paul, D. Carleton Gajdusek, C. J. Gibbs and David M. Asher, 'Potential Epidemic of Creutzfeldt-Jakob Disease from Human Growth Hormone Therapy', *New England Journal of Medicine*, 313 (1985), pp. 728–31

Brown, Paul, Jean Philippe Brandel, Michael Preece and Takeshi Sato, 'Iatrogenic Creutzfeldt-Jakob Disease: The Waning of an Era', *Neurology*, 67 (2006), pp. 389–93

Brubaker, Scott A., and Nancy Senst, 'The Gift of Tissue and Eye Donation', in Dianne La Pointe Rudow, Linda Ohler and Teresa Shafer (eds), *A Clinician's Guide to Donation and Transplantation* (Lenexa, KS: NATCO, 2006), p. 373

Bull, J. P., D. M. Jackson and C. Walton, 'Causes and Prevention of Domestic Burning Accidents', *British Medical Journal*, 2 (1964), pp. 1421–7

Burd, Andrew, 'Once Upon a Time and the Timing of Surgery in Burns', *Journal of Plastic, Reconstructive and Aesthetic Surgery*, 61 (3) (2008), pp. 237–9

Bureau of Justice Statistics, *Homicide Victimization, 1950-2005* (Washington DC: Department of Justice, 2007)

Burke, J. F., W. C. Quinby, C. C. Bondoc, A. B. Cosimi, P. S. Russell and S. K. Szyfelbein, 'Immunosuppression and Temporary Skin Transplantation in the Treatment of Massive Third Degree Burns', *Annals of Surgery*, 182 (1975), pp. 183–97

Burnham, Frank A., 'The Rendering Industry – A Historical Perspective', in Don A. Franco and Winfield Swanson (eds), *The Original Recyclers* (The Animal Protein Producers Industry, The Fats & Proteins Research Foundation, The National Renderers Association, 1996), pp. 1–16

Cadbury, Adrian, 'Corporate Social Responsibility', *Twenty-first Century Society*, 1 (2006), pp. 5–21

Caffrey, E. A., 'Lessons Learnt from Setting Up a Regional Tissue Bank', *Transfusion Medicine*, 5 (supplement 1) (1995), p. 2

'Pituitary Collection Program', *CAP Bulletin*, 15 (11) (1961), p. 8

—, 'Pituitary appeal made', *CAP Bulletin*, 19 (10) (1965), p. 259

Cartwright, Howard C., 'Pituitary Collection, Research Get New Aid', *CAP Bulletin*, 22 (4) (1968), pp. 133–5
Casey, Robert, and W.A.S. Douglas, *The World's Biggest Doers: The Story of the Lions* (Chicago: Wilcox & Follett Co., 1949)
Casey, T. A., 'Eye Banking', *British Medical Journal*, 288 (1984), p. 5
Centers for Disease Control and Prevention, 'Update: Allograft-associated Bacterial Infections – United States, 2002', *Journal of the American Medical Association*, 287 (13) (2002), pp. 1642–4
Chick, Leland R., 'Brief History and Biology of Skin Grafting', *Annals of Plastic Surgery*, 21 (4) (1988), pp. 358–65
'Children of Short Stature', *British Medical Journal*, 3 (1967), pp. 187–8.
Choyce, D. P., 'Eyes for Corneal Grafting', *British Medical Journal*, 2 (1966), p. 1593
Chu, Wing, 'The Past Twenty-five Years in Eye Banking', *Cornea*, 19(5) (2000), pp. 754–65
Clark, Michael A., 'The Value of the Hospital Autopsy: Is it Worth the Cost?', *American Journal of Forensic Medicine and Pathology*, 2 (1981), pp. 231–7
Clarkson, Patrick, 'A-bomb Casualties and Arrangements for their Care', *British Journal of Plastic Surgery*, 3 (1951), pp. 188–93
Cobbleton, Tom, 'LVII.-De Mortuis', *The Lancet*, 274 (1959), pp. 1081–2
Coe, John I., 'Collection of Pituitary Glands in Hennepin County', *Journal of Forensic Sciences*, 15 (1970), pp. 14–17
Cole, Gerald J., President, Tissue Banks International, *Slide Presentation to the Food & Drug Administration Transmissible Spongiform Encephalopathies Advisory Committee* (18 January 2001), www.fda.gov/ohrms/dockets/ac/01/transcipts/3681t1_ 04.pdf (accessed October 2016)
Colebrook, L., 'The Control of Infection in Burns', *The Lancet*, 1 (6511) (1948), pp. 893–9
—, *A New Approach to the Treatment of Burns and Scalds* (London: Fine Technical Publications, 1950)
—, 'The Prevention of Burning Accidents in England and America', *Bulletin of the New York Academy of Medicine*, 27 (7) (1951), pp. 425–38
Colebrook, L., and V. Colebrook, 'The Prevention of Burns and Scalds: Review of 1000 Cases', *The Lancet*, 254 (1949), pp. 181–8
—, 'A Suggested National Plan to Reduce Burning Accidents: Based on a Study of 2000 Cases', *The Lancet*, 258 (1951), pp. 579–84
Colebrook, L., V. Colebrook, J. P. Bull and D. M. Jackson, 'The Prevention of Burning Accidents: A Survey of the Present Position', *British Medical Journal*, 1 (1956), p. 1379
The Collaborative Corneal Transplantation Studies (CCTS), 'Effectiveness of Histocompatibility Matching in High-risk Corneal Transplantation', *Archives of Ophthalmology*, 110 (10) (1992), pp. 1392–1403
Cooper, M. H. and A. J. Culyer, *The Price of Blood: An Economic Study of the Charitable and Commercial Principle*, Hobart Paper No. 41 (London: Institute of Economic Affairs, 1968)
Cordes, Frederick C., 'The Eye-Bank Committee', *American Journal of Ophthalmology*, 43 (1957), pp. 309–11
Corner, George, *The Seven Ages of a Medical Scientist: an Autobiography* (Philadelphia, PA: University of Pennsylvania Press, 1981)
Cronin, Michael J., 'Pioneering Recombinant Growth Hormone Manufacturing: Pounds Produced Per Mile of Height', *Journal of Pediatrics*, 131 (1997), pp. S5–S7
Cuzzell, Janice Zeigler, 'Fighting City Hall: The Politics of Burn Care', *American Journal of Nursing*, 86 (2) (1986), pp. 194–5
Davies, Russell M., 'Archibald McIndoe, His Times, Society, and Hospital', *Annals of the Royal College of Surgeons*, 59 (1977), pp. 359–67
Dexter, Frank, 'Tissue Banking in England', *Transplantation Proceedings*, Supplement (1976), pp. 43–8
Diekema, Douglas S., 'Is Taller Really Better? Growth Hormone Therapy in Short Children', *Perspectives in Biology and Medicine*, 34 (1990), pp. 109–23
Dimick, Alan R., '1978 Presidential Address: American Burn Association', *Journal of Trauma*, 19 (7) (1979), pp. 520–24

Dimick, A. R., L. H. Potts, E. D. Charles Jr, J. Wayne and I. M. Reed, 'The Cost of Burn Care and Implications for the Future of Quality of Care', *Journal of Trauma*, 26 (3) (1986), pp. 260–66

Dorozynski, Alexander, 'French to Investigate Deaths from Growth Hormone', *British Medical Journal*, 307 (1993), p. 281

Doughman, Donald J., 'Tissue Storage', in J. H. Krachmer, Mark J. Mannis and Edward J. Holland (eds), *Cornea, Fundamentals of Cornea and External Disease*, volume 1 (St Louis, MO: Mosby, 1997), pp. 509–17

Douglas, S. R., A. Fleming and L. Colebrook, 'On Skin Grafting – A Plea for its More Extensive Application', *The Lancet*, 2 (1917), p. 5

Duffy, P., J. Wolf, G. Collins et al., 'Possible Person-to-person Transmission of Creutzfeldt-Jakob Disease', *New England Journal of Medicine*, 290 (1974), p. 692

Dyer, Clare, 'Growth Hormone Victims Seek Compensation', *British Medical Journal*, 306 (1993), p. 607

Easty, D. L., 'Eye Banking', *Transplantation Proceedings*, 21 (1989), pp. 3120–22

Edelson, Edward, 'The Remarkable Dr Li: Master of the Master Gland', *Family Health* (1971), pp. 15–18

Engelbach, William, 'The Growth Hormone: Report of a Case of Juvenile Hypopituitarism Treated with Evans' Growth Hormone', *Endocrinology*, 81 (1932), pp. 1–19

Escamilla, Roberto F., and Leslie Bennett, 'Pituitary Infantilism Treated with Purified Growth Hormone, Thyroid and Sublingual Methyltestosterone: A Case Report', *Journal of Clinical Endocrinology*, 11 (1951), pp. 221–8

Evans, Herbert M., and J. A. Long, 'The Effect of the Anterior Lobe of the Hypophysis Administered Intraperitoneally on Growth, Maturity and the Oestrus Cycles of the Rat', *Anatomical Record*, 21 (1921), p. 61

Fantus, Bernard, 'The Therapy of the Cook County Hospital', *Journal of the American Medical Association*, 109 (1937), pp. 128–31

Farge, Emile J., 'Eye Bank Association of America: 1982 Activities', *Cornea*, 2 (1983), pp. 83–5

—, 'Eye Banking: 1944 to the Present', *Survey of Ophthalmology*, 30 (4) (1989), 260–63

—, 'Ethics and Eye Banking', in J. H. Krachmer, M. J. Mannis and E. J. Holland (eds), *Cornea: Fundamentals, Diagnosis and Management* (Philadelphia: Elsevier Mosby, 1997), pp. 531–55

—, 'The Eye Bank Association of America', in Frederick S. Brightbill (ed.), *Corneal Surgery: Theory, Technique and Tissue* (St Louis, MO: Mosby, 1999), pp. 674–82

Feller, Irving, and Keith H. Crane, 'Classification of Burn-care Facilities in the United States', *Journal of the American Medical Association*, 215 (3) (1971), pp. 463–6

Filatov, Vladimir, 'The Grafting of Cornea', *American Braille Press* (1933), pp. 18–22

—, 'Ophthalmic Surgery in the Russian Army', *Journal of the American Medical Association*, 120 (1942), pp. 54–5

—, *My Path in Science* (Moscow: Foreign Language Publishing House, 1957)

Flosdorf, Earl W., 'Drying Penicillin by Sublimation in the United States and Canada', *British Medical Journal*, 1 (1945), pp. 216–18

—, *Freeze-drying: Drying by Sublimation* (New York: Reinhold, 1949)

Ford, Brian J., *BSE: The Facts* (London: Corgi Books, 1996)

Franco, Don A., and Winfield Swanson (eds), *The Original Recyclers* (The Animal Protein Producers Industry, The Fats & Proteins Research Foundation, The National Renderers Association, 1996)

Frasier, Douglas, 'The Not-so-good Old Days: Working with Pituitary Growth Hormone in North America, 1956–1985', *Journal of Pediatrics*, 131 (1), part 2 (1997), p. S4

Freed, S. Charles, 'Glandular Physiology and Therapy', *Journal of the American Medical Association*, 117 (1941), pp. 1175–82

Freedlander, E., S. Boyce, M. Ghosh, D. R. Ralston and S. MacNeil, 'Skin Banking in the UK: The Need for Proper Organization', *Burns*, 24 (1998), pp. 19–24

Freshwater, M. Felix, and Thomas J. Krizek, 'Skin Grafting of Burns: A Centennial', *Journal of Trauma*, 11 (10) (1971), pp. 862–5

Fuller, Richard L., 'Medical Legal Issues', in J. H. Krachmer, M. J. Mannis and E. J. Holland (eds), *Cornea: Fundamentals, Diagnosis and Management* (Philadelphia: Elsevier Mosby, 1997), pp. 537–41

Gajdusek, D. C., C. Gibbs and M. Alpers, 'Experimental Transmission of a Kuru-like Syndrome to Chimpanzees', *Nature*, 209 (1966), pp. 794–6

Gibb, C., D. Carleton Gajdusek, D. Asher, M. P. Alpers, Elizabeth Beck, P. M. Daniel and W. B. Matthews, 'Creutzfeldt-Jakob Disease (Spongiform Encephalopathy): Transmission to the Chimpanzee', *Science*, 161 (1968), pp. 388–9

Gibson, Thomas, 'Zoografting: A Curious Chapter in the History of Plastic Surgery', *British Journal of Plastic Surgery*, 8 (1955), pp. 234–42

Gibson, T., and P. B. Medawar, 'The Fate of Skin Homografts in Man', *Journal of Anatomy*, 77 (4) (1943), pp. 299–310

Gissane, W., and L. Colebrook, 'Discussion on the Prevention of Accidents', *Proceedings of the Royal Society of Medicine*, 42 (4) (1949), pp. 209–14

Graham Jr, C. R., 'Eye Banking: A Growth Story', *Transplantation Proceedings*, 17 (6) Supplement 4 (1985), pp. 105–11

Greaves, N. S., B. Benatar, M. Baguneid and A. Bayat, 'Single-stage Application of a Novel Decellularised Dermis for Treatment-resistant Lower Limb Ulcers: Positive Outcomes Assessed by SIAscopy, Laser Perfusion, and 3D Imaging, with Sequential Timed Histological Analysis', *Wound Repair and Regeneration*, 21 (6) (2013), pp. 813–22

Greenleaf, Glenn, 'Skin Transplantation: Clinical Applications and Current Issues', in Stuart J. Youngner, Martha W. Anderson and Renie Schapiro (eds), *Transplanting Human Tissue: Ethics, Policy, and Practice* (New York and Oxford: Oxford University Press, 2004), pp. 52–70

Greep, Roy, 'The Saga and the Science of Gonadotrophins', *Journal of Endocrinology*, 39 (1967), pp. ii–ix

—, 'Reproductive Endocrinology: Concepts and Perspectives, an Overview', *Recent Progress in Hormone Research*, 34 (1978), pp. 1–23

Gresham, Richard B., Vernon P. Perry and Thomas E. Wheeler, 'US Navy Tissue Bank', *Journal of the American Medical Association*, 183 (1963), pp. 99–102

Griffith, Frederick N., 'The Promise of International Eye Banking', *International Ophthalmology*, 14 (1990), pp. 205–10

Grummer, Ric R., and Terry Klopfenstein, 'Utilizing Rendered Products: Ruminants', in Don A. Franco and Winfield Swanson (eds), *The Original Recyclers* (The Animal Protein Producers Industry, The Fats & Proteins Research Foundation, The National Renderers Association, 1996), pp. 129–74

Gunby, Phil, 'Bill Introduced to Thwart Kidney Brokerage', *Journal of the American Medical Association*, 250 (17) (1983), pp. 2263–4

Guyett Jr, Philip J., *Heads, Shoulders, Knees and Bones* (Publisher: Author, 2011)

Hackett, M. E., 'Cadaver Homografts', *Proceedings of the Royal Society of Medicine*, 64 (1971), pp. 1292–4

Hadley, Jack, 'Medicaid Reimbursement of Teaching Hospitals', *Journal of Health Politics, Policy and Law*, 74 (4) (1983), pp. 911–26

Hadlow, W. J., 'Scrapie and Kuru', *The Lancet*, 274 (1959), pp. 289–90

—, 'Kuru Likened to Scrapie: The Story Remembered', *Philosophical Transactions of the Royal Society, Biological Sciences*, 363 (2008), p. 3644

Hartree, A. Stockell, 'Separation and Partial Purification of the Protein Hormones from Human Pituitary Glands', *Biochemical Journal*, 100 (1966), pp. 754–61

Hazlick, Randy, and Debra Combs, 'Medical Examiner and Coroner Systems: History and Trends', *Journal of the American Medical Association*, 279 (11) (1998), pp. 870–74

Heck, Ellen L., 'Retrospective of a Skin Bank', *Journal of Burn Care and Rehabilitation*, 20 (1999), pp. 103–7

Heggie, J. F., 'Training of Post-mortem Room Technicians', *Journal of Clinical Pathology*, 20 (5) (1967), pp. 793–4

Henneman, Philip, Paul Steinlauf, John Tullis and Robert Brill, 'Summary of the Final Report of the Ad Hoc Committee on Pituitary Collection', *CAP Bulletin* (July 1963), pp. 85–6

Herman, J. K., 'The Navy Tissue Bank: The World's First Still Sets the Standards for Others to Follow', *U.S. Navy Medicine*, 71 (5) (1980), pp. 6–17

Hickman, Matthew J., Kristen A. Hughes, Kevin J. Strom and Jeri D. Ropero-Miller, *Medical Examiners and Coroners' Offices, 2004* (Washington DC: Department of Justice, Office of Justice Programs, 2007)

Hill, Rolla B., 'The Autopsy: Past, Present, and Future', *Journal of the American Medical Association*, 242(10) (1979), pp. 1056–9

Hill, Rolla B., and Robert E. Anderson, 'The Autopsy Crisis Re-examined: The Case for a National Autopsy Policy', *Milbank Quarterly*, 69 (1991), pp. 51–77

History of the Organ Donor Register, updated May 2011, https://www.organdonation.nhs.uk/ukt/newsroom/fac (accessed November 2012)

Hogg, Penny, Paul Rooney, Eileen Ingham and John N. Kearney, 'Development of a Decellularised Dermis', *Cell and Tissue Banking*, 14 (3) (2013), pp. 465–74

Holder, Lawrence E., 'How Much Artifice in Asking for an Autopsy?', *Hospital Physician*, 7 (1971), pp. 87–91

Hoyert, Donna L., *The Changing Profile of Autopsied Deaths in the United States, 1972–2007. NCHS Data Brief, no. 67* (Hyattsville, MD: National Center for Health Statistics, 2011)

Hyatt, G. W., 'The Navy's Wonderful Tissue Bank', *Hospital Management* (1956), pp. 41–4

—, The Founding of the US Navy Tissue Bank', *Transplantation Proceedings*, 8 (2), Supplement 1 (1976), pp. 17–20

Hyatt, G. W., T. C. Turner, C.A.L. Bassett, J. W. Pate and P. N. Sawyer, 'New Methods for Preserving Bone, Skin and Blood Vessels', *Postgraduate Medicine*, 12 (1952), pp. 239–54

Iltis, Ana S., Michael A. Rie and Anji Wall, 'Organ Donation, Patients' Rights, and Medical Responsibilities at the End of Life', *Critical Care Medicine*, 37 (2009), pp. 310–16

Jackson, Douglas, 'A Clinical Study of the Use of Skin Homografts for Burns', *British Journal of Plastic Surgery*, 7 (1954), pp. 26–43

—, 'The Treatment of Burns: An Exercise in Emergency Medicine', *Annals of the Royal College of Surgeons*, 13 (4) (1953), pp. 236–57

Janes, Robert M., 'CARE and MEDICO of Canada', *Canadian Medical Association Journal*, 90 (13) (1964), p. 795

Janžekovič, Zora, 'A New Concept in the Early Excision and Immediate Grafting of Burns', *Journal of Trauma*, 10 (12) (1970), pp. 1103–8

—, 'The Burn Wound from the Surgical Point of View', *Journal of Trauma*, 15 (1975), pp. 42–62

—, 'Once Upon a Time.... How West Discovered East', *Journal of Plastic, Reconstructive and Aesthetic Surgery*, 61 (2008), pp. 240–44

Jason, Donald, 'The Role of the Medical Examiner/Coroner in Organ and Tissue Procurement for Transplantation', *American Journal of Forensic Medicine and Pathology*, 15 (3) (1994), pp. 192–202

'Early Excision of Thermal Burns – An International Round-table Discussion, Geneva, June 22, 1987', *Journal of Burns Care and Rehabilitation*, 9 (5) (1988), pp. 549–61

Joyce, Michael J., 'American Association of Tissue Banks: A Historical Reflection Upon Entering the 21st Century', *Cell and Tissue Banking*, 1 (2000), pp. 5–8

Kagan, R. J., 'Human Skin Banking: Past, Present and Future', in D. M. Strong (ed.), *Advances in Tissue Banking*, vol. 2. (Singapore: World Scientific, 1998), pp. 297–321

Kateley, John R., 'Establishing a Tissue Bank', in K. J. Fawcet and A. R. Barr (eds), *Tissue Banking* (Arlington, VA: American Association of Tissue Banks, 1987), pp. 17–27

Kaufman, Herbert E., 'Tissue Storage Systems: Short and Medium Term', in Frederick S. Brightbill (ed.), *Corneal Surgery: Theory, Technique and Tissue*, 3rd edn (St Louis, MO: Mosby, 1999), pp. 892–97

Kearney, John N., 'Yorkshire Regional Tissue Bank – Circa 50 Years of Tissue Banking', *Cell and Tissue Banking*, 7 (2006), pp. 259–64

Ketchum, Lynn D., 'An Historical Account of the Development of the Calibrated Dermatome', *Annals of Plastic Surgery*, 1 (6) (1978), pp. 608–11

Kirn, Timothy F., 'How Does Tissue Banking Work? Virginia Bank While Not Typical, May Offer Some Insights', *Journal of the American Medical Association*, 258 (3) (1987), pp. 304–5

Klasen, Henk J., *History of Free Skin Grafting: Knowledge or Empiricism?* (Berlin: Springer-Verlag, 1981)
Krachmer, Jay H., Mark J. Mannis and Edward J. Holland (eds), *Cornea, Fundamentals of Cornea and External Disease*, vol. 1 (St Louis, MO: Mosby, 1997)
Kurtz, Sheldon E., Christina Woodward Strong and David Gerasimow, 'The 2006 Revised Uniform Anatomical Gift Act – A Law to Save Lives', *Health Lawyers News* (2007), pp. 44–9
Laing, J. Ellsworth, 'Design of a New Burns Unit for the Wessex Regional Hospital Board', in A. B. Wallace and A. W. Wilkinson (eds), *Research in Burns* (Edinburgh: E & S Livingstone, 1966), pp. 542–6
'Death in the Fireplace', *The Lancet*, 248 (1946), pp. 833–4
'The Medical Services in the Korean War', *The Lancet*, 261 (1953), pp. 134–5
'Transplantation and Burns', *The Lancet*, 305 (1975), pp. 1017–18
Lawrence, J. C., 'Some Aspects of Burns and Burns Research at Birmingham Accident Hospital 1944–93: A.B. Wallace Memorial Lecture, 1994', *Burns*, 21 (1995), pp. 403–13
Layton, Thomas R., 'U.S. Fire Catastrophes of the 20th Century', *Journal of Burn Care & Research*, 3 (1982), pp. 21–8
Lee, Paul P., 'Cornea Donation Laws in the United States', *Archives of Ophthalmology*, 107 (1989), pp. 1585–9
Leveille, Albert S., Janie Benson, H. Dwight Cavanagh, Bruce I. Bodner and Robert H. Byers, 'Cost-effectiveness in Eyebanking', *American Academy of Ophthalmology*, 89 (6) (1982), pp. 51A–53A
Li, Choh Hao, and Harold Papkoff, 'Preparation and Properties of Growth Hormone from Human and Monkey Pituitary Glands', *Science*, 124 (1956), pp. 1293–4
Li, C. H., and Herbert McLean Evans, 'The Isolation of Pituitary Growth Hormone', *Science*, 99 (1944), pp. 183–4
Li, Jennifer Y., and Mark J. Mannis, 'Eye Banking and the Changing Trends in Contemporary Corneal Surgery', *International Ophthalmology Clinics*, 50 (3) (2010), pp. 101–12
Loeb, Leo, *The Biological Basis of Individuality* (Springfield, IL: C.C. Thomas, 1945)
Long-Sutehall, Tracey, Emma Winstanley, Anthony J. Clarkson and Magi Sque, 'Evaluation of the Experience of Family Members Whose Deceased Relative Donated Tissue at the NHSBT Dedicated Donation Facility, at Speke, Liverpool', *Cell and Tissue Banking*, 13 (4) (2012), pp. 537–46
Longmire Jr, W. P., J. A. Cannon and R. A. Weber, 'General Surgical Problems of Tissue Transplantation', in G.E.W. Wolstenholme and Margaret P. Cameron (eds), *Preservation and Transplantation of Normal Tissues: A CIBA Foundation Symposium* (London: J. & A. Churchill, 1954), pp. 28–43
McDowell, Frank, 'James Barrett Brown, M.D., 1899–1971: Obituary', *Plastic and Reconstructive Surgery* 48 (1971), pp. 101–4
McIndoe, A., and A. Franceschetti, 'Reciprocal Skin Homografts', *British Journal of Plastic Surgery*, 2 (1950), pp. 283–9
Malinin, T. I., 'University of Miami Tissue Bank: Collection of Post-mortem Tissue for Clinical Use and Laboratory Investigation', *Transplant Proceedings*, 8 (2 Suppl. 1) (1976), pp. 53–8
Manson, R. Hunter, *Statutory Regulation of Organ Donation in the United States* (Richmond, VA: The South-Eastern Organ Procurement Foundation, 1986)
Martin, Paul, *We Serve: A History of the Lions Clubs* (Washington DC: Regnery Gateway, 1991)
Medawar, P. B., 'General Problems of Immunity', in G.E.W. Wolstenholme and Margaret P. Cameron (eds), *Preservation and Transplantation of Normal Tissues: A CIBA Foundation Symposium* (London: J. & A. Churchill, 1954), pp. 1–20
—, *Memoirs of A Thinking Radish* (Oxford: Oxford University Press, 1988)
'Medico Expanding', *Sight-Saving Review*, 31(4) (1961), p. 239
Mesich-Brant, Jennifer L., and Lawrence J. Grossback, 'Assisting Altruism: Evaluating Legally Binding Consent in Organ Donation Policy', *Journal of Health Politics, Policy and Law*, 30 (4) (2005), pp. 687–717
Michaud, R. J., and K. J. Drabu, 'Bone Allograft Banking in the United Kingdom', *Bone and Joint Journal*, 76-B (1994), pp. 350–1

Miller, Stephen H., 'Competitive Forces and Academic Plastic Surgery', *Plastic and Reconstructive Surgery*, 101(5) (1998), pp. 1389–99
Milner, R.D.G., 'Growth Hormone 1985', *British Medical Journal*, 291 (1985), pp. 1593–4
—, T. Russell-Fraser, C.G.D. Brook, P. M. Cotes, J. W. Farquhar, J. M. Parkin, M. A. Preece, G.J.A.I. Snodgrass, A. Stuart Mason, J. M. Tanner and F. P. Vince, 'Experience with Human Growth Hormone in Great Britain: The Report of the MRC Working Party', *Clinical Endocrinology*, 11(1979), pp. 15–38
Ministry of Justice, *Coroners' Statistics 2011, England and Wales* (London: National Statistics, 2012)
Miranda, Bianca, Alessandro Nanni Costa, Jacinto Sánchez-Ibáñez and Eliana Porta, 'Gathering Donor History: Ensuring Safe Tissue for Transplant', in Ruth M. Warwick, Deirdre Fehily, Scott A. Brubaker and Ted Eastlund (eds), *Tissue and Cell Donation: An Essential Guide* (Oxford: Wiley-Blackwell, 2009), pp. 98–115
Mitchison, N. A., 'Peter Brian Medawar', *Biographical Memoirs of Fellows of the Royal Society*, 35 (1990), pp. 283–301
Murphy, A. M., 'The Development of Tissue Banking in Scotland (abstract)', *Transfusion Medicine*, 5 (supplement 1) (1995), p. 2
Murray, Joseph E., 'Organ Transplantation: The Practical Possibilities', in G.E.W. Wolstenholme and Maeve O'Connor (eds), *Ethics in Medical Progress: A CIBA Foundation Symposium* (London: J. & A. Churchill, 1966), pp. 54–77
Murray, Michael, 'Good Pharmaceutical Manufacturing Practice', in Frank Wells (ed.), *Medicines: Good Practice Guidelines* (Belfast: The Queen's University of Belfast, 1990), pp. 17–27
Myers, S. R., M. R. Macheseny, R. M. Warwick and P. D. Cussons, 'Skin Storage', *British Medical Journal*, 313 (1996), p. 439
National Burn Care Review Committee, *Standards and Strategy for Burn Care: A Review of Burn Care in the British Isles* (London: British Burn Association, 2001)
National End of Life Care Intelligence Network, *Deprivation and Death: Variation in Place and Cause of Death* (London: National End of Life Care Intelligence Network, 2012) www.endoflifecare-intelligence.org.uk
Nelson, S. D. and G. H. Tovey, 'National Organ Matching and Distribution Service', *British Medical Journal*, 1 (1974), pp. 622–4
New, Bill, Michael Solomon, Robert Dingwall and Jean McHale, *A Question of Give and Take: Improving the Supply of Organs for Transplantation* (London: King's Fund Institute, 1994)
NHS Blood and Transplant, Ocular Tissue Advisory Group, 'Ocular Tissue Pathway Service Review, Draft Report' (2012) www.odt.nhs.uk/transplantation/advisory-groups/ocular/
Niall, Hugh D., 'Revised Primary Structure for Human Growth Hormones', *Nature New Biology*, 230 (1971), pp. 90–91
Noble, W. C., *Coli: Great Healer of Men* (London: William Heinemann Medical Books, 1974)
North, J. F., 'The Development of Plastic Surgery in the West Midlands Region', *British Journal of Plastic Surgery*, 40 (1987), pp. 317–22
Orentlicher, David, 'Presumed Consent to Organ Donation: Its Rise and Fall in the United States', *Rutgers Law Review*, 61 (2009), pp. 295–331
Orsini, Lou A., 'Health Insurance and the Autopsy', *Pathologist*, 69 (7) (1978), Supplement, pp. 248–9
Overcast, Thomas D., Roger W. Evans, Lisa E. Bowen, Marilyn M. Hoe and Cynthia L. Livak, 'Problems in Identification of Potential Organ Donors', *Journal of the American Medical Association*, 251 (12) (1984), pp. 1559–62
Paegle, Roland D., 'Special Technologists for Autopsies', *Pathologist* (March 1971), pp. 79–86
Panel Appointed by the Clinical Endocrinology Committee of the Medical Research Council, 'The Effectiveness in Man of Human Growth Hormone', *The Lancet*, 273 (1959), pp. 7–12
Pappworth, Maurice, Letter to the Editor, *The Lancet*, 291 (1968), p. 419
Parkes, A. S., *Sex, Science and Society* (Newcastle-upon-Tyne: Oriel Press, 1966)
—, 'Herbert McLean Evans: An Interview', *Journal of Reproduction and Fertility*, 19 (1969), pp. 1–49
Pathologist, 26 (11) (1973), p. 456

Paton, David, 'The Founder of the First Eye Bank: R. Townley Paton, M.D.', *Refractive & Corneal Surgery*, 7 (1991), pp. 190–94
Paton, R. Townley, 'Donor Material and the Law', in Benjamin Rycroft (ed.), *Corneal Grafts* (London: Butterworth, 1955), pp. 216–30
—, 'Sight Restoration Through Corneal Grafting', *Sight-Saving Review*, 15 (1945), pp. 3–11
Payne, John W., 'New Directions in Eye Banking', *Transactions of the American Ophthalmological Society*, 78 (1980), pp. 982–1026
Perry, Vernon P., 'Description of Current Tissue Banks' Methods of Skin Preservation', *Cryobiology*, 3 (2) (1966), pp. 178–91
Philips, Glyn O., and Jorge M. Pedraza, 'The International Atomic Energy Agency (IAEA) Programme in Radiation and Tissue Banking: Past, Present and Future', *Cell and Tissue Banking*, 4 (2003), pp. 69–76
Pirnay, Jean-Paul, Alain Vanderkelen, Martin Zizi, Daniel De Vos, Thomas Rose, Geert Laire, Nadine Ectors and Gilbert Verbeken, 'Human Cells and Tissues: The Need for a Global Ethical Framework', *Bulletin of the World Health Organization*, 88 (2010), pp. 870–72
Pirnay, Jean-Paul, Alain Vanderkelen, Nadine Ectors, Christian Delloye, Denis Dufrane, Etienne Baudoux, Michel Van Brussel, Michael P. Casaer, Daniel De Vos, Jean-Pierre Draye, Thomas Rose, Serge Jennes, Pierre Neirinckx, G. Laire and M. Zizi, 'Beware of the Commercialization of Human Cells and Tissues: Situation in the European Union', *Cell and Tissue Banking*, 13 (2012), pp. 487–98
Pontius, Edwin E., 'Financing Mechanisms for Autopsy', *Pathologist*, 69 (7) (1978), Supplement, pp. 245–7
Post, Lawrence T., 'Eye Banks', *American Journal of Ophthalmology*, 30 (1947), pp. 920–22
Pruitt Jr, Basil A., Cleon W. Goodwin and Arthur D. Mason Jr, 'Epidemiological, Demographic, and Outcome Characteristics of Burn Injury', in David N. Herndon (ed.), *Total Burn Care*, 2nd edn (London: Saunders, 2002), pp. 16–30
Raben, M., 'Treatment of a Pituitary Dwarf with Human Growth Hormone', *Journal of Clinical Endocrinology Metabolism*, 18 (1958), pp. 901–3
—, 'Clinical Use of hGH', *New England Journal of Medicine*, 266 (1962), pp. 82–6
Raiti, Salvatore, 'The National Pituitary Agency', in Salvatore Raiti (ed.), *Advances in Human Growth Hormone: A Symposium Held at Baltimore, Maryland, October 9–12, 1973* (Bethesda, MD: DHEW Publication No. (NIH) 74–612)
—, 'Letter to the Editor', *College of American Pathologists Newsletter* (May 1974), p. 221
—, 'An Appreciative Word', *Pathologist*, 30 (7) (1976), p. 255
—, 'Update from the National Pituitary Agency', *Pathologist* (September 1976), p. 333
—, 'Pituitary Glands – Your Help is Needed', *Pathologist* (June 1978), p. 335
—, 'Collecting Pituitaries: An Update', *Pathologist* (March 1979), p. 112
—, 'Collecting Pituitaries: Why the Program Continues to Need Your Help', *Pathologist* (January 1985), p. 43
—, 'The National Hormone and Pituitary Program: Achievements and Current Goals', in Salvatore Raiti and Robert A. Tolman (eds), *Human Growth Hormone* (New York: Plenum Medical Book Company, 1986), p. 5
Ralston, David R., Sue G. Boyce, Sheila MacNeil and Eric Freedlander, 'Demand has Outstripped Supply in Sheffield's Skin Bank', *British Medical Journal*, 313 (1996), p. 1485
Ravage, Barbara, *Burn Unit: Saving Lives After the Flames* (Cambridge, MA: Da Capo Press, 2004)
Reals, William J., 'The National Pituitary Agency and Human Growth', *Pathologist*, 25 (8) (1972), pp. 191–2
Ridgeway, A., and B. Marcyniuk, 'The U.K. Cornea Transplant Service and the Establishment of the David Lucas Manchester Eye Bank', in P. Sourdille (ed.), *Evolution of Microsurgery. Dev Opththalmol.* (Basel: Karger, 1991), pp. 44–9
Roberts, Mohini, 'The Role of the Skin Bank', *Annals of the Royal College of Surgeons*, 58 (1976), pp. 70–74
Rockafellar, Nancy M., *Conversations with Dr Leslie Batty Bennett: The Research Tradition at UCSF* (San Francisco: UCSF Oral History Program, Department of History and Health Sciences, University of California, San Francisco, 1992)

Rogers, Blair O., 'Guide and Bibliography for Research into the Skin Homograft Problem', *Plastic and Reconstructive Surgery* 7 (1957), pp. 169–201

Rosenwasser, George O. D., and Miriam Rosenwasser, 'Vladimir Filatov 1875–1956', in Mark J. Mannis and Avi M. Mannis (eds), *Corneal Transplantation: A History in Profiles* (Ostend: Hirschberg History of Ophthalmology, The Monographs, volume 6, 1999), pp. 170–91

Rycroft, Benjamin, 'Second-hand Sight', *Medicine Illustrated* (1953), pp. 582–4

—, 'The Organization of a Regional Eye Bank', *The Lancet*, 279 (1962), pp. 147–51

—, 'Contemporary Views on the Surgery and Biology of the Corneal Graft', *Annals of the Royal College of Surgeons of England*, 36 (1965), pp. 152–71

—, 'Eyes for Corneal Grafting', *British Medical Journal*, 2 (1966), p. 1452

— (ed.), *Corneal Grafts* (London: Butterworth, 1955)

Sadler Jr, Alfred M., Blair L. Sadler and E. Blyth Stason, 'The Uniform Anatomical Gift Act', *Journal of the American Medical Association*, 206 (11) (1968), pp. 2501–6

—, 'Transplantation and the Law: Progress Towards Uniformity', *New England Journal of Medicine*, 282 (13) (1970), pp. 717–23

Sadler Jr, Alfred M., Blair L. Sadler and George E. Schreiner, 'A Uniform Card for Organ and Tissue Donation', *Modern Medicine* (29 December 1969), pp. 71–4

Servino, E. M., H. Nathan and J. S. Wolf, 'Unified Strategy for Public Education in Organ and Tissue Donation', *Transplantation Proceedings*, 29 (1997), p. 3247

Shakespeare, Peter G., 'The Role of Skin Substitutes in the Treatment of Burn Injuries', *Clinics in Dermatology*, 23 (2005), pp. 415–18

Sharp, J. R., 'Background to the New "Orange Guide"', *Pharmaceutical Journal* (25 June 1983), p. 7

Shuck, Jerry M., Basil A. Pruitt and John A. Moncrief, 'Homograft Skin for Wound Coverage: A Study in Versatility', *Archives of Surgery*, 98 (4) (1969), pp. 472–8

Siegel, Andrew, Martha W. Anderson, Tracy C. Schmidt and Stuart J. Youngner, 'Informed Consent to Tissue Donation: Policies and Practice', *Cell and Tissue Banking*, 10 (2009), pp. 235–40

Siminoff, Laura A., and Heather M. Traino, 'Consenting to Donation: An Examination of Current Practices in Informed Consent for Tissue Donation in the US', *Cell and Tissue Banking*, 14 (1) (2013), pp. 85–95

Siminoff, Laura A., Heather M. Traino and Nahida Gordon, 'Determinants of Family Consent to Tissue Donation', *Journal of Trauma, Injury, Infection, and Critical Care*, 69 (4) (2010), pp. 956–63

Simpson, S. L., 'Dwarfism, Hormones, and Dieting', *British Medical Journal*, 2 (1947), p. 977

Skolnick, Andrew A., 'Tissue Bank Expands Facilities, Efforts', *Journal of the American Medical Association*, 266 (10) (1991), pp. 1329–31

Smith, Philip E., 'General Physiology of the Anterior Hypophysis', *Journal of the American Medical Association*, 104 (7) (1935), pp. 548–53

Snyder, William H., Bettie M. Bowles and Bruce G. MacMillan, 'The Use of Expansion Meshed Grafts in the Acute and Reconstructive Management of Thermal Injury: A Clinical Evaluation', *Journal of Trauma*, 10 (9) (1970), p. 742

Sokohl, Karen, 'First Person Consent: OPOs Across the Country are Adapting to the Change', *Update* (September–October 2003), pp. 2–7

Sparkes, B. G., 'Treating Mass Burns in Warfare, Disaster or Terrorist Strikes', *Burns*, 23 (1997), pp. 238–47

Starzl, Thomas, E., *The Puzzle People: Memoirs of a Transplant Surgeon* (Pittsburgh and London: University of Pittsburgh Press, 1992)

Stelnicki, Eric J., V. Leroy Young, Tom Francel and Peter Randall, 'Vilray P. Blair, his Surgical Descendants, and their Roles in Plastic Surgical Development', *Plastic and Reconstructive Surgery*, 10 3(7) (1992), pp. 1990–99

The Stephen Kirby Skin Bank and Research Unit (London: Richmond Twickenham and Roehampton Healthcare NHS Trust, n.d.)

Stephenson, Kathryn Lyle, 'Earl C. Padgett: As I Remember', *Annals of Plastic Surgery*, 6 (2) (1981), pp. 142–53

Strobos, J., and Food and Drug Administration, 'Safety of Tissue Allografts', *American Association of Tissue Banks Newsletter*, 16 (1) (1993), pp. 3–4

Strong, D. Michael, and Naoshi Shinozaki, 'Coding and Traceability for Cells, Tissue and Organs for Transplantation', *Cell and Tissue Banking*, 11 (2010), pp. 305–23

Strong, W. Ronald, 'The Tissue Bank, its Operation and Management', in G.E.W. Wolstenholme and Margaret P. Cameron (eds), *Preservation and Transplantation of Normal Tissues: A CIBA Foundation Symposium* (London: J. & A. Churchill, 1954), pp. 220–31

Stuart, F. P., 'Need, Supply, and Legal Issues Related to Organ Transplantation in the United States', *Transplantation Proceedings*, 16 (1) (1984), pp. 87–94

Ta, K. T., John A. Persing, Henry Chauncey, H. Bradley and Stephen H. Miller, 'Effects of Managed Care on Teaching, Research, and Clinical Practice in Academic Plastic Surgery', *Annals of Plastic Surgery*, 48 (4) (2002), pp. 348–54

Tanner, J. M., 'Towards Complete Success in the Treatment of Growth Hormone Deficiency: A Plea for Earlier Ascertainment', *Health Trends*, 7 (1975), pp. 61–5

Tanner, J. M., and R. H. Whitehouse, 'Growth Response of 26 Children with Short Stature Given Human Growth Hormones', *British Medical Journal*, 2 (1967), pp. 69–75

Thomson, Lorna, 'Renal Transplants – A Sin of Omission', *World Medicine* (29 February 1972), pp. 17–22

Tolman, Robert A., 'The NIADDK Hormone Distribution Program', in Salvatore Raiti and Robert A. Tolman (eds), *Human Growth Hormone* (New York: Plenum Medical Book Company, 1986), pp. 13–18

Trier, William C., and Kenneth W. Sell, 'United States Navy Skin Bank', *Plastic and Reconstructive Surgery*, 41 (6) (1968), pp. 543–8

Truitt, Christopher, *The Dark Side of Tissue Donation* (Publisher: Author, 2009)

Truog, Robert D. T., 'Consent for Organ Donation – Balancing Conflicting Ethical Obligations', *New England Journal of Medicine*, 35 (12) (2008), pp. 1209–11

Tudor Thomas, J. W., 'The Technique of Corneal Transplantation as Applied in a Series of Cases', *Ophthalmological Society Transactions*, 55 (1935), p. 55

—, 'Corneal Transplantation: Results of a Series of 56 Operations on 48 eyes', *British Medical Journal*, 2 (1938), pp. 740–42

Underhill, Frank P., 'Changes in Blood Concentration with Special Reference to the Treatment of Extensive Superficial Burns', *Annals of Surgery*, 86 (6) (1927), pp. 840–49

Vail, A., S. M. Gore, B. A. Bradley, D. L. Easty, C. A. Rogers and W. J. Armitage, 'Conclusions of the Corneal Transplant Follow Up Study', *British Journal of Ophthalmology*, 81 (1997), pp. 631–6

Walgate, Robert, 'Pituitary Slump', *Nature*, 290 (5801) (1981), pp. 6–7

Wallace, A. B., and A. W. Wilkinson (eds), *Research in Burns* (Edinburgh: E & S Livingstone, 1966)

Wallace, Antony F., 'The First Decade of the North East Thames Regional Plastic and Surgery and Burns Unit', *British Journal of Plastic Surgery*, 38 (1985), pp. 422–5

Ward, Elizabeth D., *Timbo: A Struggle for Survival* (London: British Kidney Patients Association, 1996)

Warden, John, 'UK Blood Products are Banned', *British Medical Journal*, 316 (1998), p. 723

Watson, John, 'Experience of Burns Unit Design – Queen Victoria Hospital, East Grinstead, England', in A. B. Wallace and A. W. Wilkinson (eds), *Research in Burns* (Edinburgh: E & S Livingstone, 1966), pp. 534–41

Williams, Jason L., 'Patient Safety or Profit: What Incentives are Blood Shield Laws and FDA Regulations Creating for the Tissue Banking Industry?', *Indiana Health Law Review*, 2 (2005), pp. 295–328

Wilson, Diane, and Glenn Greenleaf, 'The Availability of Allograft Skin for Large Scale Medical Emergencies in the United States', *Cell and Tissue Banking*, 15 (2014), pp. 35–40

Winstanley, Emma, and Anthony J. Clarkson, 'The Dedicated Donation Facility. Routine Notification: A Novel Approach', *Blood Matters*, 22 (2007), pp. 13–14

Woien, Sandra, Mohamed Y. Rady, Joseph L. Verheijde and Joan McGregor, 'Organ Procurement Organizations Internet Enrollment For Organ Donation: Abandoning Informed Consent', *BMC Medical Ethics*, 7 (14) (2006)

Wolf, J. S., E. M. Servino and H. N. Nathan, 'National Strategy to Develop Public Acceptance of Organ and Tissue Donation', *Transplantation Proceedings*, 29 (1997), pp. 1477–8

Wolstenholme, G.E.W., and Margaret P. Cameron (eds), *Preservation and Transplantation of Normal Tissues: A CIBA Foundation Symposium* (London: J. &. A. Churchill, 1954)

Working Group of the British Burn Association and Hospital Infection Society, 'Report on the Principles of Design of Burns Unit', *Journal of Hospital Infection*, 19 (1991), pp. 63–6

World Health Organization Scientific Group, *Viral Hepatitis*, Technical Report Series No. 512 (Geneva: World Health Organization, 1973)

Yesner, Raymond, 'Medical Center Autopsy Costs', *American Journal of Clinical Pathology*, 69 (1978), pp. 242–4

Youngner, Stuart J., Martha W. Anderson and Renie Schapiro (eds), *Transplanting Human Tissue: Ethics, Policy, and Practice* (New York and Oxford: Oxford University Press, 2004)

Yudin, S. S., 'Transfusion of Cadaver Blood', *Journal of the American Medical Association* 106 (12) (1936), pp. 997–8

Zink, Sheldon and Stacey Wertlieb, 'A Study of the Presumptive Approach to Consent for Organ Donation: A New Solution to an Old Problem', *Critical Care Nurse*, 26 (2) (2006), pp. 129–36

Interviews with Author

Armitage John, Corneal Transplant Service (Bristol, March 2005)

Cochrane, Tom, Blond McIndoe Centre, Queen Victoria Hospital (East Grinstead, August 2005)

Easty, David, Corneal Transplant Service (Bristol, March 2005)

Fehily, Deidre, NHS Blood Service (London, January 1999)

Franks, Rollin W., United States Navy Tissue Bank (Bethesda, Maryland, December 1998)

Gallo, Nancy, New York Firefighters Skin Bank (Savannah, Georgia, April 2008)

Griffith, Frederick N., Tissue Banks International (Baltimore, April 2005).

Jacobs, Heather, New York Firefighters Skin Bank (New York, November 2006)

Kaye, Stephen, The Royal Liverpool University Hospital (Liverpool, July 2007)

Kearney, John, NHS Tissue Services (Speke, July 2007)

Keely, Pamela, East Anglian Eye Bank (Norwich, May 2007)

Larkin, Frank, Moorfields Eye Hospital (London, May 2013)

Leimkuhler, James H., Tissue Banks International (Baltimore, April 2005.)

Marchant, Bob, Blond McIndoe Centre, Queen Victoria Hospital (East Grinstead, August 2005)

Ostrander, Jim, United States Navy Tissue Bank (Bethesda, Maryland, December 1998)

Pegg, David, University of York/British Association of Tissue Banks (York, November 2007)

Rollins, Franks, United States Navy Tissue Bank (Bethesda, Maryland, December 1998)

Rowe, Andrea, Blond McIndoe Centre, Queen Victoria Hospital (East Grinstead, August 2005)

Stevenson, Robert E., United States Navy Tissue Bank staff (Savannah, Georgia, April 2008)

Secondary Sources

Published Works

Aduddell, Robert M., and Louis P. Cain, 'The Consent Decree in the Meatpacking Industry, 1920–1956', *Business History Review*, 5 (3) (1981), pp. 359–78

Agar, John, 'Modern Horrors: British Identity and Identity Cards', in Jane Caplan and John Torpey (eds), *Documenting Individual Identity: The Development of State Practices in the Modern World* (Princeton, NJ: Princeton University Press, 2001), pp. 101–20

Amoroso, E. C., and G. W. Corner, 'Herbert McLean Evans 1882-1971', *Biographical Memoirs of Fellows of the Royal Society*, 18 (1972), pp. 83-186

Andersen, Kathleen S., and Daniel M. Fox, 'The Impact of Routine Inquiry Laws on Organ Donation', *Health Affairs*, 7 (5) (1988), pp. 65-78

Anderson, Warwick, *The Collectors of Lost Souls: Turning Kuru Scientists into Whitemen* (Baltimore, MD: Johns Hopkins University Press, 2008)

Ariès, Philippe, *The Hour of our Death* (Harmondsworth: Penguin Books, 1983)

Bennett, Leslie L., 'Herbert McLean Evans: A Rare Spirit', *Perspectives in Biology and Medicine*, 22 (1978), pp. 90-103

Berman, Daniel M., *Death on the Job: Occupational Health and Safety Struggles in the United States* (New York: Monthly Review Press, 1978)

Berridge, Virginia, and Alex Mold, 'Professionalisation, New Social Movements and Voluntary Action in the 1960s and 1970s', in Matthew Hilton and James McKay (eds), *The Ages of Voluntarism: How We Got to the Big Society* (Oxford: Published for the British Academy by Oxford University Press, 2011), pp. 114-34

Bliss, Michael, *The Discovery of Insulin* (Boston, MA: Faber and Faber, 1988)

Blumstein, James F., 'Government's Role in Organ Transplantation Policy', in James F. Blumstein and Frank A. Sloan (eds), *Organ Transplantation Policy: Issues and Prospects* (Durham, NC and London: Duke University Press, 1989), pp. 5-40

Blumstein, James F. and Frank A. Sloan (eds), *Organ Transplantation Policy: Issues and Prospects* (Durham, NC and London: Duke University Press, 1989)

Bogdan, Robert, *Freak Show: Presenting Human Oddities for Amusement and Profit* (Chicago and London: Chicago University Press, 1988)

Borell, Merriley, 'Brown-Sequard's Organotherapy and its Appearance in America at the End of the Nineteenth Century', *Bulletin of the History of Medicine*, 50 (1976), pp. 309-20

—, 'Organotherapy, British Physiology, and the Discovery of the Internal Secretions', *Journal of the History of Biology*, 9 (2) (1976), pp. 232-68

Cantor, D., 'Cortisone and the Politics of Drama, 1949-55', in J. Pickstone (ed.), *Medical Innovations in Historical Perspective* (Basingstoke: Macmillan, 1992), pp. 165-84

—, 'Cortisone and the Politics of Empire: Imperialism and British Medicine, 1918-1955', *Bulletin of the History of Medicine*, 67 (1993), pp. 463-93

Cohen, Laurence, 'The Other Kidney: Biopolitics Beyond Recognition', in N. Scheper-Hughes and L. Wacquant (eds), *Commodifying Bodies* (London: Sage Publications, 2002), pp. 9-30

—, 'Operability, Bioavailability, and Exception', in A. Ong and S. J. Collier (eds), *Global Assemblages: Technology, Politics, and Ethics as Anthropological Problems* (Oxford: Blackwell Publishing, 2005), pp. 79-90

Cohen, Susan, and Christine Cosgrove, *Normal at Any Cost: Tall Girls, Short Boys, and the Medical Industry's Quest to Manipulate Height* (New York: Jeremy P. Tarcher/Penguin, 2009)

Cole, R. D., 'Choh Hao Li: April 21, 1913 - November 28, 1987', *Biographical Memoirs. National Academy of Sciences*, 70 (1996), pp. 221-39

Connor, Steven, *The Book of Skin* (London: Reaktion Books, 2004)

Cronon, William, *Nature's Metropolis: Chicago and the Great West* (New York and London: W. W. Norton, 1991)

Denise, S. H., 'Regulating the Sale of Human Organs', *Virginia Law Review*, 71 (6) (1985), p. 1022

Dunnigan, James F., and Albert A. Nofi, *Dirty Little Secrets of the Vietnam War* (New York: St. Martin's Press, 1999)

Eachter, Kenneth W., 'Secular Changes in American and British Stature and Nutrition', *Journal of Interdisciplinary History*, 14 (2) (1983), pp. 445-81

Elkins, James, *Pictures of the Body: Pain and Metamorphosis* (Stanford, CA: Stanford University Press, 1999)

Engerman, Stanley L., and Robert E. Gallman (eds), *The Cambridge Economic History of the United States, Volume II: The Long Nineteenth Century* (Cambridge: Cambridge University Press, 2000)

Farrugia, Albert, 'When do Tissues and Cells Become Products? - Regulatory Oversight of Emerging Biological Therapies', *Cell & Tissue Banking*, 7 (4) (2006), pp. 325-35

Field, Mark G., 'Soviet Medicine', in Roger Cooter and John Pickstone (eds), *Medicine in the 20th Century* (London: Harwood Academic Publishers, 2000), pp. 51–66

Fogel, Robert W., Stanley L. Engerman, Roderick Floud, Gerald Friedman, Robert A. Margo, Kenneth Sokoloff, Richard H. Steeckel, T. James Trussell, Georgia Villaflor and Kenneth W. Wachter, 'Secular Changes in American and British Stature and Nutrition', *Journal of Interdisciplinary History*, 14 (2) (1983), pp. 445–81

Fontaine, Philippe, 'Blood, Politics, and Social Science: Richard Titmuss and the Institute of Economic Affairs, 1957–1973', *Isis*, 93 (2002), pp. 401–34

Foote, Susan Bartlett, *Managing the Medical Arms Race: Innovation and Public Policy in the Medical Device Industry* (Berkeley, CA: University of California Press, 1992)

Fox, Renée C., and Judith P. Swazey, *The Courage to Fail: A Social View of Organ Transplants and Dialysis* (Chicago: University of Chicago Press, 1974)

—, *Spare Parts: Organ Replacement in American Society* (New York: Oxford University Press, 1992)

Frost, Dan, 'Farewell to the Lodge', *American Demographics*, 18 (1996), pp. 40–45

Gaul, Gilbert M., and Neill A. Borowski, *Free Ride: The Tax-exempt Economy* (Kansas City, MO: Andrews and McMeel, 1993)

Giacomini, Mita, 'A Change of Heart and a Change of Mind? Technology and the Redefinition of Death in 1968', *Social Science and Medicine*, 44 (1) (1997), pp. 1465–82

Gilman, Sander L., *Making the Body Beautiful: A Cultural History of Aesthetic Surgery* (Princeton, NJ: Princeton University Press, 1999)

Goldstein, Donald M., and Harry J. Maihafer, *The Korean War: The Story and Photographs* (Washington DC: Basseys, 2000)

Goodwin, Michele, 'Deconstructing Legislative Consent Law: Organ Taking, Racial Profiling & Distributive Justice', *Virginia Journal of Law and Technology* (2001), pp. 1522–687

—, 'Rethinking Legislative Consent Law', *De Paul Journal of Health Care Law*, 5 (2) (2002), pp. 257–318

—, *Black Markets: The Supply and Demand of Body Parts* (New York: Cambridge University Press, 2006)

Haigh, Elizabeth, 'The Roots of Vitalism of Xavier Bichat', *Bulletin of the History of Medicine*, 49 (1975), pp. 72–86

Haiken, Elizabeth, *Venus Envy: A History of Cosmetic Surgery* (Baltimore, MD and London: Johns Hopkins University Press, 1997)

Haines, Michael R., 'The Population of the United States, 1790–1920', in Stanley L. Engerman and Robert E. Gallman (eds), *The Cambridge Economic History of the United States, Volume II: The Long Nineteenth Century* (Cambridge: Cambridge University Press, 2000), pp. 143–205

Hall, Stephen S., *Size Matters: How Height Affects the Health, Happiness and Success of Boys – and the Men they Become* (Boston, MA: Houghton Mifflin Company, 2006)

Harrison, Robert Pogue, *The Dominion of the Dead* (Chicago and London: University of Chicago Press, 2003)

Hawkins, Gay, and Stephen Muecke, 'Introduction: Cultural Economies of Waste', in Gay Hawkins and Stephen Muecke (eds), *Culture and Waste: The Creation and Destruction of Value* (Lanham, MD: Rowman & Littlefield, 2003), pp. ix–xvii

Hayden, Cori, *When Nature Goes Public: The Making and Unmaking of Bioprospecting in Mexico* (Princeton, NJ and Oxford: Princeton University Press, 2003)

Healy, Kieran, *Last Best Gifts: Altruism and the Market for Human Blood and Organs* (Chicago and London: Chicago University Press, 2006)

Hochschild, Arlie, 'Emotion Work, Feeling Rules, and Social Structure', *American Journal of Sociology*, 83 (3) (1979), pp. 551–75

Hoeyer, Klaus, 'After Novelty: The Mundane Practices of Ensuring a Safe and Stable Supply of Bone', *Science as Culture*, 19 (2) (2010), pp. 123–50

Hogle, Linda F., 'Standardization Across Non-standard Domains: The Case of Organ Procurement', *Science, Technology and Human Values*, 20 (4) (1995), pp. 482–500

—, 'Transforming "Body Parts" into Therapeutic Tools: A Report from Germany', *Medical Anthropology Quarterly*, 10 (4) (1996), pp. 675–82

Horowitz, Roger, '"Where Men Will Not Work": Gender, Power, Space and the Sexual Division of Labor in America's Meatpacking Industry, 1890–1990', *Technology and Culture*, 38 (1997), pp. 187–213
—, *Putting Meat on the American Table: Taste Technology, Transformation* (Baltimore, MD: The Johns Hopkins University Press, 2006)
Höyer, Klaus, 'After Novelty: The Mundane Practices of Ensuring a Safe and Stable Supply of Bone', *Science as Culture*, 19 (2) (2010), pp. 123–50
—, *Exchanging Human Bodily Material: Rethinking Bodies and Markets* (Heidelberg, New York and London: Springer, 2013)
Ives, J. Russell, *The Livestock and Meat Economy of the United States* (Chicago: AMI Center for Continuing Education American Meat Industry, 1966)
Jackson, Percival E., *The Law of Cadavers and of Burial and Burial Places*, 2nd edn (New York: Prentice-Hall, 1950)
Jaffe, Erik S., '"She's Got Bette Davis['s] Eyes"': Assessing the Nonconsensual Removal of Cadaver Organs Under the Takings and Due Process Clauses', *Columbia Law Review* (1990), pp. 528–74
Jones, D. Gareth, *Speaking for the Dead: Cadaver in Biology and Medicine* (Aldershot: Ashgate, 2000)
Jupp, Peter C., and Clare Gittings, *Death in England: An Illustrated History* (Manchester: Manchester University Press, 1999)
Karpf, Anne, *Doctoring the Media: The Reporting of Health and Medicine* (London: Routledge, 1988)
Katz, Robert A., 'The Re-Gift of Life: Can Charity Law Prevent Non-profit Firms from Exploiting Donated Tissue and Non-profit Tissue Banks?', *De Paul Law Review*, 55 (3) (2006), p. 969
Keenan, Tiarnan D. L., Mark N. A. Jones, Sally Rushton and Fiona M. Carley, 'Trends in the Indications for Corneal Graft Surgery in the United Kingdom: 1999 through 2009', *Archives of Ophthalmology*, 130 (5) (2012), pp. 621–8
Klein, Rudolf, *The Politics of the National Health Service*, 2nd edn (London: Longman, 1989)
Králova, Jana, 'What is Social Death?', *Contemporary Social Science*, 10 (3) (2015), pp. 235–48
Laderman, Gary, *A Cultural History of Death and the Funeral Home in Twentieth-century America* (Oxford: Oxford University Press, 2003)
Lederer, Susan E., *Flesh and Blood: Organ Transplantation and Blood Transfusion in Twentieth-century America* (Oxford: Oxford University Press, 2008)
Leechman, Michael, 'Regulation of the Human Tissue Industry: A Call for Fast-track Regulation', *Louisiana Law Review*, 65 (2004), pp. 441–71
Lewis, Alan, and Martin Snell, 'Increasing Kidney Transplantation in Britain: The Importance of Donor Cards, Public Opinion and Medical Practice', *Social Science and Medicine*, 22 (10) (1986), pp. 1075–80
Lindee, Susan, *Suffering Made Real: American Science and the Survivors at Hiroshima* (Chicago and London: Chicago University Press, 1997)
Lock, Margaret, *Twice Dead: Organ Transplants and the Reinvention of Death* (Berkeley, CA: University of California Press, 2002)
Lombardo, Paul A., 'Consent and "Donations" from the Dead', *Hasting Center Report*, 11 (6) (1981), pp. 9–11
Löwy, Ilana, 'Tissue Groups and Cadaver Kidney Sharing: Socio-cultural Aspects of a Medical Controversy', *International Journal of Technology Assessment in Health Care*, 2 (1986), pp. 195–218
—, 'The Impact of Medical Practice on Biomedical Research: The Case of Human Leucocyte Antigens Studies', *Minerva*, 25 (1–2) (1987), pp. 171–200
Lynch, Thomas, *The Undertaking: Life Studies from the Dismal Trade* (London: Jonathan Cape, 1997)
Lyon, David, 'Under My Skin', in Jane Caplan and John Torpey (eds), *Documenting Individual Identity: The Development of State Practices in the Modern World* (Princeton and Oxford: Princeton University Press, 2000), pp. 291–310

MacDonald, Helen, 'Conscripting Organs: "Routine Salvaging" or Bequest? The Historical Debate in Britain, 1961–1975', *Journal of the History of Medicine and Allied Science*, 70 (3) (2014), pp. 425–61

McKelvey, Maureen, *Evolutionary Innovation: Early Industrial Uses of Genetic Engineering* (Linköping, Sweden: Department of Technology and Social Change, Linköping University, 1984)

—, 'Engineering the Biological and Political: Enabling Industrial Use of Genetic Engineering in Sweden', *Polhem: Tisskrift för Teknikhistoria*, 12 (1994), pp. 216–59

McLeave, Hugh, *McIndoe: Plastic Surgeon* (London: Frederick Muller, 1961)

Madison, James H., *Eli Lilly: A Life, 1885–1977* (Indianapolis: Indiana Historical Society, 1989)

Mahoney, Julia D., 'The Market for Human Tissue', *Virginia Law Review*, 86 (2) (2000), pp. 164–223

Mayhew, E. R., *The Reconstruction of Warriors: Archibald McIndoe, the Royal Air Force and the Guinea Pig Club* (London and Pennsylvania: Greenhill Books, 2004)

Moore, Francis D., *Give and Take: The Development of Tissue Transplantation* (Philadelphia: W.B. Saunders, 1964)

Moreno, Jonathan D., *Undue Risk: Secret State Experiments on Humans* (New York and London: Routledge, 2001)

Mulkay, Michael, and John Ernst, 'The Changing Profile of Social Death', *European Journal of Sociology*, 22 (1991), pp. 172–96

Nestle, Marion, *Safe Food: The Politics of Food Safety* (Berkeley, Los Angeles, and London: University of California Press, 2010)

O'Brien, Martin, 'Rubbish Values: Reflections on the Political Economy of Waste', *Science as Culture*, 8 (3) (1999), pp. 269–95

—, *A Crisis of Waste? Understanding the Rubbish Society* (London: Routledge, 2007)

Offer, Avner, 'Between the Gift and the Market: The Economy of Regard', *Economic History Review*, 50 (3) (1997), pp. 450–76

Pegg, D. E., 'The History and Principles of Cryopreservation', *Seminars in Reproductive Medicine*, 20 (2002), pp. 5–13

Pfeffer, Naomi, 'How Abattoir "Biotrash" Connected the Social Worlds of the University Laboratory and the Disassembly Line', in David Cantor, Christian Bonah and Matthias Dörries (eds), *Meat, Medicine and Human Health in the Twentieth Century* (London: Pickering and Chatto, 2010), pp. 63–76

Pickstone, John V., 'Bureaucracy, Liberalism and the Body in Post-revolutionary France: Bichat's Physiology and the Paris School of Medicine', *History of Science*, xix (1981), pp. 115–42

Pollock, Allyson M., *NHS plc: The Privatisation of our Health Care* (London: Verso, 2004)

Ponte, Maya L., 'Insights into the Management of Emerging Infections: Regulating Creutzfeldt-Jakob Disease Transfusion Risk in the UK and the US', *PLoS Medicine*, 3 (10) (2006), pp. 1751–64

Price, David, 'The Human Tissue Act 2004', *Modern Law Review*, 68 (5) (2005), pp. 798–821

Prothero, Stephen, *Purified by Fire: A History of Cremation in America* (Berkeley, CA: University of California Press, 2001)

Prottas, Jeffrey M., 'Obtaining Replacements: The Organizational Framework of Organ Procurement', *Journal of Health Politics, Policy and Law*, 8 (2) (1983), pp. 235–50

—, 'Organization of Organ Procurement', in Blumstein and Sloan (eds), *Organ Transplantation Policy: Issues and Prospects*, pp. 41–56

—, 'The Organization of Organ Procurement', *Journal of Health Politics, Policy and Law*, 14 (1989), pp. 41–55

—, 'Competition for Altruism: Bone and Organ Procurement in the United States', *Milbank Quarterly*, 70 (2) (1992), pp. 299–317

—, *The Most Useful Gift: Altruism and the Public Policy of Organ Transplants* (San Francisco: Jossey-Bass Publishers, 1994)

Putnam, Robert D., *Bowling Alone: The Collapse and Revival of American Community* (New York: Simon and Schuster, 2000)

Quinn, James Brian, and Frederick G. Hilmer, 'Strategic Outsourcing', *Sloan Management Review* (Summer 1994), pp. 43–55

Randle, Philip, 'Frank George Young', *Biographical Memoirs of Fellows of the Royal Society*, 36 (1990), pp. 583–99

Rasmussen, Nicolas, 'Of "Small Men", Big Science and Bigger Business: The Second World War and Biomedical Research in the United States', *Minerva*, 40 (2002), pp. 115–46

—, 'Steroids in Arms: Science, Government, Industry, and the Hormones of the Adrenal Cortex in the United States, 1930–1950', *Medical History*, 46 (2002), pp. 299–324

Rettig, Richard A., 'The Politics of Organ Transplantation: A Parable of our Time', in Blumstein and Sloan (eds), *Organ Transplantation Policy: Issues and Prospects*, pp. 191–228

—, 'Special Treatment – The Story of Medicare's ESRD Entitlement', *New England Journal of Medicine*, 364 (2011), pp. 596–8

Rhodes, Richard, *Deadly Feasts: Tracking the Secrets of a Terrifying New Plague* (New York: Simon and Schuster, 1997)

Richardson, Ruth, *Death, Dissection, and the Destitute* (Harmondsworth: Pelican, 1988)

Rose-Ackerman, Susan, 'Altruism, Nonprofits, and Economic Theory', *Journal of Economic Literature*, 34 (1996), pp. 701–28

Rothman, David J., *Strangers at the Bedside: A History of How Law and Bioethics Transformed Medical Decision-making* (New York: Basic Books, 1991)

Rugg, Julie, 'From Reason to Regulation: 1760–1850', in Peter C. Jupp and Clare Gittings (eds), *Death in England: An Illustrated History* (Manchester: Manchester University Press, 1999), pp. 202–29

Santoni-Rugiu, P., and P. J. Sykes, *A History of Plastic Surgery* (Berlin and Heidelberg: Springer-Verlag, 2007)

Sappol, Michael, *A Traffic of Dead Bodies* (Princeton, NJ: Princeton University Press, 2002)

Sawday, Jonathan, *The Body Emblazoned: Dissection and the Human Body in Renaissance Culture* (London: Routledge, 1995)

Sawin, Clark T., 'Maurice S. Raben and the Treatment of Growth Hormone Deficiency', *Endocrinologist*, 12 (2) (2002), pp. 73–6

Schneider, William H., 'Blood Transfusion Between the Wars', *Journal of the History of Medicine*, 58 (2003), pp. 187–224

Schuck, Peter H., 'Government Funding for Organ Transplants', in Blumstein and Sloan (eds), *Organ Transplantation Policy*, pp. 169–90

Sharp, Lesley A., 'Organ Transplantation as a Transformative Experience: Anthropological Insights into Restructuring of the Self', *Medical Anthropology Quarterly*, 9 (3) (1995), pp. 357–89

Shem, Samuel, *The House of God* (New York: Doubleday Dell, 1978)

Skaggs, Jimmy M., *Prime Cut: Livestock Raising and Meatpacking in the United States, 1607–1983* (College Station, TX: A & M University Press, 1986)

Solomon, Andrew, *Far from the Tree: Parents, Children, and the Search for Identity* (New York: Scribner, 2012)

Star, Susan Leigh (ed.), *Ecologies of Knowledge: Work and Politics in Science and Technology* (Albany, NY: State University of New York Press, 1995)

Stark, Tony, *Knife to the Heart: The Story of Transplant Surgery* (Basingstoke: Macmillan, 1996)

Starr, Paul, *The Social Transformation of American Medicine* (New York: Basic Books, 1984)

Stevens, Rosemary, *Medical Practice in Modern England: The Impact of Specialization and State Medicine* (New Haven, CT and London: Yale University Press, 1966)

—, *In Sickness and in Wealth: American Hospitals in the Twentieth Century* (New York: Basic Books, 1989)

Strasser, Susan, *Waste and Want: A Social History of Trash* (New York: Owl Books, 1999)

Sudnow, David, *Passing On: The Social Organization of Dying* (Englewood Cliffs, NJ: Prentice-Hall, 1967)

Swanson, Kara W., *Banking on the Body: The Market in Blood, Milk and Sperm in Modern America* (Cambridge, MA and London: Harvard University Press, 2014)

Sweeting, Helen, and Mary Gilhooly, 'Doctor, Am I Dead? A Review of Social Death in Modern Societies', *Omega*, 24 (4) (1991–2), pp. 251–69
—, 'Dementia and the Social Phenomenon of Social Death', *Sociology of Health and Illness*, 19 (1997), pp. 93–117
Tansy, E. M., 'Ergot to Ergometrine: An Obstetric Renaissance', in Anne Hardy and Lawrence Conrad (eds), *Women and Modern Medicine* (Amsterdam and New York: Editions Rodopi B.V., 2001), pp. 195–216
Tilney, Nicholas L., *Transplant: From Myth to Reality* (New Haven, CT: Yale University Press, 2003)
Titmuss, Richard, 'The Gift Relationship: From Human Blood to Social Policy', in Ann Oakley and John Ashton (eds), *The Gift Relationship: From Human Blood to Social Policy. Expanded and Updated Edition* (New York: The New Press, 1997 [1970]), pp. 57–315
Tocqueville, Alexis de, *Democracy in America* (Ware: Wordsworth, 1998)
Turney, Jon, *Frankenstein's Footsteps: Science, Genetics and Popular Culture* (New Haven, CT and London: Yale University Press, 1998)
Vialles, Noëlie, *Animal to Edible* (Cambridge: Cambridge University Press, 1994)
Webster, Charles, *The Health Services since the War, Volume I: Problems of Health Care, the National Health Service before 1957* (London: Her Majesty's Stationery Office, 1988)
—, *The National Health Service: A Political History* (Oxford: Oxford University Press, 1998)
Weisbrod, Burton A., 'The Nonprofit Mission and its Financing: Growing Links Between Nonprofits and the Rest of the Economy', in Burton A. Weisbrod (ed.), *To Profit or Not to Profit: The Commercial Transformation of the Nonprofit Sector* (Cambridge: Cambridge University Press, 1998), pp. 1–22
— (ed.), *To Profit or Not To Profit: The Commercial Transformation of the Nonprofit Sector* (Cambridge: Cambridge University Press, 1998)
Welsome, Eileen, *The Plutonium Files: America's Secret Medical Experiments in the Cold War* (New York: Delta Trade Paperbacks, 1999)
Zelizer, Viviana A., *The Social Meaning of Money: Pin Money, Pay Checks, Poor Relief, and Other Currencies* (New York: Basic Books, 1994)
Zunz, Olivier, *Philanthropy in America* (Princeton, NJ and Oxford: Princeton University Press, 2012)

FIGURE REFERENCES

1. Lions Clubs International, Chicago © *Lion Magazine*. 2. © *Chemist & Druggist*. 3. Choh Hao Li's papers, University of California San Francisco Archives and Special Collection. 4. Wellcome Library Archives, PP/COL/C.5 © Wellcome Images. 5. National Archives MH 78/370 © National Archives, London. 6. © Getty Images. 7. Richard B. Gresham, Vernon P. Perry, Thomas E. Wheeler, 'US Navy Tissue Bank', *Journal of American Medical Association*, 183 (1963), p. 99. 8. Vernon P. Perry, 'Description of current tissue banks methods of skin preservation', *Cryobiology* 3(2) (1966), p. 183 © Elsevier Inc. 9. QVHA11, Queen Victoria Hospital Collection, East Grinstead Museum © East Grinstead Museum. 10. © Getty Images. 11. [1] Salvatore Raiti, The National Pituitary Agency', in Salvatore Raiti (ed.), *Advances in Human Growth Hormone: A symposium held at Baltimore, Maryland, October 9–12, 1973*, Bethesda, MD: DHEW Publication No. (NIH) 74–612, p. 5; [2] Salvatore Raiti, 'Letter to the Editor', *College of American Pathologists Newsletter*, (May 1974), p. 221; [3] Salvatore Raiti, 'Update from the National Pituitary Agency', *Pathologist* (September 1976), p. 333; [4] Salvatore Raiti, 'Pituitary glands – your help is needed', *Pathologist* (June 1978), p. 335; [5] Salvatore Raiti, 'Collecting pituitaries: an update', *Pathologist* (March 1979), p. 112; [6] Salvatore Raiti, 'Collecting pituitaries: why the program continues to need your help', *Pathologist* (January 1985), p. 43. 12. [1] Minutes of Meeting of Growth Hormone Sub-committee, 10 November 1959, National Archives FD 23/1593; [2] Report to the Cell Board of the Steering Committee for Human Pituitary Collection: December 1972 – March 1977, May 1977, National Archives FD 23/4428; [3] Estimate based on previous year; [4] Minutes of Meeting of Human Pituitary Sub-Committee, 7 February 1966, National Archives FD 1/8379; [5] Report on Collection of Human Pituitary Glands

and Supply of Hormones, October 1969, National Archives FD 1/8388; [6] Dr Anne Stockwell Hartree, Report on Collection of Human Pituitary Glands and Supply of Hormones from 1 October 1970 to 1 October 1972, November 1972, National Archives FD 7/3542; [7] Human Pituitary Gland Collection – Discussion Paper on Arrangements for Collection: M.R.C./D.H.S.S. Meeting, 17 January 1978, National Archives FD 7/3516; [8] Robert Walgate, 'Pituitary slump', *Nature*, 5 March 1981; [9] Estimate from July to December data, In confidence: Reception of Pituitary Glands at CAMR, National Archives FD 7/3523; [10] In confidence: Reception of Pituitary Glands at CAMR, National Archives FD 7/3523. 13. Committee on Death Certification and Coroners, *Report,* Cmnd. 4810 (London: Her Majesty's Stationery Office, 1971), p. 1. 14. Home Office, Coroners' statistics, 2011, www.gov.uk (accessed August 2015). 15. Committee on Death Certification and Coroners, p. 5; Bristol Royal Infirmary Inquiry, 'Interim report: Removal and retention of human material, 2000', in Public Inquiry into children's heart surgery at the Bristol Royal Infirmary 1984–1995, *Report: Learning from Bristol*, CM 5207(1), Annex C. (London: Stationery Office, 2001), p. 5. 16. Royal National Institute for the Blind Archive, R3/3 © Royal National Institute of Blind People. 17. Blond McIndoe Research Unit Archives, Queen Victoria Hospital, East Grinstead © Queen Victoria Hospital NHS Foundation Trust. 18. National Archives BD 18/275 © National Archives, London. 19. Royal National Institute for the Blind Archive, R 3/3 © Royal National Institute of Blind People. 20. James Barrett Brown, Minot P. Fryer, and Thomas J. Zaydon, 'Establishing a skin bank: use and various methods of preservation of postmortem homografts', *Plastic and Reconstructive Surgery* 16(5) (1955), p. 339 © Wolters Kluwer Health, Inc. 21. National Archives MH 150/655 © National Archives, London. 22. © Mirrorpix. 23. 'Code that controls growth', *LIFE*, 14 October 1966, p. 96 © Getty Images. 24. Emile J. Farge, 'Eye Banking: 1944 to the present', *Survey of Ophthalmology* 30(4) (1989), p. 262. 25. © LifeNet Corporation. 26. Marion Nestle, *Safe Food: the Politics of Food Safety* (Berkeley, Los Angeles, and London: University of California Press, 2010), p. 69. 27. © *Express*. 28. Paul Ashford, 'Traceability', *Cell and Tissue Banking*, 11 (2010), p. 331.

INDEX

acellular skin, *see* skin
Acelity, 279, *see* LifeCell
advance directives, 250
African Americans, 22
  as recipient, 185
  as source, 141, 144, 164, 211
agreement to extraction, *see* kinfolk
America Burning, 131
AIDS (Acquired Immune Deficiency Syndrome), *see* Human Immunodeficiency Virus
Alder Hey Hospital (Royal Liverpool Children's Hospital), 260–2, 278
AlloDerm®, 219
AlloTech, 233
AlloSource, 219, 222, 269
American Association of Tissue Banks (AATB), 175, 186, 221, 238, 229–30, 247, 248, 252–3
  industry statistics, 214, 244, 282
  medical standards (voluntary), 175, 202, 221, 226, 233
  membership, 202, 253, 255, 269, 282
American Burn Association, 133, 184
  and burn care finances, 184, 186
  and burn care organization, 133, 186
  and cadaver skin shortages, 244
American Hospital Association, 269
American Medical Association, 43, 159, 186
American Red Cross
  blood banks, 12, 60, 216, 272
  Tissue Services, 216–17, 268–9
Anatomy Act (UK)
  1832, 70–1, 76, 81–2, 104, 141, 275
  1984, 274, 276
anatomy acts (USA), 141

Ancient Arabic Order of the Nobles of the Mystic Shrine (Shriners), 135, 137, 216, 244–5
Ares-Serono, 205
Ariès, Philippe, 70, 102, 105
Armour, 38–40, 44–5
artefacts, *see* cadaver
Association of Medical Examiners, 212
autopsy, *see* post-mortem examinations

B. Braun Melsungen, 228, *see also* dura mater; Lyodura
Baltic Tissue Bank, 231
'bank', *see* entities
Baxter, Charles, 137–8, 186
Beecher, Henry, 140
Bentham, Jeremy, 71
Bernhard Baron Memorial Fund, 29, 49
Bewick, Michael, 179, 193
Bichat, François-Xavier, x
bioavailability, 104–5
  medical criteria of, 105–6, 149, 172, 181, 203–4, 206, 226–7, 230–2, 239, 242, 253, 256, 258, 280–1
bioburden, 30, 106, 108, 175, 201, 231–2, 242, 252–3, 256, *see also* Good Tissue Practices; putrefaction
biological wound dressings, *see* skin
Biomedical Tissue Services, *see* Mastromarino, Michael
bioprospecting, 22–3, 271
biotrash, *see* waste
Birmingham Accident Hospital Burns Unit, 49–50, 54, 122
Black, D.A.K., 53
Blair, Vilray Papin, 6–7, 28, 125

Blizzard, Robert M., 96, 155, 161
Blond, Elaine and Neville, 68, 120
Blond McIndoe Burns Unit, 120
Blond McIndoe Research Unit, 120, 122, 195
blood
  banks, 12, 198, 279
  cadaver, 10, 12
  'cross-racial', 12–13
  donors-on-the-hoof, 106, 139, 158, 199, 225, 227, 243
  groups, 5
  products, 217, 242, 254, 271
  safety, 173–6, 199, 225, 227, 242
blood shield statutes, 174, 253
Blumberg, Baruch, 199–201
body philanthropy, see philanthropy
Bogdan, Robert, 42
bone (human), 57–8, 60, 62–3, 64, 82, 122, 171, 186, 187, 188, 215, 216, 219, 221, 226, 233, 238–9, 254, 255, 258, 272, 278
Bovine Spongiform Encephalopathy (BSE or 'mad cow' disease), see spongiform encephalopathies
Boyer, Herbert, 205
Breckenbridge, Aida d'Acosta, 16, 18
Bristol Eye Hospital, 197
Bristol Heart Children's Action Group, 259
British Association for Tissue Banking (BATB), 238
British Burn Association, 239
British Kidney Patient Association, 194
British Medical Association, 276
British Orthopaedic Association, 239
British Standards Institute, 51, 53
Brooke Army Hospital, Fort Sam Houston, Texas, 132–3
Brown, Harry, 7
Brown, James Barrett, 6–7, 27, 28, 31–2, 125, 129–31, 137
Brown, Paul, 203, 205–6
BSE, see spongiform encephalopathies
Burke, John, 137
Burnett, Frank Macfarlane, 121
burn care, 2, 28, 119, 125–6, 132–3, 239
  in UK, 27, 28, 49, 53–5, 119–20, 122, 126–7, 239–40
  in USA, 132–7, 184, 186, 244–5, 283, 244, 245
  low status of, 54, 135, 282
burn shock, 5–6, 53, 129
burns
  airman's, 66
  and atomic bombs, 54, 56–7
  and children, 48–53, 128, 245
  and infection, 2, 29, 53, 54, 125
  civilian, 27, 48–51, 53, 186
  incidence in UK, 49–50, 53
  incidence in USA, 128–9, 131–2
  prevention, 48–55, 131–2
  type of, 2, 245
  war, 27–8, 56, 64, 68–9, 132, 132, 186, 196
Burr, Susanna, 196–7

cadaver artefacts/products, 122, 126
  definitions, 63–4, 229–30, 247, 254
  demand, 215–16, 238, 247, 250, 269, 282–5
  import/exports, 271–2, 277–8, 284
  safety, 171–2, 202, 225–6, 229–33, 276
  types, 132, 165, 217, 218, 283–5
  see also medical devices
Calbiochem, 155
Carrel, Alexis, 59
Casey, Robert P., 213
Casey, Thomas Aquinas, 117–18, 177–8, 195
category error, see consent
cemeteries, 25–6
Centers for Disease Control and Prevention (USA, CDC), 227, 228, 252
Chain, Ernest, 30
Chambers, Ron, 123–4, 137
Chief Medical Officer of England and Wales' Census, 261–2, 263
CIBA Foundation, 60
civil society, 14, 69
clean rooms, 220, 270, 278
clinical gold rush, 124, 134
Clemetson, Mrs Gordon ('Aunt Agatha'), 69, 71, 109
Coalition on Donation (USA), 210
Cochrane, Tom, 123, 137
Cohen, Lawrence, 104
Colebrook, Leonard, 28
  and Birmingham Accident Hospital Burns Unit, 49, 122
  and Glasgow Burns Unit, 29–30
  and organization of burn care, 54–5, 121–2, 133
  and prevention of burns, 50–3
  and wound infection, 28–30, 53
Colebrook, Vera, 49
collecting sites, 22–3, 142, 224, 232, 280
  competition for, 94, 177, 183, 187, 277, 221–2, 268–70, 278
  deathbeds, 61, 88, 89, 176, 279, 280
  funeral homes, 88, 137, 171, 177, 218, 221, 266–7, 269, 271, 277, 279, 280
  hospices, 118, 218, 221, 279, 281
  intensive (critical) care units, 139, 178–9, 183, 185, 187, 250

INDEX

medical examiner/coroners' premises, 23–4, 137, 142, 160, 164, 187, 211–12, 218, 221, 269
mortuaries, 8–10, 11, 22–3, 31, 47, 58, 65, 67, 77, 97, 108, 116, 118, 123–4, 137, 149–50, 172, 177, 187, 214, 218, 233, 268, 271, 277–9, 280
overseas, 94, 231, 233, 284
College of American Pathologists, 94–5, 97
Committee on Chronic Kidney Disease (Gottschalk Committee, USA), 135
Committee on Death Certification and Coroners (Broderick Committee, UK), 103
Committee on Microbiological Safety of Blood and Tissues for Transplantation (UK), 242
Committee on Safety of Medicine, 242
Community Tissue Services, 217, 269, 283
compatibility, *see* tissue typing
Connor, Steven, 1
consent
 appropriate, 276, 282
 category error, 17, 282
 first-person, 248–9, 271–2, 275–6, 282
 legislative, 161–2, 165, 168–9, 170, 187, 207–11, 212–13, 233, 250, 256, 272
 presumed, 161, 179, 191, 276
 see also kinfolk, agreement to extraction
Cook County Hospital, Chicago, 12, 198, 278
Cooke, Alistair, 266, 268,
corneas, 8
 cadaver, 8, 9, 11, 67
 contaminated, 171, 202, 206, 225, 234
 excision of, 116, 165, 187
 exports, 272–3
 shelf life of, 10, 11, 116, 165, 176, 197
corneal blindness, 8–9, 12, 66–7, 73, 111–12, 164–5, 169
corneal grafts (keratoplasty), 9, 11–12, 19, 23, 32, 67, 74, 90, 108, 164–5, 177, 195–6, 207, 211, 258, 273, 280, 281, 285
Corneal Grafting Act (1952, UK), 71–3, 75–7, 81–3, 107, 176
Corneal Tissue Act (1986, UK), 197, 275
Corneal Transplant Service (CTS), 196, 198, 238, 278–9, 281, 285, *see also* NHS Blood and Transplant
Corner, George, 39
coroners (UK), 49
 and extraction of cadaver stuff, 81,124, 177, 179, 180–1, 197, 180–1, 278
 and post mortem examinations (autopsies), 84, 104, 180
 scope of authority, 76, 103–5, 181, 262

coroners (USA), *see* medical examiners/coroners
corporate ethics, 282–3
corpse philanthropy, *see* philanthropy
corpse
 disassembly, 59, 61–2, 209, 268, 272, *see also* bioavailability; putrefaction
 kinfolk's private interest in, 25–6, 75–6, 83, 142, 170, 207–8, 249, 262, 268, 275–6, *see also* kinfolk, agreement to extraction
 no-property (common law principle), 21–2, 83
 possession of, 24, 75–6, 82, 89, 161, 179, 180, 208, 209, 212, 224, 233, 262
 quasi-property ruling, 24–5, 76,142, 160–1, 207, 268
 unclaimed, 76, 105, 141
cosmetic enhancement and rejuvenation, 42, 215–16, 246
Council of Europe, 153, 242, 263
cottage industry, *see* entities
Crescormon®, 153, 156, 172
Creutzfeldt-Jakob disease (CJD), *see* spongiform encephalopathies
Creutzfeldt-Jakob Disease, Australian Inquiry, 203
critical care units, *see* collecting sites
Croxatto, Hector, 94
Cruickshank, Robert, 54
CryoLife, 218–19, 221, 229, 252–3
cultural work, 16
currency of regard, *see* money

Daily Express, 260, 264
Dausset, Jean, 121
Davidson, E. C., 6
dCELL®, 285
Dealler, Stephen, 241
death
 brain, 139–41, 144, 178, 179, 192–3, 222, 265
 denial of, 69–70, 102, 150, 259
 social, 105, 161, 202, 208, 233
'death-row donors', 11, 36
Department of Health and Human Services (USA, HHS)
 oversight of organ explant/transplantation industry, 188, 222, 248
Department of Health and Social Security (DHSS, UK)
 and burn care, 126
 and cadaver pituitary gland collection, 181–2,
 and eye banks, 198, 205
 and human growth hormone safety, 205

INDEX

and Human Tissue Act 1961, 151, 180–1
and mortuary staff, 150
and pledges, 190, 192
and tissue typing, 195, 196
see also Ministry of Health
Department of Health (England), 265, 274, 286
dermal fillers, 246
dermatomes, 7
Dexter, Frank, 122, 238
DHSS, *see* Department of Health and Social Security
Diall, Hugh D., 100
Dickinson, Alan, 202–4
disgust, 1, 12, 17, 81–2, 110–11, *see also* feeling rules
doctrinal tyranny, *see* skin
Doheney Eye and Tissue Transplant Bank, 211
donation conversation, *see* kinfolk
donation statutes, 141
'donor', 17, *see also* consent, category error
donor cards, *see* pledges
donor designation, *see* consent, first-person
donors-on-the-hoof, 2, *see also* blood; eyes; kidney; skin
Dooley, Thomas, 91
Duke-Elder, Stewart, 68
dura mater, 62, 64, 122, 206, 226, 228–9, 254, 255
Dutch National Skin Bank, 239–40, *see also* Euro Skin

E. coli, *see* Escherichia coli
East Anglian Tissue Bank, 239
Easty, David, 197
Eli Lilly, 35, 46–7
endocrinological gold rush, 35
End-Stage Renal Disease Program (USA), 134, 136, 166, 183, 186
Engelhardt, Tristram, 247
entities
  'agency' (USA), 188
  'bank', 12–13, *see also* blood banks; eye banks; skin banks
  conglomerate, 216, 223, 226, 268–9, 277, 283
  cottage industry, 62, 217, 220, 221, 237–8
  extractors (independent), 171, 221, 224, 255, 270–1
  for profit, 86, 156, 214, 219, 223, 246, 248, 269–70
  garage/mom-and-pop, 149, 171, 221, 271
  high-volume, 80, 94, 164, 168, 197, 234, 269, 284
  just-in-time, 108, 130, 176, 226, 277

non-profit (UK), 71, 237, 286, *see also* National Health Service
non-profit (USA), 166–8, 174, 183, 188, 215, 223, 246–8, 269–70
processors, proprietary, 217–23, 231, 232, 233, 244, 245, 246, 247, 268–71, 273
quasi-non-profit, 215–16, *see also* organ procurement organizations
service provider, 14, 166–8, 174–5, 214, 215, 223, 238, 253, 270, *see also* Lions Clubs
stash, 57–8, 64, 171, 183, 238–9, 277
enucleation, *see* eyes
Escamilla, Roberto F., 93–4, 8
Escherichia coli, 28, 205
Euro Skin, 240
European Homograft Bank, Brussels, 277
European Union
  Medical Devices Directive, 276
  Tissue and Cells Directive, 276–7, 284
Evans, Herbert Mclean, 34–6, 40, 42–4, 93
excision, *see* corneas
extirpation, *see* pituitary glands
extractors (independent), *see* entities
Eye Bank Association of America, 86, 143, 169
  and regulation, 167, 187, 248, 250, 256
  and service charges, 166
  membership, 86, 202, 255
  voluntary medical standards, 175, 202, 226, 248
Eye-Bank for Sight Restoration, 11–14, 16–18, 21, 23–4
eye banks, UK, 67–8, 116–18, 177, 264, 281, 285–6, *see also* Corneal Transplant Service; Moorfields Hospital; Queen Victoria Hospital; South East Regional Eye Bank
eye banks, USA, 11–13, 62, 86, 187, 197, 208, 214, 230
  exports from, 90–1, 169–70, 273
  high-volume, 164–5, 168–9, 234
  see also Eye Bank Association of America; Eye-Bank for Sight Restoration; Lions Clubs; Medical Eye Bank of Maryland
Eyeball Network (radio hams), 89–90, 165
eyes
  cadaver, 9
  donors-on-the-hoof, 67
  enucleation, 23, 26, 68, 88–9, 108, 115, 165, 176, 197, 275
  non-human, 9
  shelf life of, 12, 75, 116, 165,
  suitability for repurposing, 23, 164, 195

Fantus, Bernard, 12-13
Federal Trade Commission, 38, 174
feeling rules, 17-18, 73, 110, 143, 227, 243, 246, 249, 268, 286
fees, *see* money
Festenstein, Hilliard, 195
Filatov, Vladimir Petrovich, 8-11, 13, 231, 233
first-person consent, *see* consent
Fisher, Russell S., 164
food manufacturing, 37-8, 57, 60, 137, 194, 201, 217, 218, 230-1, 232, 257, 284-6
Food and Drug Administration, 38
  and safety of body/cadaver stuff/artefacts/products, 172, 204-5, 221, 225, 229, 232-3, 243, 252-8, 267, 270-1, 272, 276, 285
  and safety of medical devices (Medical Devices Amendment), 172-8, 226-7, 229-30
  and safety of medicines (Kefauver-Harris Amendment), 155, 204-5
  and scope of/limitations to authority, 172-3, 204-5, 226-7, 229
Flosdorf, Earl, 60
France-Hypophyse, 203, 205
freeze drying (lyophilization), *see* shelf life
freezing, *see* shelf life
Frenchay Skin Bank, 240
funeral industry (USA), 26, 70, 84, 88-90, 102, 150
  and extraction of cadaver stuff, 88, 99, 129, 142, 171, 177, 187, 266-7, 271, 277, 278, 280
funeral grants, 105, 141, 267
funeral homes, *see* collecting sites

Gajdusek, Carleton, 199-200, 202-4, 206, 228
Genentech, 205, 218
Gibbs, Douglas, 177, 195
Gibson, Thomas, 31
Gillies, Harold, 27, 50, 126
Girdner, John H., 1-2
Glasgow Royal Infirmary Burns Unit, 28, 30-1, 49, 54
Glaxo, 79
golden age of immunology, 120
Good Manufacturing Practices, 239, 242
Good Pharmaceutical Manufacturing Practice (UK), 148
Good Tissue Practices, 226, 230, 253, 256-8, 277
Goodwin, Michele, 211-13
Gore, Albert Jr, 186-7, 216
Gorer, Geoffrey, 69

Gorer, Peter, 123
Grafton®, 219
Great Ormond Street Hospital for Children, London, 80, 97, 259
Griffith, Frederick N., 163-6, 168-9, *see also* Medical Eye Bank of Maryland
growth hormone, carcass, 34
growth hormone, human (hGH), 45
  and safety of recipients, 148-9, 157, 172, 202-6, 229
  commercial, 148, 151, 153, 155-6, 172, 182, 205
  dosage, 80, 98
  price, 148, 153, 156, 205
  production costs, 147, 156
  recombinant (rhGH), 205, *see also* Genentech
  soup recipes, 34, 92-3, 147-8, 149, 156-7, 158, 203, 229
  species specificity of, 45-7
  synthetic, 92, 99-100, 147, 156, 180
Guide Dogs for the Blind, 75, 196
Guild of Mortuary Administration (UK), 150
Guinea Pig Club, 67
Gulf War (1990-1), 219, 271
Gummer, John, 236
Guyett, Philip, 271

Hazard Analysis and Critical Control Point (HACCP), 257-8
Hadlow, William, 200
Hagens, Gunther von, 275
Haiken, Elizabeth, 216
Hartree, Anne Stockell, 146-9, 182, 203
Hayden, Cori, 22-3
Healy, Kieran, 16
heart transplants, 178, 194, 238
heart valves, 172, 218-19, 221, 229, 239, 253, 254, 255, 269, 278, 283, *see also* CryoLife
Heck, Ellen L., 137
Heggie, James, 150
Hennepin Medical Examiner's Office and County Hospital, 161
hepatitis (serum), 106, 148, 149, 201, 203
  post-transfusion, 158, 173-4, 199, 228
  see also bioavailability, medical criteria
hGH, *see* growth hormone, human
Hill-Burton program, 87
Hillingdon Hospital, London, 177
Hintz, Raymond, 204
history of sources (medical and social), 199, 227, 242, 249-50, 256, 270, 272
HIV, *see* Human Immunodeficiency Virus

INDEX

Hochschild, Arlie, 17
homostatic stuff, 32
homovital stuff, 32
Hood, George J., 7
Human Bodies, Human Choices, 274
Human Fertilisation and Embryology Act, 274
Human Growth Inc., 96
Human Immunodeficiency Virus (HIV), 225, 227, 230
    see also bioavailability, medical criteria
Human Organs Transplants Act (UK, 1989), 237, 275–6
Human Pituitary Bank, 96
Human Rights Act (UK, 1998), 263, 278, 281
Human Tissue Act (UK, 1961), 82–3, 107, 113, 176, 187, 239, 274
    criticisms of, 83–4, 123, 179–82, 262–3, 274
Human Tissue Act (England, Wales & Northern Ireland, 2004), 274–6, 278, 281
Human Tissue Authority (England, Wales & Northern Ireland), 277, 280
Hyatt, George W., 58, 60–2
hypophysectomy, 35

iatrogenesis, 199
immunosuppression, 122, 124, 137, 142, 185, 198
Informed Consent in Tissue Donation: Expectations and Realities, 248
insulin, 35, 40, 45, 79, 218, 231
intensive (critical) care units, 133, 139, 140, 178, see also collecting sites
International Atomic Energy Agency, 278
Iris Fund for the Prevention of Blindness (UK), 195–7
Isaacs Report, 262–3

Jackson, Douglas MacGregor, 122
Jacobs, Harvey Barry, 184–5
Janžekovič, Zora, 125, 137, 239
Joint Panel on Medical Aspects of Atomic Warfare (US), 56–7
Jordan, Michael, 210–11
Joseph, Keith, 190
J. Walter Thompson Advertising Agency, 16, 18

KabiVitrum (KABI), 155–6, 172, 205
Kagan, Richard, 247
Kaufman, Herbert E., 165
Kearney, John, 238
Keller, Helen, 14

KeraLink International, 283
keratoplasty, see corneal grafts
Khvatov, Valery, 233
kidney, 32
    cadaver, 142, 145, 178–9, 183–4, 187
    donors-on-the-hoof, 104, 121, 142, 184–5, 237, 275
    explant/transplants, 121, 124, 140, 142, 145, 178, 185
    see also Medicare, and End-Stage Renal Disease Program; tissue typing; UK Transplant Service
Kinetic Concepts, 270
kinfolk,
    agreement to extraction of, 11, 23, 24, 36, 61, 75–6, 81, 83, 107, 108, 123, 129, 135, 137, 143, 151–2, 163–4, 176–7, 180–1, 208, 210–12, 224, 233, 249, 260–4, 266
    'donation conversation' with, 23, 61, 83, 88, 124, 129, 163, 168, 176–8, 180–1, 187, 189, 191, 197, 213, 222, 248–51, 256, 281, 286, see also history of sources; Required Request; Routine Inquiry statutes; Routine Referral; transplant coordinators
Kiwanis, 14
Klatzo, Igor, 200
Knight, Bernard, 180–1
Korean War casualties, 54, 64, 128, 132
Krause, Fedor, 3
kuru, see spongiform encephalopathies

Lackam, William H. Von, 57
Landsteiner, Karl, 5
Larkin, Frank, 264
leather, 1–2, 7, 37
Lederer, Susan, 4, 16
Leeds Tissue Bank, 122
legislative consent, see consent
Li, Choh Hao, 43, 46–7, 78, 94, 156
    and human growth hormone molecule, 92, 99–100, 156
LifeCell, 219, 221, 245–6, 270
LifeNet, 216–17, 225–7, 282–3
LifePoint, 269
Lions Clubs, 14–15, 16
    and cadaver eye exports, 90–1, 273
    and communism, 170
    and enucleation, 88–9
    and eye banks, 13–14, 86–90, 137, 163, 217
    as lobbyists, 141, 187
    and pledges, 19–20, 141, 143
    Lions (UK), 108, 279
Loeb, Leo, 32

Longmore, Donald, 194
Los Angeles County Coroner's Office, 211–12
Los Angeles Times, 211–12
Lowry, Philip, 149–50, 203
Luft, Rolf, 47, 156
lunch-hour facelift, 282
Lurie, Leonard, 176
Lykins, Brian, 252–3
Lyodura®, 228

McIndoe, Archibald, 27–8, 50, 66–9, 71–3, 120
McIndoe Burns Unit, East Grinstead, 120, 122–3
'mad cow' disease, *see* spongiform encephalopathies
Malinin, Theodore, 171
Manchester Royal Eye Bank, 198
Mande, Jerold, 187
Marchioness pleasure boat tragedy, 262
Masters, Colin, 203
Mastromarino, Michael, 266–8, 270, 271, 278
maximizing the gift, 223, 268, 281
meat industry, UK, 78–9
meatpackers, USA, 35, 37–8, 243
  and carcass pituitary glands, 38–45
Medawar, Peter, 30–2, 50, 60, 120–2, 123
Medicaid, 133, 160, 166, 188, 209, 213, 221, *see also* Medicare
medical devices, 172–3, 229–30, 254, 276
  regulation (UK), 276
  regulation (USA), *see* Food and Drug Administration
medical examiners/coroners (USA)
  and extraction of cadaver stuff, 142–3, 158, 160–2, 208–9, 211–12, 224, 232, 233
  and post mortem examinations (autopsies), 160–1
  scope of authority, 24, 143, 160–2, 209, 211
  see also collecting sites; consent, legislative
Medical Eye Bank of Maryland, 163–5, 168, *see also* Tissue Banks International
medical negligence, 175, 202, 206, 228
Medical Research Council, 28, 59
  and cadaver pituitary gland collection, 79–80, 84, 85, 95, 98, 101–2, 105, 146, 151
  and human growth hormone clinical trial, 80, 97, 147
  and human growth hormone production, 146, 149

  and human growth hormone safety, 106, 149, 203
  and Human Tissue Act 1961, 81, 84, 124, 150–1
  and Strontium-90 research, 83, 262
  and tissue typing, 123
  and war burns, 28, 31, 49
Medicare, 133, 166, 215
  and autopsies, 160
  and burn care, 133, 184
  and End-Stage Renal Disease Program, 134–5, 166, 183–4, 186, 188, 221–2, 248
  Condition of Participation Rule (COP), 209, 213–14, 248, 268
Medicines Act (UK, 1968), 148
MEDICO, 91
Medtronic, 270
Mile High Transplant Bank, Denver, 216, 219
Mims, Cedric, 204
Ministry of Health (UK), 77, 110, 117, 176, *see also* Department of Health and Social Security
Mitford, Jessica, 88–9
money (as consideration) 167, 215
  cost recovery, 40, 95, 134, 137, 165–7, 175, 180, 184, 188, 215, 221–2, 233, 238, 244–5, 247, 270, 277, 279, 282, 285
  currency of regard, 18, 67, 110, 185, 286
  debt repayment, 12
  'dirty', 167, 212, 233, 270–1
  fees, 84, 126, 104, 124, 138, 166, 167, 214, 218
  gifts, 98, 177
  profit, 37, 88, 89, 136, 153, 202, 216, 218, 231, 245, 246–8, 282, 286
  service charges, 137, 166–7, 175, 185, 198, 215, 217, 238, 247
  tips (incentives etc.), 84–5, 97, 115, 124, 146, 151–3, 156, 158, 163–4, 171, 181–2, 224
  see also philanthropy
Moorfields Hospital, London, 68, 115, 117, 195, 197–8, 238, 264, 279–80, 285
mortuaries, 26, 70, 80–1, 102, 106, 149–50
  ethos and practices, 82, 89, 92, 96, 129, 180, 152–3, 159, 180, 183, 199, 218, 239, 259, 261–2, 263, 280
  staff, 82, 84, 89, 95, 98, 101, 104, 106, 124, 129, 146, 150, 153, 177, 180, 181–2, 202, 211, 277
  see also collecting sites, mortuaries; money; no-place
Murray, Joseph, 6, 121, 142

INDEX

Musculoskeletal Transplant Foundation, 221, 223, 268–9, 283

Nader, Ralph, 172
nucleic acid amplification (NAT) tests, 256
National Academy of Sciences, 155
National Association of Medical Examiners (USA), 212
National Cattlemen's Beef Association (USA), 243
National Coalition of Burn Center Hospitals (USA), 184
National Commission on Fire Prevention and Control (USA), 131, 134
National Conference of Commissioners on Uniform State Laws, 142–3, 249–50
National Health Service (NHS, UK), 68, 69, 77, 80, 115
 and burn care, 55, 126, 239–40
 and death denial, 68, 70, 109
 ethos, 102, 157, 195, 215, 279
 finances, 119, 236–7, 259, 274
National Institutes of Health
 and burn care, 134
 and spongiform encephalopathies, 200, 203
 see also Gajdusek; National Pituitary Agency
National Kidney Foundation (USA), 143, 183, 185
National Organ Transplant Act (NOTA) (USA, 1984), 187–9, 215, 237, 246, 247
National Pituitary Agency, 94–5, 98
 and cadaver pituitary gland collection, 95–8, 154–5, 202, *see also* pledges
National Research Council Review, 155–7
National Uniform Law Commission (USA), 209
Naval Regional Medical Center, San Diego, California, 65
Navy Tissue Bank (USA), 58, 60–2, 122, 172, 175, 216, 217
Nazi concentration camps, 2, 155
Nestle, Marion, 201–2
New York Firefighters Skin Bank, *see* skin banks (USA)
New York Presbyterian Hospital Burn Center, 136, 138
NHS Blood and Transplant Tissue Services (England), 278, 285
NHS Blood and Transplant (UK), 278, 279, 281, 286
NHS Blood Service (England), 240–1, 242, 259, 277–8
NHS and Community Care Act (UK, 1990), 236–7, 242

Nixon, Richard, 65, 131
'no-place'
 in abattoir, 81
 in mortuary, 81, 84, 151, 228, 235, 280
 in operating theatre, 140, 178, 183
Nordisk, 151, 153
North Wales & Oswestry Tissue Bank, Wrexham, 277
North London Blood Centre, Edgware, 239
notification of potentially bioavailable corpse, 61, 108, 114, 124, 177, 178–9, 185, 189, 213–14

O'Brien, Martin, 37
Occupational Safety and Health Act (USA, 1970), 165
Odstock Hospital, Salisbury, 119–20
Ollier, Louis, 3
O'Neill, Mary Jane, 208
Orange County Register, 244–6, 248, 275, 282, 286
organ procurement agencies, 136, 183, 185–7
Organ Procurement and Transplantation Network (USA), 246
organ procurement organizations (OPO), 188–9, 208, 215
 diversification, 209, 213–14, 216, 219, 221–3, 248, 268–9
Osteotech, 219, 223, 268–70
Ostrander, James E., 175

Padgett, Earl C., 7
Pappworth, Maurice, 140
Parkes, Alan S., 35, 59
Parkland Hospital, Dallas, Texas, 137
Parlow, Albert F., 148, 158
Patient Self-Determination Act (USA, 1991), 250
Paton, David, 16
Paton, R. Townley, 11–14, 18, 22
Peanut Club, 68–9, 71
Pegg, David, 239
Pennsylvania Act 102 (1994), 213, 249
Peter Brent Brigham Hospital, Boston, Mass., 121, 142
philanthropy
 big-money, 14–15, 68, 71, 120, 123
 body, 16, 227
 corpse, 16, 71, 246, 264, 282, 286
 discretionary, 15, 71, 109
 marketing of, 16–20, 69, 98, 109–10, 143–4, 194, 210–11, 246, 251, 265, 268
 mass, 15, 135, 137
Pillsbury Company, 257–8
Pinderfields General Hospital, Wakefield, 119, 122

pituitary gland, 34–6
  adrenocorticotrophic hormone (ACTH), 45
  and disorders of growth, 34, 47, 80
  anterior lobe hormones, 36, 43, 45–6
  cadaver, 79–80, 84, 94–5, 106, 124, 149, 151, 157–8, 161, 180–2, 206
  carcass, 37, 39–40, 43–5, 46, 78, 84–5
  donor-on-the-hoof, 47
  extirpation of, 38–9, 84–5
  gonadotropins, 156
  imports, 94, 151
  pituitrin, 40
  monkey, 79
  soups, 34, 36, 43, 92–3
pituitary pirates, 92, 158
pledges, 17, 250
  cards, 17, 72, 98, 112–13, 143, 191, 193
  in/utility of, 22, 68, 77, 107–8, 110, 113–15, 117, 145, 191, 193, 208, 265, 273, 281
  legal authority of, 21, 70–2, 141–3, 161, 179, 190–1, 249, 286
  of corpses, 71, 141
  of eyes, 17–20, 67–8, 71, 73–5, 107–9, 110–11, 191–2
  of kidneys, 139, 190–1
  of 'multi-organs', 192–4, 265
  of pituitary glands, 81, 98–9
  of skin, 265
  of tissue, 286
  on driving licences, 144–5, 192, 193, 249
  registers of, 113, 192, 193, 249, 265, 275–6, 286
  see also philanthropy, corpse
poke and sniff test, 78, 106, 234, 252, 256, 267, 280, see also putrefaction
polio vaccine, 47
Pollock, George David, 3
post-mortem examinations, UK, 104, 263,
post-mortem examinations, USA, 95, 96, 98, 159–60
presumed consent, see consent
prions, 228, 236, 241, 242–3, 257, 259
processors, proprietary, see entities
product liability, 173–5, 229, 242, 253
Protein Hut, 79, 148, 149
Prottas, Jeffrey, 185, 189
Prusnier, Stanley B., 228
Putnam, Robert, 168
putrefaction, 9, 12, 26, 216, 252, 280

Queen Victoria Hospital, East Grinstead, 66
  and burns, 66, 69, 120, see also McIndoe Burns Unit

eye bank, 67–9, 71–2, 108, 110–11, 115–17, 177–8, 198, 279–80, 285
skin bank, 123, 137

Raben, Maurice, 45, 78, 92, 94, 146, 148, 156–7
Raiti, Salvatore, 97, 203
Ravenis v. Detroit General Hospital, 171–2, 175, 202, 226
Regeneration Technologies, 219, 223
  RTI Biologics, 270–1
  RTI Donor Services, 267
Regional Organ Bank of Illinois, Chicago, 219, 222
rendering, 235–6
Repliform®, 245
requesters, 248–51, 270, see also history of sources; kinfolk
Required Request statutes, 208–9, 213
retained organs scandal, 259–63
Retained Organs Commission (UK), 264, 276
Rettig, Richard A., 134–5
Richardson, Ruth, 76, 105
Ross, Donald, 238
Rotary Clubs, 14
RNIB, see Royal National Institute for the Blind
Routine Inquiry statutes, 208–9
Routine Referral, 213–14; see also, kinfolk; Medicare, Condition of Participation Rule; notification
Royal Brompton Hospital, London, Heart Valve Bank, 238
Royal Eye Hospital, London, 74, 196
Royal Liverpool Children's Hospital, see Alder Hey Hospital
Royal National Institute for the Blind (RNIB), 74–5, 107–15, 117, 191–4, 196, see also pledges, of eyes
Royal Society for the Prevention of Accidents, 52
Rumsfeld, Donald, 199, 255
Rycroft, Benjamin, 66–8, 73–5, 107–11, 116–17, 176, 177, 195

St Andrews Hospital, Billericay, 127
San Francisco Pituitary Bank
  collecting sites, 93–4, 97–8
Sappol, Michael, 141
Saxena, Brij E., 157
scalds, see burns
Scarry, Elaine, 73
Scottish Blood Transfusion Service, 238
Scottish Executive, 84, 262
scrapie, see spongiform encephalopathies

INDEX

Sell, Kenneth W., 175
service charges, *see* money
Sheffield Skin Bank, 239–40
shelf life, techniques for extending,
    freeze drying (lyophilization), 60, 63, 122, 123, 147, 171, 219
    freezing (cryopreservation), 57, 59–60, 116, 123, 137–8, 150, 171, 218
    organ culture, 197, 279
    refrigeration, 58, 116, 130, 171, 239–40
    storage media, 10, 85, 116, 165, 197
Shem, Samuel, 163
Shipman, Harold, 103
short stature, 41–3, 80, 99
Shriners, *see* Ancient Arabic Order of the Nobles of the Mystic Shrine
Simmel, George, 167
Simonsen, Morten, 123
Sinclair, Upton, 38
skin
    accellular artefacts/products, 219, 245–6, 283, 285
    as biological wound dressings, 31–2, 33, 63, 82, 131, 132, 244, 283
    biosynthetic, 245
    cadaver, 1–2, 31, 62–3, 129, 221, 231, 270
    cosmetic use of, 244
    doctrinal tyranny of, 5, 30–1, 32, 120–3, 124–5, 129, 219
    donors-on-the-hoof, 2, 4, 7, 122, 125, 126, 129
    extraction, 124, 129, 244
    shelf life, 123, 130
    supply controversies, 186–7, 239–40, 244–5, 283
    see also leather
skin banks UK, 121, 122–3, 239–41, 278,
skin banks USA, 130–1, 136–8, 186, 216, 244–5, 254, 282–3
skin grafts, 1, 2–4, 7, 28, 30, 54, 125–6
    autograft, 3
    homograft (later allograft), 4
    zoograft, 5, 132
Sklifosovsky Institute, Moscow, 10, 233
soups, *see* growth hormone, human
Smith, Audrey, 59
Smith, Wilson, 106
South East Regional Eye Bank (UK), 116–17, 177–8
South Thames Regional Burns and Plastic Surgery Unit, Roehampton, 240
Southwest Medical Foundation, Crystal Charity, Texas, 137
Spembly Technical Products Ltd, 123
SpinalGraft Technologies, Philadelphia, 270
spongiform encephalopathies, 199, 228
    bovine (BSE; 'mad cow' disease), 235, 243, *see also* rendering
    Creutzfeldt-Jakob disease (CJD), 200, 202–6, 228, 235, 241–3
    kuru, 199–200, 203
    scrapie, 200, 204, 236
    variant (v) Creutzfeldt-Jakob disease (CJD), 241, 243
    see also Gajdusek, prions
Stanislaw, Francis, 73–5
Starzl, Thomas, 124, 142
Stephen Kirby Skin Bank, Queen Mary's Hospital, Roehampton, 240, 265, 278
Stevenson, Robert E., 186–7
Stroever, Bruce W., 221, 268
Sunday Herald, 83, 262
Sunday People, 152–3, 176, 180, 203
supply chains, 10, 18, 23, 37, 85, 89–90, 95, 96–7, 117, 123–4, 138, 146, 151, 165, 177–8, 194–6, 228, 231, 267, 271–3, 284, 285

Tanner, James Mourilyan, 80
tanning (tannic acid), 2, 6, 28
Terasaki, Paul, 194
terminal sterilization, 106, 122, 148, 201, 204, 218, 220, 232, 236, 253, 254, 278
Thiersch, Karl, 3, 7
Thompson, J. Walter (advertising agency), 16
'tissue', ix, 263
Tissue Banks International, 211, 221, 273, 283
Tissue Regenix Group, 285
tissue typing (compatibility), 121, 123, 137, 185, 188, 194, 195, 197, 198, 278
Titmuss, Richard, 157–8, 199, 243
Tocqueville, Alexis de, 14, 20
transplant coordinators
    UK, 197, 240
    USA, 189, 222, 281
trephines, 11
Trevor-Roper, Patrick, 115–18, 192
Tudor Thomas, James William, 67
Tutogen Medical, Inc., New Jersey, 270
Tyrrell, David, 149

Service, 195–7, 237, 278, *see also* NHS Blood and Transplant
Underhill, Frank Pell, 6
Uniform Anatomical Gift Act (USA)
    of 1968, 139, 142–3, 161, 190, 209, 249
    of 1987, 209, 249
    of 2006, 249–50
United Network for Organ Sharing (USA), 189, 210, 246

United States Army Institute of Surgical Research, 132
United States Navy Tissue Bank, 58–65, 131, 132, 143, 217–18, 278
University of Florida Tissue Bank, 219
University of Miami Tissue Bank, 171
University of Texas Medical Branch, Galveston (UTMB Tissue Bank), 216, 244
Unrelated Live Transplant Regulatory Authority (ULTRA, UK), 237

Valley Forge General Hospital, Pennsylvania, 32
Van Velsen, Dick, 261–2
Veterans Administration (USA), 96, 134
Vialles, Noëlie, 81
Vietnam War casualties, 64–5, 132
Virginia Tissue Bank, 186
Vision Share, 273

Ward, Elizabeth, 190–1, 194
waste, 37
  abattoir, 40, 94, 235, 243
  biotrash, 37–8, 264
  surgical, 4, 9, 31, 47, 57, 58, 67, 122, 238–9, 264

Watson, John, 120, 122–4, 137
Watson, Thomas J., 11
Werne, Jacob, 23–4
Westminster Hospital, London, 117–18
Westminster/Moorfields Regional Eye Bank, 117
Weston, Simon, 196, 240
Wildy, Peter, 204
Wilhelmi, Alfred E., 94, 148, 156, 158
Wilmer Eye Institute, Johns Hopkins Hospital, Baltimore, 16
Wolverhampton Eye Infirmary, 77, 108
World Health Organization, 109, 169, 201, 271

yellow jaundice, *see* hepatitis
Yorkshire Regional Tissue Bank, 122, 238, 277
Young, Frank George, 78–9, 151
Your Life in their Hands, 110
Yudin, Serge, 10

Zelizer, Viviana, 84, 167
Zirm, Eduard, 9
Zunz, Oliver, 14